To Man Yuk, Jim Yuan, and Joan Man
for their encouragement, understanding, and assistance

加拿大華埠發展史

黎全恩

Chinatowns

This book is a definitive history of Chinatowns in Canada. From instant Chinatowns in gold- and coal-mining communities to new Chinatowns which have sprung up in city neighbourhoods and suburbs since World War II, it portrays the changing landscapes and images of Chinatowns from the late nineteenth century to the present. It also includes a detailed case study of Victoria's Chinatown, the earliest such settlement in Canada.

David Chuenyan Lai begins by outlining the history of Chinese immigration to Canada and provides an overview of the demographic changes of Canada's population of ethnic Chinese, including regional origins, from the period of free entry (1858–84) to the period of selective entry (1948 and on). He then classifies Chinatowns into four groups—Old, New, Replaced, and Reconstructed Historic Chinatowns—and analyses their origins, locations, viability, and socio-economic and image changes.

Focusing on Victoria's Chinatown as a typical example of a Historic Chinatown, Lai employs a stage-development model consisting of four phases (budding, blooming, withering, dying or reviving) which can be used to study Chinatowns across North America. He also describes a public involvement model of community planning which is responsible for the successful rehabilitation project in Victoria—one which has received international recognition and in which Lai has been extensively involved.

The culmination of twenty years of research which has included detailed surveys of over fifty Chinatowns in North America and interviews with numerous community leaders and city planners in all major Chinatowns in Canada, this book explains why Historic Chinatowns are seen as important by Chinese today and why they may survive despite the competing attractions of New Chinatowns. It also sheds new light on the characteristics of these communities and provides useful insights for geographers, historians, sociologists, and anthropologists.

DAVID CHUENYAN LAI is an associate professor of geography at the University of Victoria. In 1983 he was made a Member of the Order of Canada (CM) in recognition of his work on Victoria's Chinatown.

Chinatowns

TOWNS WITHIN CITIES IN CANADA

David Chuenyan Lai

UNIVERSITY OF BRITISH COLUMBIA PRESS
VANCOUVER 1988

Canadian Cataloguing in Publication Data
Lai, Chuen-yan David, 1937-
Chinatowns

Bibliography: p.
Includes index.
ISBN 0-7748-0309-6

1. Chinese - Canada - History. 2. Chinese Canadians - History. 3. Chinatown (Victoria, B.C.) - History. 4. Chinese - British Columbia - Victoria - History. 5. Chinese Canadians - British Columbia - Victoria - History. I. Title.
FC106.C5L34 1988 971′.004951 C88--91555-2
F1035.C5L34 1988

This book has been published with the help of grants from the Social Science Federation of Canada, using funds provided by the Social Sciences and Humanities Research Council of Canada; from the B.C. Heritage Trust; and from the Canada Council.

Printed in Canada
ISBN 0-7748-0309-6

Contents

Figures

Tables

Plates

CREDITS

Bill Chi-yun Li: Plate 23
City of Toronto Archives: Plate 9
Eric Joe: Plate 37
Nanaimo Museum, Plate 6
Provincial Archives of British Columbia: Plates 1, 2, 5, 29, 30, 31, 32, 33, 34, 35
Vancouver Public Library: Plates 7, 8

Preface

This book is an urban history of Chinatowns in Canada. It traces the transformation of the physical and cultural landscapes of Canadian Chinatowns from 1858 to 1988 and analyzes their origins, locations, viability, and socio-economic and image changes. The book is also concerned with Chinese immigration to Canada and the changing demographic characteristics of Chinese people in the country. It attempts to shed new light on the characteristics of Chinatowns across Canada, which have been rarely examined by sociologists, historians, or geographers.

In the past, a "Chinatown" was a self-contained urban enclave where nearly all Chinese people, their businesses, and their social institutions were confined. Physically and functionally, it was "a town within a town." In the old days, Chinatown was regarded by the white community as a segregated, mysterious ghetto of prostitution, gambling, opium-smoking, and other vices, but it was considered by the Chinese people as a home where they could find pleasure, comfort, and companionship; it was a sanctuary where they were secure from threats and discrimination. Since the end of the Second World War, Chinatowns throughout Canada have undergone substantial changes. Usually one of the earliest settled inner-city neighbourhoods, Chinatowns are often part of the depressed downtown residential and commercial districts of a modern city. In the course of slum clearance or urban renewal programs, some Chinatowns have been destroyed and others relocated and rebuilt. A few old Chinatowns still survive *in situ,* but they are depopulated: most of their residents have moved to other parts of the city for better accommodation. Since 1967, new types of Chinese immigrants have come to Canada, mostly from Hong Kong, Taiwan, and Southeast Asian countries. The entrepreneur immigrants revitalized the economy of Old Chinatowns and created New Chinatowns. The thrust of new investment and new pride is felt not only in Chinatowns of large metropolitan cities, such as Toronto and Vancouver, but also in Chinatowns of smaller cities such as Edmonton and Calgary. During the past twenty years, I have surveyed one

Chinatown after another across Canada, observing that most of the Old Chinatowns have evolved through cycles of growth and decay, each cycle being differentiated by successive stages. A growth model, consisting of a four-stage development cycle, was devised to trace the evolution of a Chinatown through its budding, blooming, withering, and dying or reviving phases. I have also collected and studied the archives of various Chinese associations in Victoria, British Columbia, documents rarely used as sources of information about the Chinese people in Canada. The study of the Chinese population and of Chinatowns in this book is based not only on official sources and newspaper reports but also on certain kinds of data derived from unusual sources of information such as hospital donation receipts, burial records, and the reminiscences of old-timers. Unless specified in notes, information is based on interviews with old-timers and local community leaders in each Chinatown. Some of these informants do not wish to be identified; thus promises of anonymity prevent me from revealing them. Since different factions exist in some Chinatowns, I have tried my utmost to present the views and opinions of all sides and refrain from siding with any group within the Chinese community.

There are many variations in the romanization and transliteration of Chinese names. For the sake of consistency, the Pinyin system of romanization has been used in this book wherever possible. For example, Zhou is used instead of Chow, Chou, or Joe, although a Cantonese will be more familiar with the latter spellings of the surname than the former. Spellings of placenames also follow the Pinyin system, such as Guangdong instead of Kwangtung or Kuang-t'ung. Exceptions have been made, and well-established or more familiar spellings used, for some Chinese voluntary associations, personal names, and cities, because changes in spelling may cause confusion. For example, Chee Kung Tong is used instead of Zhigongtang; Dr. Sun Yat-Sen Garden instead of Dr. Sun Yixian Garden; and Hong Kong instead of Xianggang. In such cases, the Pinyin system follows the first reference in parentheses, such as Sun Yat Sen (Sun Yixian), Hakka (Kejia), and Hook Sin Tong (Fushantang). However, if a Chinese association or a Chinese person does not have a conventional English name, the Pinyin system is used, such as Yuqingtang, the predecessor of the Hoy Sun Association in Victoria. In this book, the first reference of an old street name is followed by the new street name in brackets, such as Dupont Street (East Pender Street) in Vancouver.

Acknowledgments

I would like to thank the Chinese Consolidated Benevolent Association of Victoria for permission to study its meeting minutes, which provided valuable new information about the Chinese community. Thanks are also due to many old-timers, community leaders, and other people—too numerous to mention individually—who agreed to have interviews and who contributed their reminiscences. Thanks are also extended to the city town planners, librarians, and archivists who provided access to valuable reference materials.

I am also grateful to Dr. Terry McGee for comments on the first outline of the book, Dr. Peter Murphy on the introductory chapter, and Dr. Doug Porteous on the book's title; Dr. Samuel Wong, Mr. Paul Molyski, and Mr. Paul Sales for writing the computer programs; Mrs. Laura Coles for editing earlier drafts of the book, Dr. Jane C. Fredeman for helpful suggestions, and Ms. Jean Wilson for the final editing; Messrs. Ken Josephson and Ole Heggen and Mrs. Diane MacDonald for doing cartographic work; Mr. Ian Norie for advice on the format, Miss Stella Chan for typing; Messrs. Sam Lum and Tom Ma for writing the Chinese characters on the title page; Dr. Michael P.Y. Lai and Miss Averil Harrison for proofreading and assistance in library research; Messrs. Dennis and Derek Chao for assistance in field surveys; and Mr. Norman Micklewright for taking photographs of Winnipeg's Chinese garden and gate. Some of the illustrations were prepared by Messrs. Gary M. Richardson and Steven Wynne, who were hired under the Federal-Provincial Summer Employment Program Challenge '87, and by Timothy C. Chan, who was hired under the Work-Study Program, 1986–7, funded by the Province of British Columbia. Grateful acknowledgment is also given to the Secretary of State, Department of Multiculturalism; the Social Sciences and Humanities Research Council of Canada; and the University of Victoria for supporting the research for this volume with their grants. Thanks are extended to the three anonymous referees for their valuable comments on the manuscript.

Chinatowns

1

Introduction

Many city governments in North America recently have become interested in the rehabilitation and preservation of their historical Chinatowns. In order to understand today's Chinatowns, it is necessary to study the history of Chinese migration to North America and to examine the evolution of Chinese communities. This book studies the migration of Chinese people to Canada and the development of Canadian Chinatowns from 1858 to the present.

The research for this book focused on several central questions. What are Chinatowns? What variables account for their birth, growth, decay, or revival? How did the townscapes and images of Chinatowns change before and after the Second World War? To what extent have Canadian immigration policies affected the development of Chinatowns and the demographic structures of the Chinese population in Canada? How is the survival of Chinatowns related to the more encompassing process of downtown urban renewal? Why do the Chinese community leaders want to preserve Chinatowns? Do they have the same aspirations for the future of their Chinatowns? It is hoped that this book will provide answers to these questions.

DEFINITIONS OF CHINATOWNS

There is no precise definition of a "Chinatown"; usually it is perceived as a Chinese quarter of any city outside China. During the nineteenth century, San Francisco, Victoria, and Vancouver were the major Pacific points of entry to North America from China. After the Chinese immigrants arrived in these port cities, they confined themselves to one or two streets, which the Chinese people called *Tangren Jie* (Chinese street), and the white public called "Chinamen's quarters," "Chinese community," or "Chinatown." The Chinese living quarters in the gold-mining settlements were also known as "Chinatowns." Eventually, the word "Chinatown" was so commonly used that it became a standard term. As populations increased and economic activities diversified, Chinatowns in port cities expanded, occupying several city blocks and functioning like self-contained towns. These large coastal Chinatowns became known to the Chinese people as *Huabu* (Chinese ports). Today Chinatowns are still called *Tangren Jie* colloquially, but they are also commonly referred to as *Huabu*.

In this book, Chinatowns are classified into four groups: Old Chinatowns, New Chinatowns, Replaced Chinatowns, and Reconstructed Historic Chinatowns. Old Chinatowns are defined as Chinatowns established before the Second World War. An Old Chinatown is a Chinese residential, institutional, and commercial inner-city neighbourhood that is physically discernible by its land uses, commercial façades, demographic structures, and socioeconomic activities.

New Chinatowns, Replaced Chinatowns, and Reconstructed Historic Chinatowns were established after the Second World War. New Chinatowns are basically commercial areas characterized by the concentration of Chinese business concerns along a section of a street or in nearby shopping plazas. Unlike an Old Chinatown, a New Chinatown is not a well-defined Chinese residential area although it is usually located close to neighbourhoods with a Chinese population.

A Replaced Chinatown is a planned Chinatown which replaces an Old Chinatown. Like an Old Chinatown, a Replaced Chinatown is a Chinese residential, commercial, and institutional area but unlike an Old Chinatown, it is an attractive inner-city neighbourhood. Today, the only example of a Canadian Replaced Chinatown is found in Edmonton. In the United States, Replaced Chinatowns were built in cities such as Stockton and Sacramento.

Old, New, or Replaced Chinatowns are identifiable by the brightly coloured commercial façades of Chinese business concerns, the activity of many Chinese pedestrians, the odours of Chinese merchandise and food, and the sounds of various Chinese dialects. These bustling Chinese activities are not found in a Reconstructed Historic Chinatown, which is an extinct Chinatown restored to its original streetscape and designated as a heritage site. Today, there is only one Reconstructed Historic Chinatown in Canada, in Barkerville, British Columbia. The New, Replaced, and Reconstructed Historic Chinatowns are just beginning in Canada and do not have a long evolutionary history like the Old Chinatowns.

STAGE-DEVELOPMENT MODEL

We cannot understand an Old Chinatown in Canada unless we see its whole evolution. Like a living organism, an Old Chinatown is constantly evolving and being transformed. Although Old Chinatowns change in different ways and at varying rates, they tend to follow a common pattern in their course of development. A stage-development model therefore was devised to explain this evolution (Figure 1).

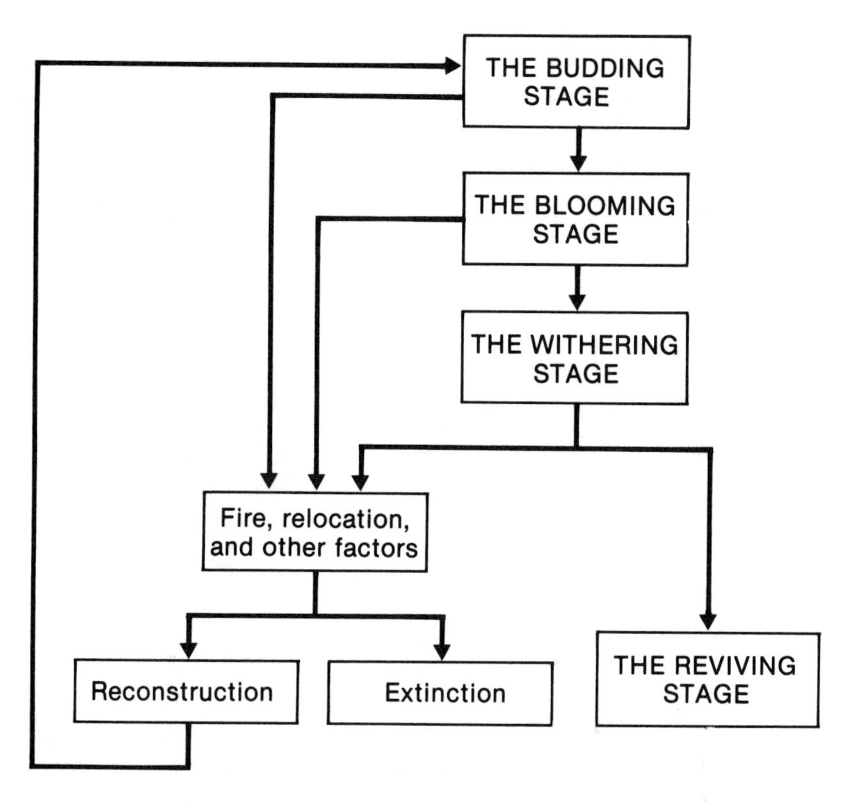

1 Chinatown Stage-Development Model: Life Cycle

Each stage of development has its own characteristics. In the budding stage, an Old Chinatown usually has few Chinese residents, predominantly male, who represent nearly the entire Chinese population of a city. Thus, a city's "Chinatown" is identical with its "Chinese community." The Chinatown's economy is likely controlled by a few wealthy merchants and the Chinese society is often polarized into two classes: a small merchant class at one end, and a large labouring class at the other end. Morphologically, a budding Chinatown is characterized by a linear pattern or a cross-shaped pattern formed by two intersecting streets (Figure 2). The streetscape is dominated by rows of closely packed wooden shacks and cabins. To the white society, a Chinatown is a filthy slum.

During the blooming stage, the Chinese population increases rapidly by in-migration. Although the number of married couples is increasing, the popu-

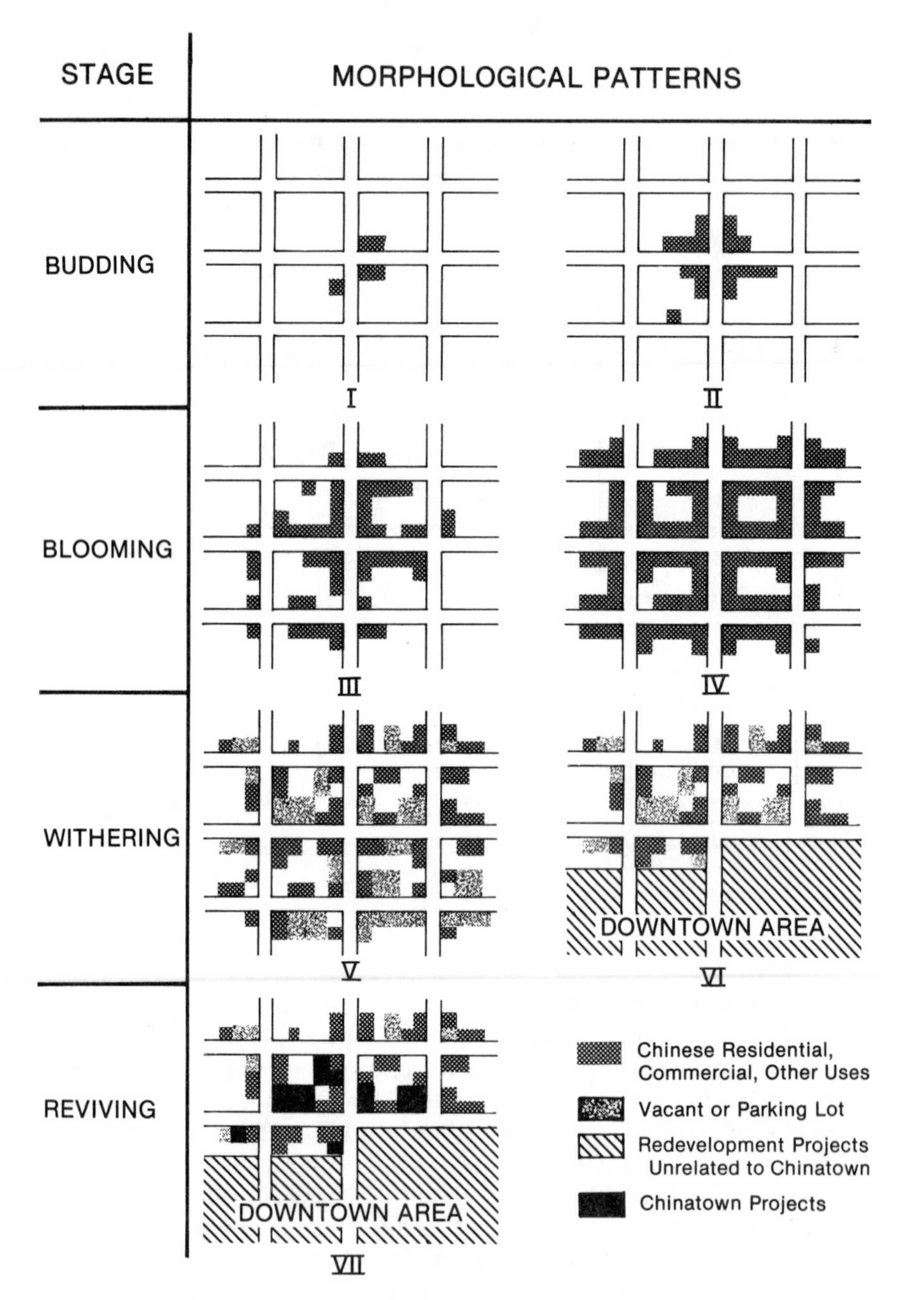

2 Chinatown Stage-Development Model: Morphological Patterns

lation is still dominated by single male labourers. Most of them live in Chinatown, but many Chinese domestic servants, market gardeners, laundrymen, and the like gradually spread out into other parts of the city. Thus, the Chinese community of a city extends beyond the boundary of its Chinatown, and the city's Chinatown is no longer a synonym for its Chinese community. A merchant class of small businesses is beginning to emerge as Chinatown's economy is becoming more diversified and prosperous. There are gambling dens and brothels meeting the needs of a "bachelor" society. (A blooming Chinatown before the 1910s abounded with opium dens as well.) Various types of Chinese associations, temples, churches, schools, and other institutions have also been built. Usually a blooming Chinatown functions like a self-contained town and has its "government" led by an umbrella organization under different names in different cities.[1] This organization helps resolve the internal conflicts within the Chinese community and deal with the discriminatory measures and treatments by the host society. An increasing number of properties in the Chinatown are owned by Chinese organizations or individuals. Covering several city blocks, the expanding Chinatown has a reticulated pattern formed by parallel streets crossing one another. The townscape is likely dominated by a mixture of wooden shacks, log cabins, and two- or three-storey tenement buildings built of wood or brick. Some outstanding association buildings have a distinctive architecture that combines Chinese symbolism and motifs with Western-style structures. The host society's image of Chinatown is as a dangerous, mysterious, exotic district, consisting of underground tunnels and trap doors.

An Old Chinatown enters its withering stage when its population decreases, its economy declines, and the ownership of properties by the Chinese diminishes. Chinese businesses are closed one after another as non-Chinese businesses such as low-class bars, second-hand shops, and pornographic bookstores move in. Chinatown residents of moderate means gradually move out of Chinatown, leaving behind the poor, elderly bachelors. Community organization weakens, and there is a declining level of participation in social activities. Many traditional associations fail to recruit young members and become defunct after their aging members die. The withering Chinatown is diminishing in size because of the encroachment of new redevelopment projects which do not conform to the traditional Chinatown land use. The townscape of Chinatown is increasingly dominated by dilapidated structures, vacant sites, parking lots, and a mixture of Chinese and non-Chinese businesses. To both the Chinese community and the host society, Chinatown is a skid row district, and its days are numbered.

The final stage of an Old Chinatown is either extinction of rehabilitation. If a withering Chinatown is not revitalized, it will be wiped out or reduced in size by fire, relocation, gentrification (or inner-city revitalization), and other factors. If there are infusions of urban renewal funds from the municipal, provincial, and / or federal governments, the withering Chinatown will be revived. Its old buildings will be renovated and new construction projects such as care facilities, cultural centres, and subsidized housing for elderly people or low-income families will be built. A revitalized Chinatown will attract new businesses and investments, and its property values will rise rapidly. It will consist of several congested streets full of restaurants and stores, and sidewalk displays of foodstuffs, dry goods, and other merchandise, catering not only to the Chinese community but also to people of other ethnic groups. The image of Chinatown then varies, as it is simultaneously considered a tourist attraction, a vibrant inner-city neighbourhood, a historic district, an emblem of Chinese heritage, and / or the root of Chinese Canadians in the multi-ethnic society of Canada.

Since an Old Chinatown is a perceptual area existing in the eyes and minds of both the Chinese community and the host society, each stage of development cannot be defined by quantitative measures such as the population size, the volume of business, and the amount of space for residential or commercial land uses. Instead, it is the comparative differences in physical and socioeconomic features that distinguish one stage from another; each stage of development passes gradually into a subsequent stage via a transitional phase. At any stage, a Chinatown may be destroyed by fire, relocation, or other factors. If it is rebuilt immediately on the same site or at another location, a second Chinatown will be born.

PERIODS OF MIGRATION

In this book, the history of Chinese migration and the development of Chinatowns are divided into four periods, based on Canadian immigration policies towards the Chinese people. These four periods are: the period of free entry, 1858–84; the period of restricted entry, 1885–1923; the period of exclusion, 1924–47; and the period of selective entry, 1948 to the present.

Captain John Meares, a retired British Naval officer (1756?–1809), noted the presence of Chinese people in Canada as early as 1788, but the migration of Chinese people to Canada did not occur until the discovery of gold in British Columbia in 1858.[2] During the first period, Chinese immigrants were permitted to enter and leave Canada without restriction, and all Canadian Chinatowns were in British Columbia. In 1885, the federal government began to discourage and restrict Chinese immigration by means of a head tax. The various restrictive measures and regulations failed to curb the influx of Chinese immigrants, however, and Chinatowns grew in many towns and cities across Canada. This period of restricted entry ended with the Immigration Act of 1923 (the Exclusion Act), which virtually prohibited any Chinese immigrant from entering Canada. During the period of exclusion, the Chinese population across Canada declined; gradually and subsequently all Canadian Chinatowns were depopulated, some of them disappearing forever. After the Exclusion Act was repealed in 1947, the admission of Chinese immigrants was resumed with certain restrictions. The Immigration Act of 1967 induced new types of Chinese immigrants and new investors to come to Canada. Some surviving Chinatowns were revitalized, and a new type of Chinatown was created.

CANADIAN CHINATOWNS

This book consists of two parts. Part One examines the stages of development of Canadian Chinatowns according to the four periods of Chinese immigration. During the period of free entry, the migration of Chinese to Canada was, in fact, a migration to the province of British Columbia. Chinatowns were mostly associated with gold-mining, coal-mining, or fish-canning in the province. The Chinatowns developed in the port cities of Victoria, New Westminster, and Nanaimo, and in some large gold-mining towns such as Barkerville and Yale, closely followed the pattern of the stage-development model, particularly in the budding stage.

During the restricted and exclusion periods, Chinese immigrants followed the transcontinental Canadian Pacific Railway (CPR) to the prairies and the eastern provinces; subsequently Chinatowns were set up across Canada in different locations, with different rates of growth, and in different stages of development. In the early 1900s, for example, the Chinatowns in Victoria, Vancouver, and New Westminster were blooming but the Chinatowns in many gold-mining towns in British Columbia were withering or dying, whereas the Chinatowns in Moose Jaw and Montreal were still in their formative years (Figure 3). Although the Chinatowns outside British Columbia were different from those in British Columbia during the budding stage, they shared many common features with the latter in the blooming and subsequent stages. After the 1930s, all Canadian Chinatowns were in the withering stage as their population and economy continued to decline.

During the period of selective entry, new types of Chinese immigrants from Hong Kong, Taiwan, and Southeast Asian countries came to Canada; these new immigrants recast the Chinese-Canadian society. Some Old Chinatowns completed the four-stage life-cycle and passed away, and some Old Chinatowns were encroached upon by redevelopment and transportation projects, but some other Old Chinatowns were revitalized, their characteristics conforming to the reviving stage of the model. New Chinatowns were created in the suburbs of metropolitan cities, and in cities where no Chinatowns had ever been formed, or Replaced Chinatowns were built to replace

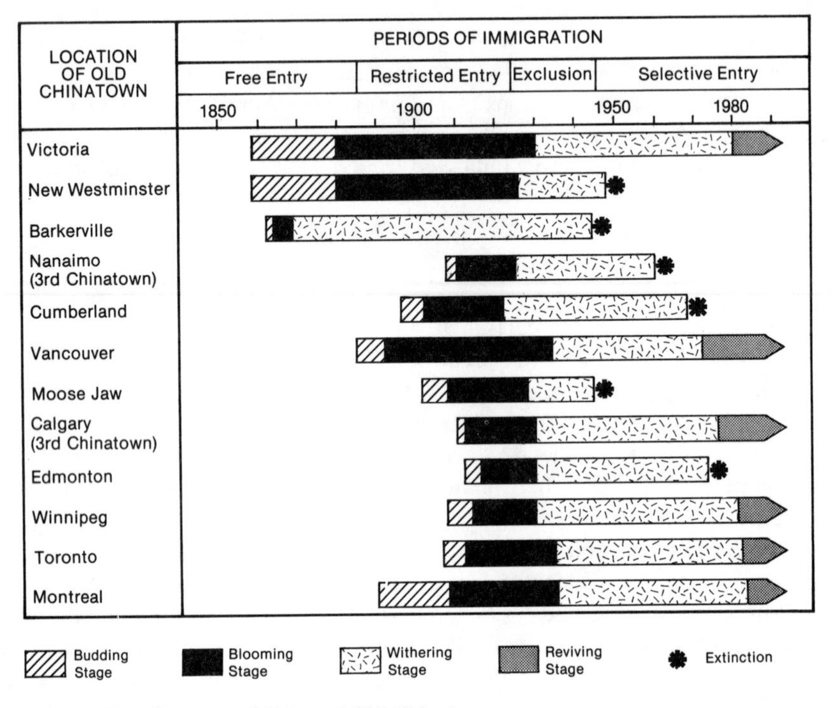

3 Stage-Development of Selected Old Chinatowns

Old Chinatowns. The stage-development model is not applicable to New or Replaced Chinatowns because their development is completely different from the historical growth of Old Chinatowns.

VICTORIA'S CHINATOWN

Part Two of this book focuses on the Chinese community as it developed in the city of Victoria, examining its Chinatown more thoroughly as a role model in comparison with other Canadian Chinatowns. Victoria's Chinatown, established in 1858, was chosen for a detailed study for three main reasons. First, it is the oldest Chinatown in Canada. Second, it is the only Canadian Chinatown that has a complete history of a functioning Chinatown from the first date of Chinese arrival in Canada to the present, illustrating many features of the evolutionary process of the stage-development model. The last reason is that I live in Victoria and have studied its Chinese community and Chinatown for nearly two decades and am deeply involved in Chinatown community projects and revitalization programs. My experience, close contacts with Chinatown community leaders, and access to the archives of the Chinese Consolidated Benevolent Association of Victoria enabled me to glean much information about Victoria's Chinatown hitherto unknown to outsiders.

Like other Chinatowns, the common evolutionary pattern of Victoria's Chinatown began with a budding stage in the late 1850s and 1860s, during which the few Chinese residents and businesses were confined to Cormorant Street, an unpaved street on the northern fringe of Victoria. The blooming stage was marked by the increase of population and socioeconomic activities after the 1880s. As Victoria's Chinatown expanded, its morphology gradually changed from a linear to a reticulated pattern, and by 1901 its Chinese population had reached nearly 3,000. For fifty years, Victoria had the largest Chinatown in Canada. Its Chinatown reached the zenith of its development in the 1910s when it was surpassed in size by the much more rapidly growing Chinatown in Vancouver. After the 1920s Victoria's Chinatown entered the withering stage as its residential population began to decline and its socioeconomic activities to abate. Nevertheless, Victoria still had the second largest Chinese population in Canada until it was surpassed by Toronto during the 1940s, then by Montreal in the 1950s. By the 1960s, other cities such as Calgary and Edmonton had begun to attract many new Chinese immigrants; soon they also surpassed Victoria in the number of Chinese residents. Like other Old Chinatowns, Victoria's Chinatown was depopulated by suburbanization and reduced in size by downtown urban renewal; it was approaching extinction. In the early 1980s the City of Victoria adopted my public involvement model of community planning, which since has been applied successfully to rehabilitation of Victoria's Chinatown. This model of community planning may also be applied to revitalization of other Old Chinatowns in North America.

The evolution of Old Chinatowns in Canada varied in time and space, but they tend to have followed a common growth cycle, experiencing a budding, blooming, withering and dying, or reviving stage. This "stage hypothesis" is tested by the application of a four-stage development model to Victoria's Chinatown. Similar detailed studies of the Chinatowns in Vancouver, Toronto, Montreal, and some other cities will show that most major historic Chinatowns in Canada can be charted in a growth pattern similar to that of Victoria's Chinatown, and in many ways conform to the stage-development model.

PART ONE

Canadian Chinatowns

2

Entry without Restriction

The first wave of Chinese migration to the two British colonies of Vancouver Island and British Columbia was associated with the Fraser River gold rushes of the 1850s and 1860s. After the rushes were over, economic recession set in and both colonies were in debt. In an attempt to reduce government expenses, they were amalgamated in 1866 as the united Colony of British Columbia. As unemployment rose, Chinese labourers became scapegoats, and demands for their restricted entry increased and discriminatory laws and regulations against them were instituted. In 1871 the colony entered Confederation as a province of Canada, on the condition that a transcontinental railway be built to link it with the rest of the country, and this led to the second wave of Chinese immigrants to British Columbia: labourers for railway construction. The simmering anti-Chinese feeling began to boil when thousands of Chinese labourers entered the province.

ORGANIZED MIGRATION

After the Chinese merchants in San Francisco heard rumours of the gold discovery in the lower Fraser River, they sent a scout to the Fraser region to ascertain the truth. He returned in May 1858 and assured them of the "marvellous richness of the gold mines of that region," starting the Chinese migration from California to British Columbia.[1] On 24 June 1858, Hop Kee & Company of San Francisco signed a contract with Allan Lowe & Company, a shipping agent, agreeing to pay the latter $3,500 to ship 300 Chinese and fifty tons of merchandise to Victoria, with twenty dollars for each additional passenger.[2] In those days, it took about thirteen days for a boat to travel from San Francisco to Victoria. Throughout the summer and fall, scores of Chinese continued to arrive at Victoria to join the Fraser River gold rush. Rumours about Chinese immigrants abounded, including one that a Chinese company in San Francisco would bring 2,000 Chinese immigrants to British Columbia in the spring of 1859, and another that Chinese companies in San

Francisco were planning to bring 7,000 Chinese into British Columbia.[3] These rumours tended to exaggerate the number of Chinese people migrating to British Columbia. Most of the early Chinese immigrants were employees of large Chinese companies in San Francisco and came as an organized team.

CHAIN MIGRATION

Chain migration is a process by which "prospective migrants learn of opportunities, are provided with transportation, and have initial accommodation and employment arranged by means of primary special relationships with previous migrants."[4] When some Chinese miners struck pay dirt in British Columbia, they immediately returned to their home villages in China, bringing with them the exciting news of another *Gim Shan* (Gold Mountains) in North America. In 1859, for example, a local newspaper correspondent interviewed some returning Chinese and wrote:

> The American bark *Sea Nymph Stege* sailed yesterday for Hong Kong... with a cargo of lumber and twelve Chinamen, homeward bound passengers from this port. These passengers are stated to have had each from $400 to $500, the proceeds of their labour in the mines of B.C. One of the above-mentioned happy-go-lucky Johns, informed us that with the $500 in cash, he could live in China for five or six years. . . . Two of the Johns intended returning in the spring, bringing their families of female Johns and demi-Johns with them and they all speak of a rush from China to this country is sure to occur, during the months of February, March and April of next year.[5]

When the returned Chinese arrived home, they would help their relatives and friends migrate to *Gim Shan*. Once the latter had settled down in British Columbia, they then would help their own relatives and friends to emigrate. Thus, throughout the early 1860s, clipper ships—notably *Hebe, Lawson, Leonidas,* and *Frigate Bird*—brought hundreds of Chinese immigrants directly to Victoria from Hong Kong or China. In those days, shipping agents usually arranged large group sailings so that the Chinese passengers would pay a lower fare and at the same time would have the benefit of companionship and mutual help during the long voyage. The fare was about thirty dollars, and one meal per day was provided on board.

The trip across the Pacific to Canada in those days was quite hazardous. In December 1859, for example, the brig *Esna* was attacked by pirates in the South China Sea while it was en route to Victoria, and the crew was imprisoned.[6] In the same month, *Lady Inglis* was lost on her voyage from China to Canada. In April 1860, after sixty-two days at sea, the Norwegian ship *Hebe* brought the first shipload of 265 Chinese passengers from Hong Kong to Victoria, including one female.[7] However, two months later, *Lawson* from China arrived at Victoria with only sixty-eight passengers, although its list called for 280; the missing passengers could not be accounted for.[8] The captain was later charged with having no clearance papers and no medicine chest on the ship. In another case, 316 Chinese labourers recruited by a Chinese employment agency were placed on board *Maria;* each labourer had to pay thirty-four dollars to the agent for his passage from Hong Kong to Victoria, although the fare charged by the ship was only twenty dollars.[9] During the entire sixty-day journey across the Pacific, the labourers were given tea only twice. One pound of rice per day was supposed to be shared by four people, but it was in fact divided among ten. Many of the labourers were neither fed nor clothed properly, and they suffered from malnutrition and starvation on their way to Canada. On one occasion, a local newspaper reporter observed that "considerable distress is manifesting itself among the recently arrived Chinese immigrants, some of whom are reported to be actually starving."[10] In spite of the hazardous voyage, however, many Chinese men travelled across the Pacific to *Gim Shan* to seek their fortunes.

SOURCE AREAS OF CHINESE IMMIGRANTS

A study of the home county origins of 5,000 Chinese in British Columbia in 1884–5 reveals that most of them came from fourteen counties in Guangdong Province of South China.[11] Nearly 64 per cent of them came from the counties of Taishan, Kaiping, Xinhui, and Enping, collectively known as Siyi (the Four Counties) because their dialects are similar to one another (Figure 4). About 18 per cent came from the counties of Panyu, Shunde, and Nanhai. They are known as Sanyi (the Three Counties), and their dialects are very similar to the Cantonese dialect. About 8 per cent came from Heshan and Zhongshan counties and 6 per cent from the counties of Zengcheng, Dongguan, and Baoan on the eastern side of the Zhujiang (Pearl River). The remaining 4 per cent came from other counties in Guangdong Province. From within Siyi, the Taishan people outnumbered those from all other counties, constituting over one-fifth of the 5,000 people.

The 5,000 Chinese can be classified into 129 clans according to their surnames. Classification reveals that eight large clans with 170 persons each accounted for half the total population in British Columbia (Table 1). These clans had come from one or two counties. For example, 76 per cent of those bearing the surname of Zhou were from Kaiping County. Similarly, most of those with the surnames Li, Huang, Lin, or Ma were from Taishan County,

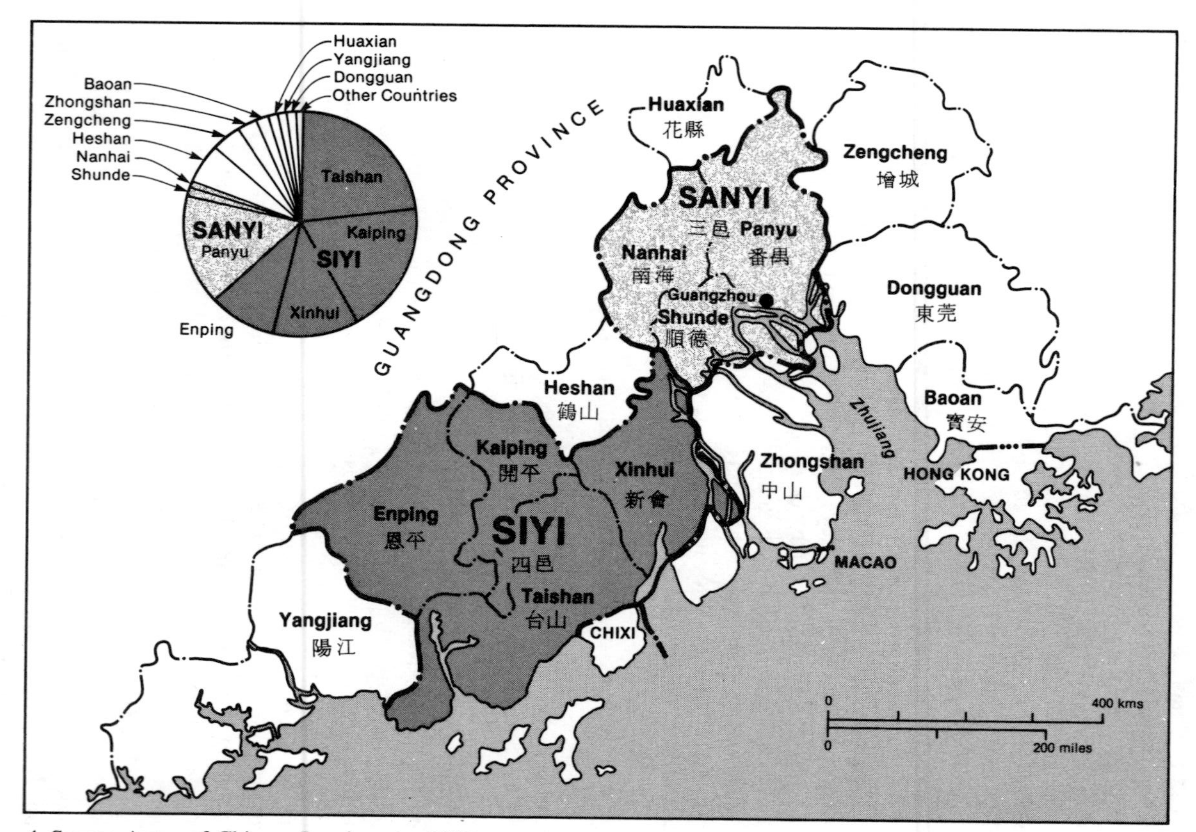

4 Source Areas of Chinese Immigrants, 1880s

TABLE 1: Home county and clan origins of Chinese in British Columbia, 1884–5

County	Zhou	Li	Huang	Chen	Lin	Liang	Xie	Ma	Others	Total
	Clan									
Taishan	0	219	119	64	122	25	0	163	446	1,158
Kaiping	408	24	56	0	0	35	82	0	344	949
Xinhui	38	123	61	33	40	21	0	0	299	615
Enping	0	22	19	39	0	32	18	0	361	491
Panyu	55	39	56	34	13	22	66	0	513	798
Heshan	0	48	41	12	45	31	0	0	125	302
Zengcheng	0	0	27	14	0	0	0	0	154	195
Zhongshan	0	13	0	0	0	0	0	0	98	111
Others	33	35	36	54	15	39	27	8	190	437
Total	534	523	415	250	235	205	193	171	2,530	5,056

Sources: stubs of donation receipts dated 1884 and 1885, Chinese Consolidated Benevolent Association of Victoria

and over two-thirds of the Xies originated from Kaiping and Panyu counties. It can be deduced that many early Chinese immigrants in Canada were part of extended family groups. This reflects the process of a chain migration.

THE ANTI-EMIGRATION POLICY

Early Chinese emigrants had to face opposition from two governments: that of the receiving country and their home government. After the Manchu overthrew the Ming Dynasty and set up the Qing Dynasty in China in 1644, many Ming royalists fled from North China to South China and thence to Formosa or Southeast Asia. Emperor Kangxi decreed that any Chinese who left China must either be disaffected with the Manchu government or be communicating with the anti-Manchu forces led by Koxinga. Accordingly, in 1712 he issued an edict to prohibit emigration. Anyone who had left for foreign islands would be beheaded upon his return to China.[12] Officials allowing emigration also would be punished by decapitation. This anti-emigration policy was relaxed slightly by Emperor Kangxi's successor, Emperor Yongzheng, who permitted coastal inhabitants to go abroad to trade provided that they first obtain a licence and subsequently return to China. Those who left China without a licence would not be permitted to return.

The Manchu government regarded emigrants as deserters from their motherland and did not care about their safety and livelihood abroad. For example, there was little concern about the massacre of over 10,000 Chinese in Java by the Dutch in 1740 because these emigrants "had quitted their native country and abandoned the tombs of their ancestors."[13] The Manchu government continued to oppose emigration until it signed the Burlingame Treaty with the United States in 1868. This treaty recognized the right of the Chinese people to leave their homeland and change their allegiance.[14] Similar agreements were signed later with Britain, France, and other European powers who were anxious to obtain cheap Chinese labour for their colonies. Eventually, in 1893, the imperial edict of 1712 was repealed; six years later an edict was issued by Emperor Guangxu ordering Manchu officials to grant emigrants protection wherever possible.[15]

PRESSURE FOR EMIGRATION

The anti-emigration policy of the Manchu government could not deter the out-migration; there was too much incentive to leave the unemployment, hunger, and banditry at home and to take advantage of economic incentives and opportunities abroad. In China, particularly in rural areas, life was truly miserable—there was little food and there were far too many people. During the early nineteenth century, for example, Taishan County in Guangdong Province had a population of over half a million. Only 400 sq. km of the county's total area of 4,600 sq. km were suitable for cultivation; the rest of the county consisted of sandy hillocks. There were 1,300 persons for each square kilometre of arable land. Food production was insufficient and many people were hungry and unemployed. They had little choice but to become land bandits or pirates or leave their home villages and live elsewhere. With the help of relatives and friends abroad, lucky ones could go to the United States and Canada.

Most early Chinese emigrants to North America had little or no knowledge of this new land, except that gold nuggets were easily found, jobs were plentiful, and wages were high in *Gim Shan*. The Chinese knew little about the unpleasant trans-Pacific voyage, nor did they know about Western people and their culture. Many Chinese emigrants had to borrow money from relatives or friends to pay for their passage to British Columbia. Upon arriving at *Gim Shan,* they would work hard and save their money to pay off their loans and support their families in China. If their passages were paid by an employment company, they would work for the company for low wages, from which they would slowly repay their fare and family loans and pay for room and board. During the 1860s, their monthly wage amounted to only thirty dollars or less. It could, therefore, take more than a decade to pay off their debts. Burdened with debts and separated from their loved ones in China, the early immigrants had only two goals: to save as much as possible, and to return to China as soon as possible. Therefore they were regarded by the host community as sojourners, not settlers, although they might have spent their entire youth in British Columbia, contributing their labour to its development.

MIGRATION UP THE FRASER, 1850s–1860s

In the summer of 1858, thousands of miners, including a few hundred Chinese, joined in the Fraser River gold rush by boat from San Francisco through Victoria or by overland routes from the Washington Territory. Soon more Chinese gold-seekers came directly by ship from Hong Kong. Generally speaking, this Chinese migration on the mainland followed two main directions. One stream of migration ascended the Fraser River and dispersed to the northern frontier region. The other stream followed the Dewdney Trail from Hope and spread out eastward to the valleys of the Similkameen, Rock Creek, and Kootenay in the border country (Figure 5). The Chinese often followed on the heels of the white miners, partly because they were not as adventurous but mostly because they dared not compete with the hostile white miners.

By the fall of 1858, thousands of gold-seekers had entered British Columbia and were panning for gold in river bars and bottom flats in the lower Fraser Valley between Hope and Yale. Some prospectors discovered that the further they went up the Fraser, the coarser the gold particles they found; many immediately abandoned their claims and rushed upstream. The British colonial government saw the need to cut trails through the dense forests to develop the upper Fraser Valley, but there were not enough labourers available to do this kind of job. Most of the white workers were interested in seeking quick fortunes in gold mining; unless driven by sheer necessity, they would spurn slow wage-earning jobs in the public service. Accordingly, the colonial government had to rely on Chinese labourers for such back-breaking jobs as clearing the forest, building trails, and digging. In 1858, for example, the government employed many Chinese labourers to build the Douglas-Lillooet trail, which was an important route from the Fraser delta to Lillooet in the middle Fraser Valley. While the trail was still under construction, many white gold-seekers had taken the shorter route by ascending the formidable Fraser Canyon beyond Yale and going upstream to Lillooet. In 1860, the colonial government employed many Chinese labourers and native Indians to widen the Douglas-Lillooet trail so it could be used by wagons; Chinese labourers also built a trail from Yale up the formidable Fraser Canyon. An undetermined number of them were killed during the construction, usually by falling off the cliffs which plunged down to the treacherous Fraser River.

By 1859, most of the white miners had deserted the lower Fraser Valley between Hope and Yale; the mining population there dropped from 5,000 to 600, of whom 500 were Chinese working on bars abandoned by the white miners.[16] By 1861, most miners in the Fraser Valley were Chinese; they were making only a few dollars a day on the abandoned bars.[17] For example, nearly 90 per cent of the 530 miners at Yale and about 60 per cent of the 210 miners

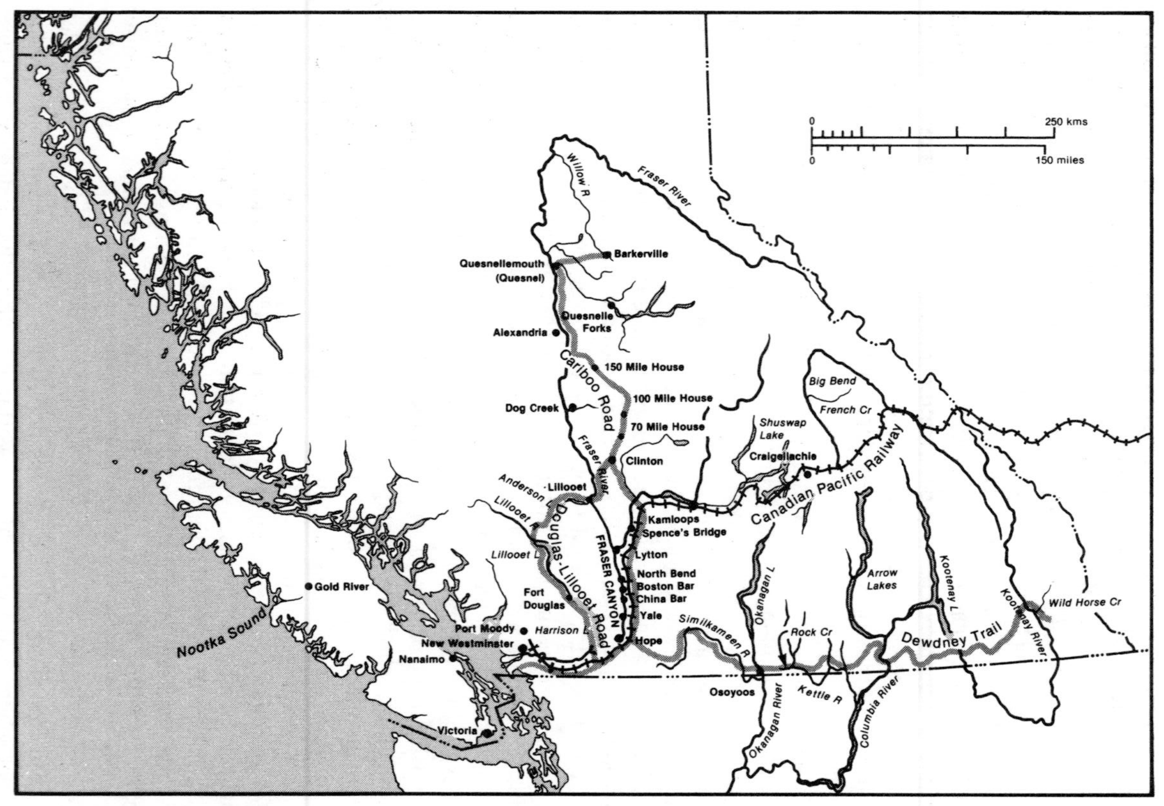

5 Migration Routes of Chinese Immigrants during the Gold Rushes

at Lytton were Chinese.[18] About 300 Chinese miners were concentrated on Tranquille Creek, which drains into Kamloops Lake. The white miners, on the other hand, had gone up the Quesnel River and discovered the fabulously rich deposits of the Keithley, Cunningham, Antler, Lightning, Williams, and other Cariboo creeks.[19] These finds sparked off the Cariboo Rush of 1862. Within two years, the Cariboo region's mining population soared to 5,000, of which over one-third was concentrated at Williams Creek, where Barkerville, the hub city of the Cariboo, was established.[20] To improve the transportation between the Fraser delta and the Cariboo region, the government let out a contract in May 1862 to construct the 393-km (244-mile) wagon road from Lillooet to Alexandria. G.B. Wright, the contractor, found it nearly impossible to obtain white labourers; most of them had gone to the Cariboo to look for gold. The native Indians who might have supplied some labour were suffering from a severe epidemic of smallpox, which eventually claimed the lives of about 20,000 people—one-third of the total native population of British Columbia.[21] Having failed to get enough native and white labour, Wright had to rely on his 1,500 Chinese labourers to finish the work.

While the Lillooet-Alexandria wagon road was being built, the government decided on another route from Yale up the Fraser to Lytton, then up the Thompson River, a tributary of the Fraser. This route later came to be known as the Cariboo Road. Once again the colonial government had to rely on Chinese labourers to do the work. The widening of the old trail into a wagon road through the Fraser Canyon was the most hazardous job; near Chapman Bar north of Yale, for example, the road had to be sliced into the cliffs and supported by wooden bridges spanning deep gullies. Landslides frequently occurred after heavy rains, and casualties were not uncommon. At one place named China Bluff after the large Chinese population there, the road had to be strung out on a wooden trestle to get around the bluff.

By the summer of 1863, the wagon road had been pushed through to Clinton, where it joined the Lillooet-Alexandria wagon road. In the following two years, the Cariboo Road was extended from Alexandria to Barkerville. By this time, shallow diggings in the Cariboo region had been exhausted and the returns were low. Many white labourers quit mining and joined Chinese labourers in road construction. By 1866, the Cariboo region had been deserted by most of the white miners, and only a few hundred Chinese miners still remained to pan for whatever small amount of leavings they could find in the creeks. Meanwhile, a large number of Chinese labourers had been recruited by the Collins Telegraph Line Company to cut a trail from New Westminster via Quesnellemouth (Quesnel) as far north as Fort Stager, a supply depot near Hazelton on the Skeena River. These workers were laid off in 1866 when the company abandoned the project.[22] The Chinese labourers did not return to the Skeena region until the salmon fishing industry developed there in the early 1880s.

MIGRATION ALONG THE BORDER, 1850s–1860s

The other direction of Chinese diffusion during the gold rushes was along the southern border of British Columbia. In 1860, some white miners discovered gold in the Similkameen River and Rock Creek, and the colonial government immediately employed Chinese labourers to open the Dewdney Trail over the forested mountains east of Hope to the Similkameen River.[23] The rushes to these places were short-lived though, as miners hurried to the Cariboo Rush in 1861. When profitable strikes were made in Wild Horse Creek in 1864, Chinese labourers were again employed to extend the trail to the Kootenay River. The following year, the mining population of Wild Horse Creek dropped abruptly from several hundred to barely one hundred because of the glowing reports of gold discoveries in French Creek near Big Bend on the Columbia River. Many Chinese miners who had been working on Tranquille Creek and the North Thompson River also joined the Big Bend rush. But again, in less than a year, the Big Bend area was deserted after its surface diggings were exhausted. However, the rushes to the Columbia and Kootenay rivers had brought many Chinese into the southeastern part of British Columbia. In early May 1865, for example, three vessels from China brought a total of 1,185 Chinese to Victoria; by the end of the month, hundreds of them were on their way to the Kootenay region.[24] Although the border area had been deserted by the white miners in the late 1860s, several hundred Chinese miners remained behind to work patiently on the abandoned claims.

POPULATION DISTRIBUTION DURING THE RUSHES

The colonial government had no way of making a census of the population in British Columbia because of the miners' great mobility.[25] Estimates of the Chinese population fluctuated not only annually but also seasonally and even monthly. In the summer of 1861, for example, nearly 4,000 Chinese miners were estimated to be working on the mainland, but towards the end of the year nearly half of them, dreading the severe winter, had left for California or China.[26] As many Chinese labourers worked on trails and wagon roads, their whereabouts were determined by where construction work was located. Similarly, those Chinese who were employed to operate mule and horse teams up and down the Cariboo Road were impossible to enumerate. Nevertheless, a computation of the estimates in the *Blue Books of Statistics* revealed that the Chinese people tended to concentrate in a few regions. Unlike white miners, who rushed from one place to another to seek bigger deposits, Chinese miners did not desert a place immediately at the next rumour of better diggings. During the second half of the 1860s, about 86 per

TABLE 2: Annual average of Chinese population in British Columbia, 1867–70

Location	Population
VANCOUVER ISLAND	
Victoria City	201
Victoria District	23
Esquimalt	4
Nanaimo	27
MAINLAND	
Cariboo Region	740
Lower Fraser Region (Hope to Lytton)	430
Lillooet	198
Kootenay River	173
Osoyoos	57
New Westminster	45
Columbia River	2
Total	1,900

Source: United Kingdom, *Blue Books of Statistics: Colony of British Columbia* (London: Colonial Office 1867), 140; (1868), 142; (1869), 150; (1870), 135–6.

cent of the 1,900 Chinese lived on the mainland; most of them were found in the Cariboo region and the lower Fraser between Hope and Lytton (Table 2). However, towards the end of the 1870s only 33 per cent of the Chinese people still remained in various towns and gold-bearing creeks on the mainland, whereas 57 per cent had settled down in the capital city of Victoria and 11 per cent in the coal-mining town of Nanaimo (Table 3). Although over half the population were engaged in gold-mining, coal-mining, and fishing, many had taken up other occupations as cooks, domestic servants, store employees, farm labourers, and the like (Table 4).

INITIAL REACTIONS OF THE HOST COMMUNITY

During the early years of the gold rush, there was little obvious discrimination against Chinese people. As far as the British government was concerned, Chinese immigrants were allowed the same rights, liberties, and privileges as other immigrants. But when they arrived in Victoria in 1858, they were received with mixed feelings by the predominantly English community. Some whites looked upon the Chinese with curiosity, some with prejudice, some with condescension, and some with spite. Leonard McClure, who had been a journalist in the Californian mining districts, came to Victoria and started the *Victoria Gazette* in June 1858. In March 1859, his editorial alerted the public for the first time to the threat of the Chinese influx into British Columbia, and

TABLE 3: Distribution of Chinese in British Columbia, 1879

Location	Number of persons			
	Male	Female	Total	% of total
Victoria	2,300	70	2,370	57
Nanaimo	435	2	437	11
Quesnelle Forks	375	10	385	9
New Westminster	300	10	310	7
Quesnellemouth (Quesnel)	224	10	234	6
Barkerville	149	10	159	4
Lightning Creek	93	5	98	2
Wild Horse Creek	80	1	81	2
Yale	41	4	45	1
Kamloops	33	2	35	1
Total	4,030	124	4,154	100

Sources: compiled from reports submitted to T.B. Humphries, Provincial Secretary of British Columbia, by various polling district officers, Oct. 1979–Feb. 1880

TABLE 4: Occupations of Chinese in British Columbia, 1879

Occupation	Number of persons	% of total
Gold miners, fishermen	1,100	26.5
Gold miners	874	21.0
Cooks, servants	434	10.5
Storekeepers, employees	383	9.2
Coal miners	347	8.4
Gardeners, farm labourers	125	3.0
Washermen	121	2.9
Tailors	61	1.5
Barbers	29	0.7
Butchers	10	0.2
Females, various occupations	44	1.0
Other labourers	286	6.9
Unspecified	340	8.2
Total	4,154	100.0

Sources: compiled from reports submitted to T.B. Humphries, Provincial Secretary of British Columbia, by various polling district officers, Oct. 1879–Feb. 1880 he questioned the British government's attitude about "this peculiar class of newcomers."[27] The editorial received little public attention, however, because most people were too occupied with the gold rush and felt that the fears of the Chinese influx were merely the phantoms of a few nervous and ill-informed individuals. At a public meeting on 5 March 1860, for example, there was a suggestion to impose a head tax on all Chinese entering the colony of Vancouver Island.[28] After the Chinese merchants in Victoria learned of the discussion at the meeting, they immediately called upon Governor James Douglas, who assured them that he had no plan to impose a tax on the Chinese.[29] Amor De Cosmos, editor of the *British Colonist,* another newspaper in Victoria, opposed the head tax; he felt that it would work against Victoria's interests since trade would suffer from a lack of customers. Chinese merchants in Victoria would help promote the city's economic prosperity and Chinese labourers would solve the labour problem; thus they should "have the same immunities from taxation as subjects of any other nation."[30] The viewpoint of De Cosmos should not be interpreted as support for egalitarianism; he also clearly stated that a heavy head tax would repel the Chinese labourers, whom "we cannot at present afford to dispense with. When the time arrives that we can dispense with them, we will heartily second a check to their immigration."[31] Indeed in later years De Cosmos did fight against Chinese immigration.

ECONOMIC RECESSION AFTER THE RUSHES

Prejudice against an ethnic minority by an ethnic majority usually does not develop in isolation from the socioeconomic circumstances of the time.[32] When British Columbia was prosperous during the height of the gold rushes, Chinese labourers were much in demand; prejudice against them was minimal. After 1864, gold-mining with a sluice and rockers was gradually replaced by placer-mining, which required more capital investment. Many free miners found it difficult to raise funds for the heavy capital expenditure, particularly after the failure in November 1864 of Macdonald's Bank, one of the largest banks in British Columbia during the gold rush era.[33] Victoria's economy was stagnant; it benefited little from the short-lived Leech and Sooke gold rushes in the fall of 1864. The economic recession was aggravated further the following year when the usual spring rush from San Francisco did not materialize, and many merchants with overextended credit and unsaleable merchandise went bankrupt. By 1866, the United Colony of British Columbia had a debt of nearly £1.3 million.[34]

The adverse economic conditions in British Columbia continued to deteriorate after the second half of the 1860s. Unable to obtain employment in restaurants, laundries, and market-gardens, Chinese labourers began to look for work on ranches, farms, or in coal mines. As a result, white workers soon felt that their livelihood was being threatened. Although there were two short-lived gold rushes on the mainland during the 1870s (the Omineca Rush in 1870–1 and the Cassiar Rush in 1874–5), the thousands of miners, hundreds of them Chinese, left within a year after the returns from diggings plummeted. By the late 1870s, most Chinese labourers and miners had drifted back to the Fraser delta and to Vancouver Island, where they looked for other occupations, accepting lower wages and longer hours. Those who relied on cheap labour felt that Chinese labourers were beneficial to the economy of British Columbia; however, this view was not shared by the white labouring class, which regarded Chinese workers as a detriment to their well-being.

DISCRIMINATION AGAINST THE CHINESE

After British Columbia entered Confederation in 1871, the Chinese question became a political issue for the province. Editorials, feature articles, and letters to the editor were published in the two Victoria newspapers, the *Colonist* and the *Standard*.[35] As well as providing international, national, and local news, these two newspapers also recorded discussions at public meetings and details of the debates of the city council and the Legislative Assembly. Since the two newspapers were widely read in the province, their comments on the issue of Chinese labour had a profound effect on the formation of public opinion about the Chinese people themselves. John Robson, editor of the *Colonist,* had a strong prejudice against the Chinese labourers, whereas Amor De Cosmos, editor of the *Standard,* considered cheap Chinese labour as an asset to the City of Victoria as well as to the province. For example, after the city council passed a bylaw in 1875 to include, in all contracts for city works, a clause that white labour only could be employed, the *Standard* immediately criticized the bylaw, the *Colonist* congratulated the council for lessening the Chinese evil and setting an example to other municipalities.[36] Three years later, the Legislative Assembly passed a resolution to exclude Chinese people from provincial works as well.[37] From the 1870s onwards, no politician in British Columbia received support from white workers if he did not speak against Chinese labour.

In 1872, the Legislative Assembly passed an Act to amend the Qualification and Registration of Voters Act, which disenfranchised both the Chinese and the native Indians.[38] The amendment act was replaced later by an Act to Make Better Provision for the Qualification and Registration of Voters, under which no Chinese or Indian would be eligible to vote in any Legislative Assembly election.[39] The act received royal assent on 22 April 1875. On 1 June 1875 the Collectors of Votes removed all Chinese from their voters' lists.[40] The following year, an Act to Amend the Municipal Act of 1872 was passed by which "no Chinese or Indians shall be entitled to vote at any Municipal Election for the election of a Mayor or Councillor."[41] From the 1870s onward, discrimination against the Chinese became a hallmark of the white citizens of British Columbia.

ANTI-CHINESE ORGANIZATIONS

The first mention of a formal anti-Chinese organization in Victoria appeared in the *Colonist* of 16 May 1873. It reported that the Anti-Chinese Society had passed resolutions demanding revision of the Sino-British Treaty of 1860, which permitted the immigration of Chinese to Canada, as well as the withdrawal of the subsidy from the Chinese steamships and the introduction of a law that would forbid men to work more than eight hours a day. After this first report, no further news about the society was published. It was probably short-lived. However, British Columbians were always bitter that the federal government did not fully understand their Chinese problem. Throughout the so-called "Fight Ottawa" period between 1874 and 1883, the province pressured the federal government to address the Chinese immigration issue.[42]

During the federal election in 1878, a better organized anti-Chinese organization known as the Workingmen's Protective Association emerged. Its objectives were "the mutual protection of the working classes of British Columbia against the influx of Chinese, and the use of legitimate means for the suppression of their immigration."[43] All persons, male or female, of any nationality, creed, or colour, except Chinese, might become a member of the association; each member solemnly pledged not to aid, abet, or patronize the Chinese or those employing them, and to use all the legitimate means possible to prevent their patronage. Noah Shakespeare was president of the association and probably the driving force behind its establishment. Playing upon the racial prejudice of white workers, he gained publicity by attacking Chinese labourers. Portraying himself as the arch-enemy of the Chinese, Shakespeare gained the votes of local white workers and eventually achieved his political aspirations. He became a city councillor in 1874, then mayor of Victoria in 1882, and eventually an MP for Victoria in 1887.[44]

One of the association's tactics was to put pressure on political candidates during municipal, provincial, and federal elections. For example, on 30 September 1878, the association organized a public meeting for the candidates to express their views on the Chinese question and made it clear at the meeting that only anti-Chinese politicians would receive its support.[45] At another meeting, the names of firms and families employing Chinese were read in order to "disgrace the Chinese employers" and housewives were advised not to employ or patronize the Chinese.[46] As a result, may businessmen and politicians publicly denounced Chinese workers, although they hired them secretly.[47]

The Workingmen's Protective Association was also very active in protesting against the employment of Chinese in construction of the CPR. In 1879, the association was reorganized and renamed the Anti-Chinese Association, with "No Surrender" as its motto; the association would not yield to Prime Minister John A. Macdonald's decision to permit Chinese labourers to do the railway work.[48] Shakespeare as president of the association declared that "he would rather vote against the construction of the transcontinental railway than see Chinese employed on it."[49] In June 1885, the association sent a letter to the Chinese merchants in the city, threatening them to send the Chinese labourers back to China:

> You have imported thousands of coolies, virtually slaves to you. . . . With these men you flood our labour market, cutting down wages in every department in which our men, women and children seek to earn a living.
>
> . . . Ship your coolies hence now, whilst you are given the chance. [You are] inimical to our institutions, laws and form of government, defiant of our courts, subtle at evading taxation, evading our sanitary laws, [and] unwilling to assist in protecting our property from fire. . . . The time has come, and now is for such an undesirable people to be banished from our land.
>
> . . . Take our advice and leave whilst you have the chance of doing so peacefully before an outraged people treat you as your conduct so deserves.
>
> . . . If you oblige us to use force to rid ourselves of this scourge, the blame will rest on your own heads.[50]

Although the forceful evacuation did not happen, the Chinese people felt that the anti-Chinese movement was gaining momentum and that the restriction of Chinese immigration would be imposed sooner or later.

CHINESE TAX ACT OF 1878

The legislature of Queensland, Australia, passed an act which obtained royal assent in 1877 to impose a £10 head tax on each Chinese immigrant.[51] This approval immediately prompted the British Columbia provincial government to follow suit. Accordingly, on 2 September 1878, the Legislative Assembly passed the bill known as the Chinese Tax Act, by which every Chinese person over twelve years of age would pay ten dollars every three months to take out a licence to reside in British Columbia. In other words, every Chinese in the province had to pay forty dollars head tax a year. Any Chinese without the licence would be liable to a $100 fine and, if in default of immediate payment of the fine, the tax would be levied by sale and distress of his goods and chattels.[52] The following month, Mr. Justice Gray of the Supreme Court of British Columbia declared the act unconstitutional and void.[53] He explained that Queensland legislated solely for itself and was constitutionally on the same footing as the Dominion of Canada in its relative position towards the British Empire. On the contrary, British Columbia was not autonomous and was restrained by the federal compact which governed the Dominion. Only Ottawa had the power to pass an act concerning immigration to the country.

After the Chinese Tax Act was declared unconstitutional, the British Columbia Legislature established a committee to consider how to stop Chinese immigration. On 28 March 1879, the committee recommended that since the provincial legislature had no power to impose a tax exclusively on Chinese people, a grievance should be sent to the Dominion government, setting forth the baneful effects of the Chinese presence and the necessity of adopting effective measures to prevent their further immigration to the province.[54] This grievance was subsequently drafted and sent to Ottawa. In May 1879, the House of Commons established a Select Committee to study the Chinese question. It suggested that Chinese immigration should not be encouraged and that Chinese labour should not be employed on Dominion public works. Although no specific action was taken on its recommendations, white workers and politicians in British Columbia were at least temporarily appeased. The Chinese received no further publicity in the press until the call for tenders for construction of the CPR was announced in October 1879.

INFLUX OF RAILWAY LABOURERS, 1880–4

The second wave of Chinese immigration to Canada began with the construction of the CPR in British Columbia. Andrew Onderdonk, who had most of the contracts to build various sections of the line in the province, could not get enough white workers. In 1880, the white population of the province was around 35,000; since he would need at least 10,000 strong men to build the railways, Onderdonk had to look elsewhere for much of his labour force.[55] At first, he recruited only white workers from San Francisco, but he was very disappointed in their performance—most were unemployed clerks or bartenders, and some had never handled a spade or a pick before.[56] Furthermore, Onderdonk accepted contracts with the lowest bids; by employing Chinese labourers, he reduced his labour cost by about four million dollars. Without the Chinese, he would have been bankrupt.[57] In 1880, he asked the Lee Chuck Company to recruit 1,500 Chinese labourers from the United States who had worked on railway construction in California and Oregon. He found them cheaper, more reliable, and harder working than white labourers. As he required more Chinese labourers, and the Lee Chuck Company could not find enough in North America, Onderdonk asked the company to recruit about 2,000 Chinese labourers from Hong Kong. In the spring of 1881, he chartered two ships to bring them to Canada. The passage across the Pacific was long and rough, and the men were kept below decks with the hatches closed. They received no fresh vegetables or fruit, and a large number developed scurvy. Nearly 10 per cent died soon after their arrival in British Columbia, and additional labourers had to be recruited in order to meet the labour shortage.[58] By 1882, the total labour force on the railway had reached about 9,000 men, of whom 6,500 were Chinese and 2,500 white. Many of the new Chinese labourers could not adjust to the weather in Canada and were sick or dying. Accordingly, in June 1883, a group of concerned Chinese merchants in Victoria sent a despatch to the employment agents in China, advising them to stop bringing Chinese people to British Columbia.[59] But the despatch was disregarded. Between January 1881 and July 1884, vessels from China, San Francisco, and Puget Sound ports had brought a total of 15,701 Chinese passengers to Victoria, and Chinese labourers continued to enter in large numbers.[60]

Nearly the entire railway line in British Columbia, from Port Moody to Craigellachie, was graded by Chinese labourers with white foremen. Hundreds of labourers died from accidents or illness.[61] In August 1880, for example, nine Chinese were blown up by a blast near Yale, and in 1882, over a hundred Chinese labourers had to climb and walk along the steep and slippery cliffs to pull a steamer on tow lines up the Fraser Canyon; many were injured.[62] Many deaths were caused not only by accidents but also by winter cold, illness, and malnutrition; an undetermined number of Chinese railway workers died from scurvy because of a lack of vegetables in the winter.[63] When a worker was ill, there was no doctor to treat him; when dead, he would be left by the railway and covered up with piles of rocks and earth. No coroner's inquest was held, as these deaths were never reported. Andrew Onderdonk estimated that between five and six hundred Chinese workers died during construction of the line—three Chinese workers for every kilometre of track. As far as the white contractor was concerned, the welfare of the Chinese workers was the responsibility of the Lee Chuck Company. However, Lee Chuck had no concern for his Chinese labourers, forcing them to purchase all the supplies from his company at prices higher than in other stores. If a Chinese worker patronized other suppliers, his wages were cut. This exploitation of the workers was made public only when they refused to work. In May 1881, for instance, the workers went on strike after the company reduced their rice ration by 5 per cent.

The use of Chinese labourers in construction of the CPR was openly opposed by the government of British Columbia, although they were virtually indispensable in the project. On 12 May 1882, Amor De Cosmos, then an MP for Victoria, repeated his request in the House of Commons that the employment of Chinese labour should be prohibited. He argued that because so many Chinese workers were entering Canada, they would soon outnumber white people in British Columbia.[64] In reply to his complaint, Prime Minister Macdonald said: "if you wish to have the railway finished within any reasonable time, there must be no such step against Chinese labour. At present it is simply a question of alternatives—either you must have this labour or you cannot have the railway."[65] John Rochester, MP for Carleton, added that British Columbia should be thankful to be able to obtain Chinese labour in building the railway. Easterners could not understand why British Columbia was so much against Chinese workers when it needed them desperately in all sorts of labour-intensive jobs. On the other hand, British Columbians felt that Ottawa did not understand their Chinese problem. The national census of 1881, which for the first time included British Columbia, revealed that 4,350 of the 4,383 Chinese in Canada lived in British Columbia: the province had over 99 per cent of the country's Chinese population.[66] The census listed only four Chinese in Manitoba, seven in Quebec, and twenty-two in Ontario. Thus, British Columbia found it hard to convince the federal government and other provincial governments that Chinese immigration to Canada was an urgent issue for the country.

3

Chinatowns in British Columbia

During the period of free entry, the Chinese were free to enter and leave Canada but they were not free in many other respects: there were restrictions on their right to vote, to seek employment, and to choose where they could live and work. These restrictions resulted from the racial discrimination and prejudice which developed initially in the mining towns of California and later spread to Oregon and Washington.[1] When the white miners of the American Pacific coast came to British Columbia during the Fraser gold rushes, they carried their prejudices with them, along with their shovels and gold pans.[2] Similarly, the Chinese gold-seekers followed the white miners to the goldfields and took with them the memory of white hostility. They arrived in British Columbia in groups and lived together in a Chinatown physically and culturally separated from the white community. An analysis of Chinatowns must begin by considering their origins, since they were the result of a complex process of voluntary and involuntary factors.[3]

BIRTH OF CHINATOWNS

White racism was one factor in the creation of a Chinatown. White landlords would not sell or lease their properties to the Chinese unless the lands were on the fringe of the town and thus unattractive to the white community. For example, the Chinese in Victoria and Vancouver established their living quarters on mudflats, the cheapest districts of the cities, where the low rent also attracted low-class saloons and brothels. In Nanaimo and other coal-mining towns, mining companies housed Chinese miners in isolated areas, separating them from white miners. Thus, the Chinese people had little or no choice of residence.

Racism, hatred, and violence had resulted in residential segregation. Many pioneer Chinese immigrants in British Columbia came from California, where they had experienced physical abuse and other violent action against them. In British Columbia, sporadic incidents of violence were recorded from the start. The Chinese were treated like dogs—bullied, scoffed at, kicked, and cuffed about on all occasions—and violence was used to keep them out of the Cariboo gold-mining areas until 1862.[4] Mobs of irate whites forced Chinese people out of town, first in Vancouver in 1887, then in the Slocan Valley, next at Atlin in the northwest, later at Salmo in the Kootenays, and finally in Penticton.[5] Chinese old-timers recalled that white youngsters often caused all kinds of mischief, such as throwing stones at them and tipping over their vegetable baskets. Sometimes the assailants were more violent: once two boys waylaid a Chinese ranch labourer and shot him with a revolver.[6] A local newspaper in Vancouver once remarked that unprovoked assaults on the Chinese people by ill-mannered boys and youths were so common that they were hardly worth reporting.[7] As a result, the Chinese tended to travel in groups and to isolate themselves from the white community in order to avoid abuse.[8] Thus, voluntary segregation resulted in the birth of a Chinatown, which was a kind of self-defence measure used by the Chinese to avoid open discrimination and hostility. They confined themselves, whenever possible, to the boundaries of Chinatowns, where they felt safe and secure.

Cultural barriers were another factor in the formation of a Chinatown. Most early Chinese immigrants could not speak English and so found it difficult to communicate with white people. As well, they were ignorant of Western customs. However, when they lived together, they could speak their own dialects, eat their own food, and worship their own gods as they had done in China. Thus, it was out of necessity that the Chinese voluntarily segregated themselves from the host society.

Economic factors also shaped the origin and growth of a Chinatown. At first, many Chinese immigrants were recruited as labourers by Chinese merchants or contractors to come to Canada. For economy's sake, the sponsors built or leased wooden shacks in the cheapest district of a town and operated stores and restaurants to serve the labourers' needs. This cluster of Chinese labourers and stores in one location constituted the nucleus of a budding Chinatown. When new Chinese immigrants joined their relatives or friends in Canada, they tended to live in the Chinatown where their sponsors were living, and shared the cost of room and board. To serve the needs of Chinatown residents, Chinese grocery stores, restaurants, and other businesses were set up, and theatres, schools, and various types of associations were established. Thus, the Chinatown began to grow.

TOWNSCAPES AND IMAGES

There were two types of Chinatowns in British Columbia. The first included the "instant Chinatowns," which were the homes for Chinese miners in small gold-mining settlements in the valleys throughout the Cariboo and other mining districts during the gold rushes. These instant Chinatowns had no stores or permanent institutional buildings; they were just clusters of crude shacks where Chinese miners lived. Many of these temporary structures fell into ruin after their inhabitants moved out or died. The stage-development model for Old Chinatowns was not applicable to these instant Chinatowns because they did not meet the basic requirements of a recognized Chinatown. The other type of Chinatown was established in large gold-mining towns such as Yale and Barkerville, or in the port cities of Victoria, New Westminster, and Nanaimo. They conformed to the growth patterns of the stage-development model.

During the gold rushes in British Columbia, all the mining towns had only one main street; a Chinatown was usually situated at its end on the edge of town. Chinatowns in Barkerville, Yale, and Kamloops were good examples. All had a very short budding stage and were physically similar; each consisted of clapboard shacks or wooden cabins on both sides of the town's sole street, much like the other parts of the town. However, it was the sights, the odours, and the sounds of Chinatown which shaped its image for whites and Chinese. Within the precincts of Chinatown, there was a concentration of Chinese males who dressed, looked, spoke, and behaved differently from the white community. Signs written in Chinese characters outside the small stores, and exotic Chinese merchandise inside them made the streetscape of a Chinatown visually different from other parts of the town. The smells from Chinese restaurants, the aroma from grocery and herb stores, and the odour from opium dens pervaded the crowded Chinatown. Sometimes, funeral processions and festival celebrations brightened the streets with the sound of firecrackers, drums or gongs, and the shouting and talking of Chinese dialects. All these activities constituted the essential components of a blooming Chinatown in Victoria, Barkerville, and other cities or towns.

The Western image of a Chinatown was shaped by its own stereotypes. Whites perceived Chinatown as a filthy, unsanitary, overcrowded, sinister, and insidious slum. All its residents were downtrodden slaves and victims of gambling, opium-smoking, and prostitution. Chinatown was conceived as a godforsaken place to be kept as far away as possible from the white community. These images usually had their roots in the sensational reports in local newspapers or the biased accounts of policemen or church people who judged the Chinese people within the context of Christian white society. One clergyman described the Chinese immigrants thus:

> The Chinese immigrant comes as an Alien and a Stranger. . . . he comes from the coolies class, the despised and oppressed of his own land. Downtrodden and abused, and inured to thoughts, practices, customs and laws vitiated by a pagan atmosphere that has for centuries been slowly

> sapping the mental, moral and spiritual vitality of his own class, he comes without any definite moral conviction of right or wrong. . . . As a coolie or a serf, he comes with one qualification, he can slave and toil.[9]

The same man depicted Chinatown as

> a miniature Chinese town built by Chinese carpenters, without any regard for beauty, regularity, sanitation or comfort; a segregated group of individuals who realized that they were unloved and separated from their neighbours by an almost impassable gulf of race, colour, language and thought. . . . Within the unshapely structures of Chinatown were the parasites of the Chinese race—professional gamblers, opium eaters, and men of impurity. . . . Chinatown became the carcass to attract the foul birds of Western vices, the dumping ground of those evils which the white man wishes removed from his own door.[10]

Frances MacNab, the literary pseudonym of Agnes Fraser, a writer living in British Columbia, wrote:

> Chinatown is an offence to at least two senses—sight and smell. It reeks of opium, and is suggestive of low gambling-halls. There sit the fat merchants, who are probably deep in usurious practices of the most blood-sucking description. It is impossible not to suspect that the hard toil of many a poor John goes to increase the paunch of some of these fat tyrants who sit lurking like spiders in their dark and silent dens, concocting in their minds webs for the unwary.[11]

These were some of the stereotyped images of the Chinese and Chinatowns shared by many white people, people who might never talk to a Chinese person or see a Chinatown.

CHINATOWNS ON VANCOUVER ISLAND, 1860s–1870s

During the gold rushes, there were four Chinatowns on Vancouver Island. Victoria's Chinatown, established in 1858, was the first one in Canada. (It is discussed in detail in Part Two.) The other Chinatowns were established in the three coal-mining areas: Nanaimo, Wellington, and South Wellington. Nanaimo, about 110 km (70 miles) north of Victoria, was built on a rocky promontory extending into the Commercial Inlet (Figure 6). The village site and thousands of acres of nearby land were owned by the Vancouver Coal Mining and Land Company (VCC). During the 1870s, the VCC initiated the development of Nanaimo and extended the workings of its Nanaimo Colliery. Settlement began to spread up the slopes from the waterfront of Com-

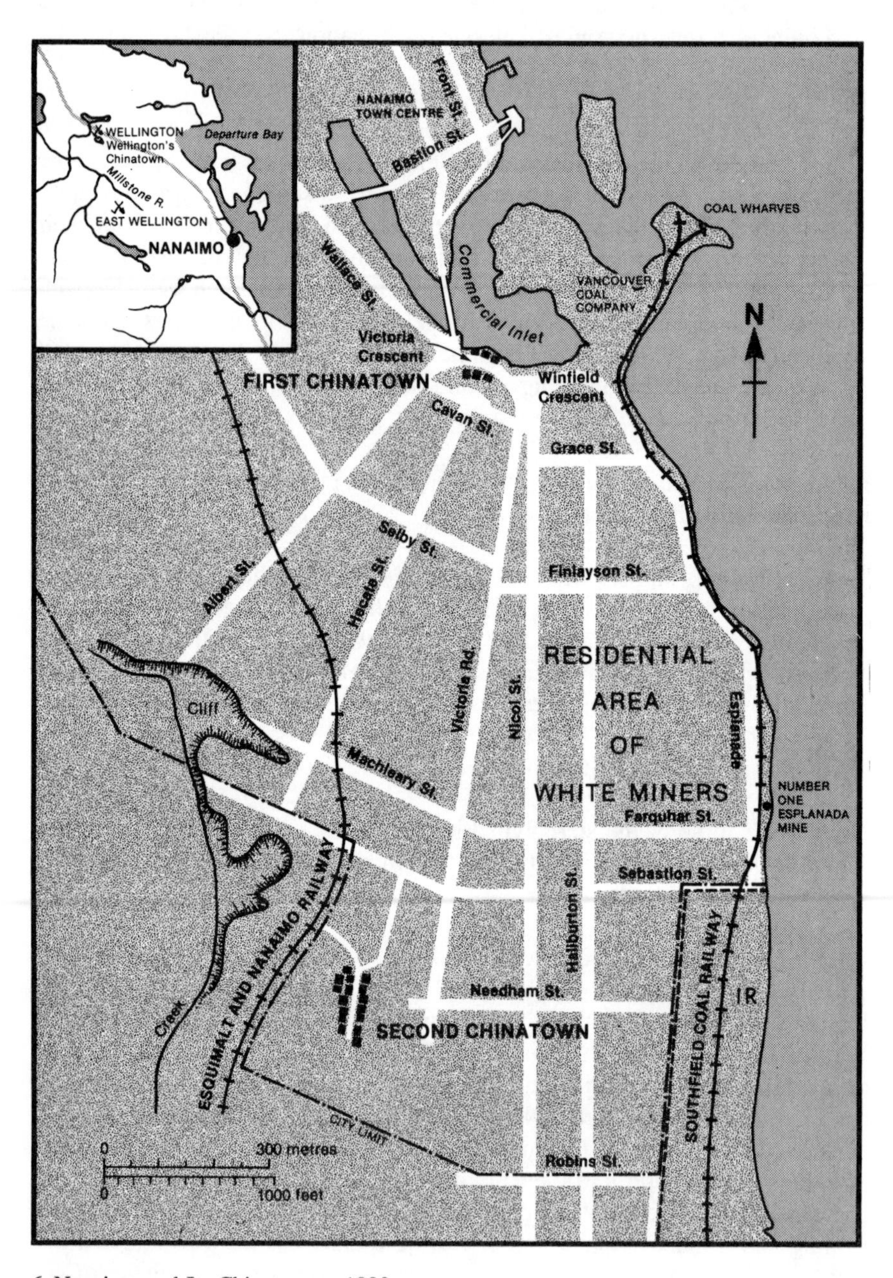

6 Nanaimo and Its Chinatowns, 1880s

mercial Inlet; the main streets radiated uphill and concentric cross-streets followed the contours, making a basin-like formation.

As early as the 1860s, a few Chinese labourers were recruited by the VCC to load coal at the pit mouth and at the wharves. In 1867, there were only thirteen Chinese among 580 white residents in the small mining community.[12] Their presence was unknown to the outside world until 24 July 1867, when the *Victoria Colonist* reported that "John has caught the 'striking fever' from the white colliers and struck for a dollar and a half a day instead of one dollar."[13] As there was a shortage of labour in Nanaimo, more Chinese labourers were recruited, and by 1871, the Chinese population in the small mining village had increased to thirty-six.[14] They lived in company-built shacks on Victoria Crescent, where an embyronic Chinatown was begun. At the beginning, there was apprehension that some white miners might react with violence to Chinese labourers working in the mines, but since Chinatown was separated from the centre of Nanaimo by Commercial Inlet, "their presence, though naturally distasteful to the white miners, has not caused any ill-feeling."[15] During its budding stage, Nanaimo's Chinatown had only two or three small stores, which provided the daily needs of the small Chinese mining population; Yee Kee & Company was probably the largest shop.[16] When Nanaimo was incorporated in 1874, the city's population had risen to 1,884, and its Chinatown was in its blooming stage and had 200 Chinese residents, nearly one-fifth of the city's population.[17]

In 1870, Dunsmuir, Diggle & Company (DDC) began the Wellington Colliery, about seven kilometres (four miles) north of Nanaimo, and established the mining village of Wellington. Like the VCC, the DDC also recruited Chinese labourers to work in its mines, building scores of shacks for them some distance away from the main village. As an ever-increasing number of Chinese workers were recruited, more shanties were built. A small Chinatown thus emerged in Wellington.

By 1877, Chinatowns in the mining villages of Nanaimo and Wellington had a total population of about 300.[18] Nearly all of them were coal miners: 206 Chinese were employed by the VCC in Nanaimo and ninety by the DDC in Wellington.[19] Two years later, the DDC purchased the coal mine at South Wellington, where the Chinese were employed not only in the mines but also on the railway and in the active logging operations.[20] A small Chinatown thus came into existence in South Wellington, providing essential services to Chinese miners.

CHINATOWNS ON THE MAINLAND, 1860s–1870s

On the mainland of British Columbia, Chinatowns were established in New Westminster, Barkerville, Quesnellemouth, Keithley, Yale, Lytton, and a few other gold-mining towns. During the 1860s, New Westminster was still an undeveloped frontier town on the northern bank of the Fraser River. It consisted of a small group of wooden huts clinging to the hillside among fallen trees and blackened stumps.[21] In 1867, the town had a population of about 1,000, of which one-tenth was Chinese.[22] The budding Chinatown was established on the western end of Front Street, which ran east-west along the Fraser River. With a residential population of sixty-six Chinese males and thirty-seven Chinese females, New Westminster's Chinatown functioned mainly as a "resort" for Chinese miners, who came down by river boat from Yale, Hope, and other goldfields to spend the winter. Many of them indulged in gambling, opium-smoking, and prostitution in order to alleviate the loneliness and monotony of their idle hours.

Barkerville, established in 1862, consisted of a series of wooden structures erected on log posts on either side of a single muddy street running parallel to Williams Creek.[23] As a main distribution centre, Barkerville provided supplies to other mining settlements in the Cariboo region. Chinatown, at the southern end of the town, had a very short budding stage, and flourished virtually overnight, containing the largest Chinese community in British Columbia (Plate 1). Except for the Chinese signs hanging on storefronts and the concentration of Chinese residents, the streetscape of Chinatown was little different from that in other parts of town. In the early 1860s, at the zenith of its development, Barkerville's Chinatown had a population of about 5,000 and several large stores, the largest of which was the Kwong Lee Company, owned by Loo Chuck Fan and his brother.[24] The only organization in the Chinese community was probably Chee Kung Tong (Zhigongtang), which was set up in 1862, the first secret Hongmen organization in Canada.[25] Although the society's objectives were to maintain friendly relationships among its fellow countrymen and to accumulate wealth "through proper business methods for the benefit of all members," it also controlled all brothels, opium dens, and gambling clubs in Chinatown.[26] For more than a decade, Chee Kung Tong controlled the socioeconomic activities of most Chinese miners not only in Barkerville but also in other parts of the Cariboo region.

The local white community did not have a good image of Barkerville's Chinatown, always complaining that it was dirty and untidy. The Chinese residents fed pigs in the street in front of their buildings as they had done in China and, as a result, the drains were always choked with filth.[27] After the frozen creek melted, its water often flooded the street, making it muddy, smelly, sticky, and slippery. Pedestrians had to use the raised sidewalks on both sides of the street.

Barkerville's Chinatown prospered for only a few years; on 16 September 1868 a fire wiped out Barkerville and its Chinatown, and many Chinese miners left. Thus, Barkerville's Chinatown completed its life-cycle and became extinct. Although a second Chinatown was rebuilt immediately, most Chinese miners did not return since Barkerville's success was over. The second Chinatown in Barkerville remained in the budding stage and grew very slowly; in 1879 it had only 159 Chinese residents, of whom 142 were gold miners, two were washermen, five were storekeepers, and ten were women.[28]

Yale, situated at the entrance to the interior goldfields, was not only a significant mining town during the gold rushes but also an important supply depot for miners. Its Chinatown, at the eastern end of Front Street, was also very busy; it accommodated Chinese miners heading for the upper Fraser River or returning to New Westminster or Victoria. Yale's Chinatown population fluctuated between 100 and 500 from 1861 to 1865; during the 1870s, it still had about 300 Chinese residents.[29]

In addition to these Chinatowns, there were many small short-lived Chinese settlements in the valleys throughout the Cariboo and other mining districts during the gold rushes. These instant Chinatowns died soon after their transient population disappeared. Traces of the locations of these instant Chinatowns may be found in the place-names of streams, lakes, and mountains in the mining district. For example, at least six streams (in Princeton, Lillooet, Robson, Osooyos, Quesnel, and Alberni) are called China Creek for the Chinese miners who were once concentrated there.[30] Similarly, north of Big Bar Creek were China Lake and China Gulch. China Cabin Creek and China Cabin Lake are located on the southwestern side of Horsefly Lake. Some place-names such as Ahbau, Ahbau Creek, Ahbau Lake, and Ahbau Landing, all near Quesnel, are named after one Chinese called Ah Bau, who had lived there as a miner and trapper for many years.

DEMOGRAPHIC STRUCTURE OF CHINATOWNS IN THE 1880s

During the 1880s Chinatowns in British Columbia were at varying stages of development; some Chinatowns were budding and some were still thriving, but other Chinatowns had entered the withering stage, some approaching extinction. The 1881 census revealed that the 4,350 Chinese people in British Columbia were the largest and most visible ethnic minority in the province, although they accounted for only 9 per cent of its total population. Chinese residents were widely dispersed in various mainland mining towns. For example, 402 Chinese lived in Quesnellemouth and 413 in Keithley, and about 500 lived in Richfield, Barkerville, and other small mining settlements in the Cariboo region. Nearly 540 Chinese were concentrated in the Fraser Valley between Hope and Yale, and almost 500 around Lytton, Cache Creek, and nearby mining settlements. On Vancouver Island, Chinese were found mainly in the capital city of Victoria and the coal-mining town of Nanaimo,

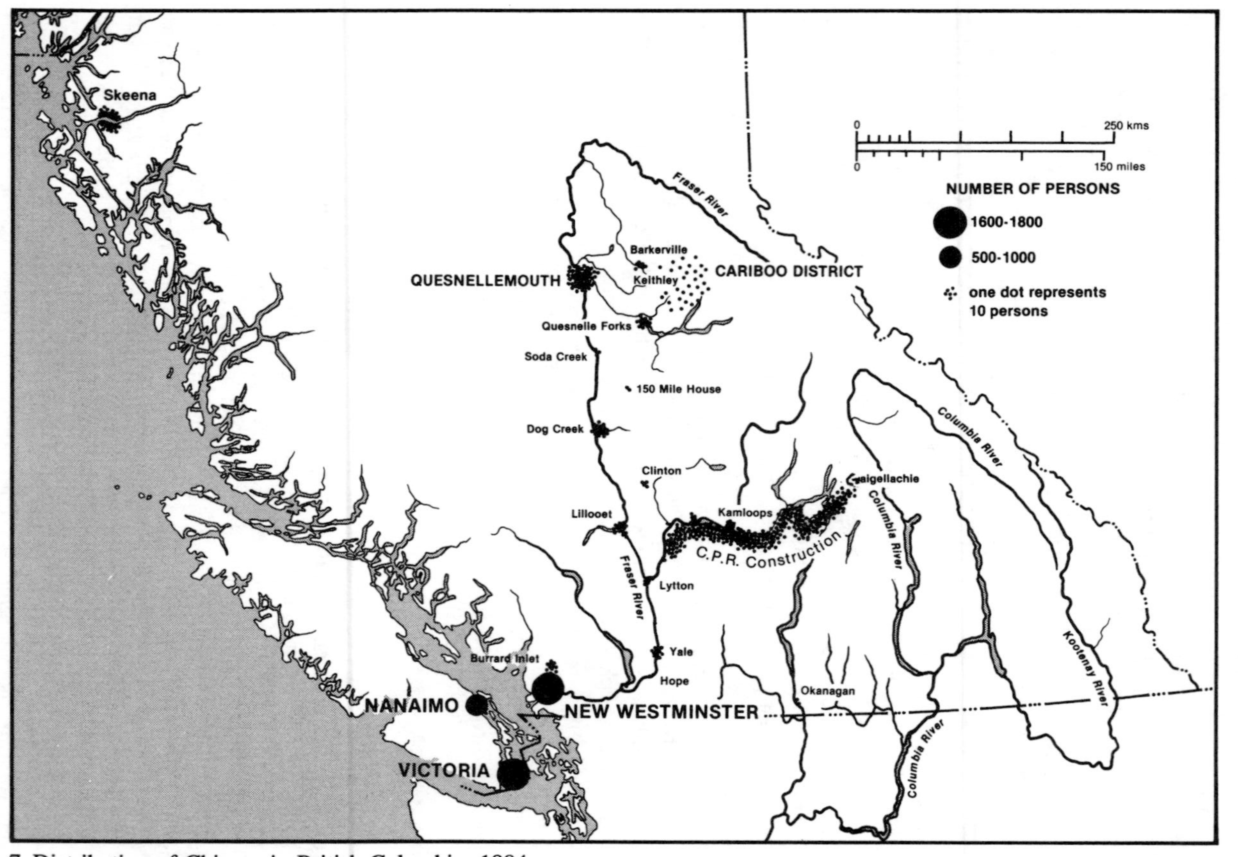

7 Distribution of Chinese in British Columbia, 1884

with 492 persons in the former and 287 in the latter.

During the first few years of railway construction, thousands of Chinese labourers entered British Columbia. By 1884, there were over 10,000 Chinese people living in the province. The outcry against Chinese immigration in British Columbia prompted the federal government to set up a Royal Commission in 1884 to investigate the whole issue of Chinese immigration to Canada. The report was the first and most accurate study of the demographic structure of the Chinese in British Columbia; it revealed for the first time the distribution of the 10,492 Chinese by towns and cities in the province.[31] Nearly 4,000 Chinese railway workers lived in makeshift tents along the route of the railway line from the Fraser Canyon to Craigellachie, and small Chinese communities were found in various interior towns, such as Quesnellemouth, Quesnelle Forks, Yale, Dog Creek, and Lillooet (Figure 7). Elsewhere, pockets of Chinese miners and farmers were found throughout the Cariboo region, and about 300 Chinese worked in canneries along the Skeena River. The largest Chinese communities were in Victoria, with a Chinese population of 1,767; New Westminster, with 1,680; and Nanaimo, with 969, and their flourishing Chinatowns displayed similar features of the blooming stage of the development model.

The report also classified the Chinese according to their occupations. Eighty-six per cent were labourers; less than 2 per cent belonged to the merchant and professional classes; about 6 per cent were young boys, girls, and women; and the remaining 6 per cent were new arrivals (Table 5). Half the labourers were railway workers and gold miners, and the other half were coal miners, fish hands, farm labourers, store employees, cooks, servants, and the like. Chinese society was dominated by single male labourers—over ninety-two males for every female. Nearly all the married women and young girls lived in the Chinatowns of Victoria, New Westminster, or Nanaimo. Chinese communities in Quesnellemouth, Quesnelle Forks, Lillooet, Yale, Lytton, and other small mining and farming towns did not have any Chinese females, except for a few prostitutes.

A study of the home county and clan origins of 5,000 Chinese in British Columbia in 1884–5 reveals that Chinese immigrants had specific places of origin and destination.[32] For example, over half the 400 Chinese in Quesnellemouth were Zhou people from Kaiping County. Similarly, nearly 45 per cent of the Chinese in Nanaimo were Ma people from Taishan County. Such a phenomenon is simply an extension of life in South China, where many villages are inhabited predominantly by people with a single surname. A stream of migration came from a few villages in China to specific towns in British Columbia, where lineage groups were maintained. These patterns of residence, with great concentration of people according to clanship, were also noticeable in the United States. For example, the Mei people from Taishan

TABLE 5: Occupations of Chinese in British Columbia, 1884

Occupation	No. of persons	% of total
LABOURERS		
railway workers	2,900	27.6
gold miners	1,709	16.3
coal miners	727	6.9
fish hands	700	6.7
farm labourers	686	6.5
store employees	302	2.9
cooks and servants	279	2.7
sawmill workers	267	2.5
wood-cutters	230	2.2
washermen	156	1.5
ditch diggers	156	1.5
fuel cutters	147	1.2
boot-makers	130	1.1
vegetable gardener	114	0.8
other labourers	519	4.9
NON-LABOURERS		
merchants	120	1.1
restaurant keepers	11	0.1
doctors	42	0.4
teachers	8	0.1
OTHERS		
prostitutes	70	0.7
married women and girls	88	0.8
boys under 17	529	5.0
new arrivals	602	5.7
Total	10,492	100.0

Source: Royal Commission report, 1885, 363–5

County were the dominant clan in Chicago, and the Chen people from Taishan were prominent in Seattle.[33]

Another demographic feature of Chinese communities was that certain occupations were monopolized by people from certain counties. For example, most Chinese cooks and servants were Taishan people, and gardeners and farm labourers were Zhongshan people. This kind of occupational monopoly is similar to that in the United States. In San Francisco, for example, most drygoods stores were owned by people from Zhongshan County, laundries by Yu people from Taishan County, and fruit and candy stands by Xie clansmen from Kaiping County.[34] Common dialect was one factor in employment preference. In those days, the Siyi dialect was the main tongue of the Chinese in North America. A Chinese unable to speak the dialect found it difficult to get a job. As a result, those people who spoke the Sanyi dialect or

Zhonghshan dialect, for example, tended to employ others who spoke the same language. Nepotism also resulted in the occupational monopoly of certain clans. In China, the villagers always favoured people of the same lineage, excluding outsiders or neighbours with other surnames. When they came to Canada, the Chinese retained this attitude towards occupations and trade. For example, during the late 1880s, the 400 Chinese residents in Winnipeg had come from Heshan County; most were surnamed Li.[35] For many years, they had look-outs posted at roads and railways entering Winnipeg in order to prevent other Chinese from coming to compete with their laundry business in the city, a practice which ceased only after the Li Association in Vancouver persuaded their clansmen in Winnipeg to change their attitude. County and clan origins therefore are important as a basis of residence and occupation in the Chinese community. Their significance in the formation of Chinese voluntary associations will be dealt with in Chapter 9.

CHINATOWNS ON VANCOUVER ISLAND, 1880s

The Royal Commission of 1884 revealed that Chinese communities were found in over twenty cities, towns, or villages in British Columbia.[36] Victoria's Chinatown was the largest in the province, followed by the Chinatowns of New Westminster, Nanaimo, and Quesnellemouth. In addition, there were numerous small Chinatowns, with populations ranging from twenty to two hundred, and in several interior gold-mining towns such as Dog Creek, Quesnelle Forks, and Lytton.

On Vancouver Island, Nanaimo's Chinatown on Victoria Crescent was the largest Chinatown outside Victoria (see Figure 6). Because of its central location in the coal-mining districts, Nanaimo became a regional centre for surrounding towns and villages and maintained direct shipping traffic with Victoria. Its Chinatown also functioned as a regional centre for satellite Chinatowns in Wellington and South Wellington. In 1882, there were eight Chinese businesses in Nanaimo's Chinatown: five stores, two hand laundries, and one tailor shop.[37] Of all the Chinese business concerns, Hong Hing & Company was the largest company in Nanaimo's Chinatown. It imported Chinese commodities directly from Victoria's Chinatown and distributed them to smaller stores in Chinatowns in Wellington and South Wellington. The Royal Commission Report revealed that only Nanaimo's Chinatown had Chinese female and married residents (Table 6). About 80 per cent of the Chinese people in Nanaimo and Wellington (including South Wellington) were miners and cooks. The VCC and DDC were the major employers of Chinese labourers, paying them only $1 to $1.25 a day, whereas an Indian labourer was paid $1 to $1.50 and a white labourer $2 to $3.75 a day.[38] As in other Chinatowns, the sex ratio in Nanaimo's Chinatown was highly unbal-

TABLE 6: Occupations of Chinese in Nanaimo and Wellington, 1884

Occupation	No. of persons			
	Nanaimo	Wellington	Total	% of total
LABOURERS				
miners and cooks	64	727	791	81.6
servants and cooks	18	21	39	4.0
store employees	6	8	14	1.5
farm labourers	13	–	13	1.4
washermen	8	4	12	1.2
barbers	4	8	12	1.2
NON-LABOURERS				
merchants	6	12	18	1.9
doctors	3	5	8	0.8
teachers	1	1	2	0.2
OTHERS				
prostitutes	2	–	2	0.2
married women	4	–	4	0.4
boys	15	15	30	3.1
children	2	–	2	0.2
new arrivals	22	–	22	2.3
Total	168	801	969	100.0

Source: Royal Commission report, 1885, 363

anced, with one female for 160 males. Nearly half the Chinese people in Nanaimo were surnamed Ma and had originated from Taishan County (Table 7), another indication of Chinese chain migration to Canada

The stereotyped images of Chinatown emerged in testimony to the Royal Commission. The Knights of Labour described Chinatown as an unsanitary place consisting of wretched hovels which were dark, poorly ventilated, filthy, and unwholesome. Packed with people, Chinatown was "an abomination to the eyes and nostrils, and a constant source of danger to the health and life of the community."[39] Although this description of Chinatown was given by an anti-Chinese secret society, it represented the image many white people held about Chinatowns in those days. Nanaimo's Chinatown was undoubtedly filthy and overcrowded; its unsanitary condition, however, was exaggerated.

The development of Nanaimo's Chinatown followed a different path of growth from that of Barkerville's Chinatown. When it was still blooming, Nanaimo's Chinatown entered the stage of extinction soon after it faced eviction in November 1884. The VCC, owner of Chinatown's land, set aside three hectares (eight acres) of its undeveloped land south of Pine Street at a remote corner of the city and ordered the Chinese merchants and residents to move

TABLE 7: Home county and clan origins of Chinese in Nanaimo, 1884–5

Clan	Home county			
	Taishan	Enping	Others	Total
Ma	109	0	1	110
Huang	11	0	2	13
Feng	0	12	0	12
Wu	0	12	0	12
Zhen	9	0	0	9
Zheng	3	6	0	9
Liang	0	4	5	9
Cen	0	7	0	7
Others	17	16	31	64
Total	149	57	39	245

Sources: stubs of donation receipts dated 1884 and 1885, Chinese Consolidated Benevolent Association of Victoria

there.[40] There were three possible reasons for the decision to relocate Chinatown. First, as the white business district of Nanaimo began to expand southward along the waterfront, the land on Victoria Crescent became too valuable for a Chinatown. It had to make way for the expansion of the town centre. Second, white miners in Wellington vented their rage upon Chinese labourers after Robert Dunsmuir broke their strike in the autumn of 1883 by using the Chinese as scabs.[41] To avoid racial trouble in its mines, the VCC did not hesitate to remove its Chinese labours and Chinatown to the southwestern corner of the city. The third reason for the relocation of Chinatown was as a precaution against the arrival of many Chinese labourers to work on the Esquimalt and Nanaimo Railway. It was felt that they would be more noticeable if they lived on Victoria Crescent. The new site, next to the railway line, was an ideal place to isolate Chinese labourers from the white community. And the VCC only leased the land to the Chinese, who had to clear the forest, level the site, and erect the buildings for their homes and businesses; thus, without paying the cost of labour, the company developed a small townsite on its own property.

By 1885, almost all Chinese businesses and residents had been moved to the new site, about ten city blocks south of downtown Nanaimo. Local white people were pleased with the VCC's decision since the second site for Chinatown was situated in remote wilderness and separated from their residential areas. The second Chinatown was thriving soon after it was begun. It was particularly busy between 1884 and 1886 when the E & N Railway was being built. Railway workers were reluctant to go to downtown Nanaimo since some remembered being stoned upon their arrival; when they were not working, they spent most of their time in their camps or in Chinatown.[42]

CHINATOWNS ON THE MAINLAND, 1880s

By the late 1880s, most of the Chinatowns in the gold-mining districts on the mainland had reached their withering stage or disappeared. Barkerville's Chinatown, for example, had been reduced to a small Chinese settlement of less than 200 residents. Quesnellemouth, which became "Quesnelle" for Post Office purposes, was still flourishing.[43] Quesnelle's Chinatown, with a Chinese population of about 500, replaced Barkerville's Chinatown as the main supply centre for Chinese miners, and emerged as the largest Chinatown in the Cariboo region. In the lower Fraser River, Yale's Chinatown maintained its dominant position. It covered a triangular city block bounded by Front Street, Rail Road (Douglas Street), and Regent Street and had spilled over to the adjacent Indian reserve. In 1882, Yale's Chinatown was heavily damaged by fire and suffered an estimated loss of $10,000.[44] Several cabins were rebuilt immediately, and On Lee, the wealthiest Chinese merchant, constructed four buildings on his property.[45] Nevertheless, Yale's Chinatown continued to decline and by 1884, its Chinese population had dropped to about a hundred.[46]

After the gold rushes, an undetermined number of Chinese had left British Columbia, heading for the United States or back to China. But construction of the CPR brought Chinese labourers back to the interior and reactivated some stagnant Chinatowns along the railway line. In 1884, for example, there were only sixty-two Chinese (three merchants, three store employees, and fifty-six miners and farmers) in a tiny Chinatown on the western end of Kamloops.[47] When construction of the CPR reached the town in 1885, hundreds of Chinese railway workers came from Savona's Ferry and set up their base camps in what is now Riverside Park. Kamloops was booming: land was cleared along Kamloops Lake and up the South Thompson River, and the town's main road, Main Street (Victoria Street), was graded for the railway. A budding Chinatown, established at the western end of Main Street, was bisected by the railway track. Its northern section was sandwiched between the Thompson River and the railway, and its southern section fell between the railway and a high cliff (Plate 2). On its western side was Overlanders Bridge and on its eastern side was the white business district. Although it was still in the budding stage, Kamloops's Chinatown was busy and had a few stores to provide some daily necessities for railway workers.

Another impact of the CPR was the release of a large number of Chinese railway workers after the line was completed. Only a few had saved enough money to return to Victoria and thence to China. Those who had no means to return to the coast were found loitering in towns and villages along the railway, suffering from hunger. Those who were stranded along the line in the lower Fraser River Valley were seen eating decayed vegetables and dead salmon. Those who found employment worked as domestic servants in white families, as labourers on farms, or as cooks on cattle ranches. Some people opened hand laundries and others began truck-farming, labouring long hours to grow vegetables for settlers who yearned for something more than sourdough and bacon.[48] Therefore, after the CPR line was completed, an undetermined number of former railway workers remained in the interior.

In 1884, about a hundred Chinese were living around Burrard Inlet on the coast; half worked in the sawmills near Granville on the south shore of the inlet.[49] The city of Vancouver did not yet exist and its predecessor, Granville, was just a small town geared to the needs of local lumber mills and logging camps. Gin Tei Hing Grocery Store, Wah Chong Wash House and General Merchandise, and Yune Chong Washing were the only three Chinese stores in Granville, where less than fifty Chinese lived.[50]

New Westminster's Chinatown was the second largest Chinatown in British Columbia, after Victoria's, so the Chinese people called New Westminster *Erbu* (the Second Port) and Victoria *Dabu* (the First Port or the Big Port). New Westminster's Chinatown was thriving in the early 1880s mainly because of the rapid development of the city's farming, lumbering, and salmon-canning industries. These industries required a great deal of cheap labour, and their operators welcomed Chinese workers. As a result, the Chinese population in New Westminster, which stood at only 485 in 1881, had more than tripled three years later. About 66 per cent were workers in canneries, sawmill workers, farm labourers, and ditch-diggers (Table 8).

Canneries in particular had to rely on the Chinese. The industry required plenty of labour, but it was only seasonal work, lasting for two or three months a year. Few white workers were interested in seasonal employment, and native women were not a reliable source of labour because they often had to return to their own communities to prepare for winter. During the fishing season, a cannery would contract a Chinese company to supply labourers, who would work as a team to cut and clean the fish and seal the cans. After the season was over, the workers would be laid off for the rest of the year, staying in Chinatown and waiting for the next fishing season. Often they had to do other general labouring work in order to make ends meet.

Because of its population increase, New Westminster's Chinatown began to expand uphill along McInnes and McNeely streets to Columbia Street, covering a square block bounded by Front Street on the south, Columbia Street on the north, McInnes street on the west, and McNeely Street on the east. The main commercial area was still centred on Front Street. In 1882, for example, there were thirteen Chinese businesses in Chinatown, including six grocery stores, four hand laundries, one restaurant, one butcher shop, and

TABLE 8: Occupations of Chinese in New Westminster, 1884

Occupation	No. of persons	% of total
LABOURERS		
farm labourers	400	23.8
fish hands	390	23.2
sawmill workers	190	11.3
ditch diggers	156	9.3
fuel cutters	82	4.9
cooks and servants	50	3.0
washermen	20	1.2
charcoal burners	18	1.1
store employees	18	1.1
barbers	15	0.8
vegetable sellers	9	0.5
sewing-machine workers	6	0.4
carpenters	3	0.2
NON-LABOURERS		
merchants	12	0.7
doctors	6	0.4
teachers	2	0.1
OTHERS		
prostitutes	7	0.4
married women	4	0.2
girls	2	0.1
boys under 17	90	5.4
new arrivals	200	11.9
Total	1,680	100.0

Source: Royal Commission report, 1885, 363

one tailor shop.[51] Except for two laundries on Columbia Street, all the Chinese businesses were found on Front Street (see Figure 13, p.78).

Conforming to other features of the stage-development model, New Westminster's Chinatown had a very unbalanced sex ratio: 113 males to one female in 1884.[52] Most of its Chinese people had come from Enping and Taishan counties (Table 9). A few clans originated mainly from one county. For example, nearly all people surnamed Tan came from Kaiping County, and those surnamed Xu from Panyu County. This is another example of the tendency of Chinese immigrants of particular home counties to go to certain places in Canada.

OTHER CHINESE SETTLEMENTS, 1880s

Before the 1880s, no Chinatowns were found outside British Columbia. The census of 1881 listed only ten Chinese in Toronto, eight in Barrie Town, and

TABLE 9: Home county and clan origins of Chinese in New Westminster, 1884–5

	Home county						
Clan	Enping	Taishan	Kaiping	Xinhui	Panyu	Others	Total
Li	1	9	1	15	5	6	37
Tan	0	0	28	0	0	1	29
Zhen	7	3	0	2	3	5	20
Lu	5	8	0	3	1	5	19
He	11	0	0	0	2	1	14
Liang	3	0	0	0	7	4	14
Huang	0	6	1	1	0	5	13
Feng	8	4	0	0	0	0	12
Xu	0	0	0	0	10	2	12
Others	39	21	14	23	13	26	139
Total	74	51	44	44	41	55	309

Sources: stubs of donation receipts dated 1884 and 1885, Chinese Consolidated Benevolent Association of Victoria

seven in Montreal.[53] Other towns such as Winnipeg, Hamilton, and Emerson Town had only one or two Chinese residents. These few Chinese pioneers had crossed the border from the United States. No Chinese immigrants went to other provinces from British Columbia because of the formidable barrier of the Rocky Mountains. But after completion of the CPR, Chinese immigrants began to move along the line to the prairies and other parts of Canada.

4

From Restriction to Exclusion

During the period of restricted entry, from 1885 to 1923, every Chinese immigrant had to pay a head tax to the Canadian government before he was permitted to land in the country; the tax rose from fifty dollars to one hundred dollars, and eventually to five hundred dollars. However, this monetary deterrent proved ineffective, and an increasing number of immigrants from China continued to enter British Columbia. Many of them left the province, taking the train east. And the prejudice and discrimination against them, hitherto only in British Columbia, spread to other parts of Canada. The anti-Chinese movement became so widespread that in 1923 the federal government was compelled to pass an act to exclude the Chinese from entering the country. During this period of exclusion, from 1924 to 1947, only a few Chinese immigrants were allowed into Canada.

IMMIGRATION ACTS AND HEAD TAXES

Throughout the late nineteenth and early twentieth centuries, European powers had been competing to expand their African and Asian empires. The concept of white supremacy was rooted in the minds of the Caucasians, and in one battle after another they defeated China, once a powerful dragon.[1] The Manchu government was so feeble militarily that it could not protect its own country against foreign exploitation, much less care for Chinese people overseas. The discriminatory immigration acts passed in the United States and Canada were but a reflection of the international climate of racism and white superiority at the beginning of the twentieth century.

On 8 May 1882, the American government passed the Chinese Exclusion Act, which suspended the admission of Chinese labourers to the United States for ten years; Chinese residing in the United States before 5 August 1882 were to be given identification certificates.[2] As a result, many Chinese labourers who had lived in the United States but were working on the railway in British Columbia could not get these American identity certificates, and when their Canadian jobs ended, they could not return to the United States. The whites in British Columbia were furious when many stranded Chinese labourers thus were forced to stay in the province, and they were enraged further when new Chinese immigrants continued to come by the hundreds. Noah Shakespeare and others rallied workers to parade in protest and sent petitions to the provincial legislature and Parliament to complain about Chinese competition in the labour market. On 18 February 1884, the Legislative Assembly of British Columbia passed three acts: one prohibiting Chinese immigrants to land in the province; one preventing them from acquiring Crown lands; and one requiring every Chinese in the province over fourteen years of age to pay a ten-dollar annual fee for a residential licence.[3] These acts were soon disallowed by a federal Order-in-Council, but to appease the anger of whites in British Columbia the federal government in July 1884 set up the Royal Commission referred to earlier to investigate the "Chinese question" in the province. The result was the introduction of Canada's first anti-Chinese immigration law in 1885.

The Chinese Immigration Act of 1885 stipulated that every Chinese person who entered Canada for the first time as a landed immigrant had to pay a head tax of fifty dollars for a certificate of entry.[4] The intention of the act was to impose a financial burden on Chinese labourers, in order to discourage them from coming. According to the Commission Report, the average Chinese labourer could earn only $225 a year. After deducting $130 for food and clothing, twenty-four dollars for rent, and twenty-eight dollars for road taxes, medicine, and other expenses, he could save only forty-three dollars a year.[5] In other words, the fifty-dollar head tax used up the rest of a year's savings. Meanwhile, those Chinese already in British Columbia were unable to find work; in 1885, as the CPR neared completion, numerous Chinese labourers were laid off. They drifted back to Victoria's Chinatown, unable to afford the fare back to China. Many were found loitering in the streets, suffering from cold and hunger. In January 1886, for example, soup kitchens were set up to help some 300 hungry Chinese through the bitterly cold winter.[6] And while the unemployed labourers were struggling to survive, new Chinese immigrants continued to arrive.

The tempo of the anti-Chinese movement in British Columbia accelerated. The Knights of Labour, founded in Philadelphia in 1869, had developed into a powerful labour union by the early 1880s.[7] Instead of promoting unionism, it concentrated on attacking Oriental labourers and establishing anti-Oriental organizations in Victoria and other cities. The Provincial Federated Labour Congress, formed by various labour unions in British Columbia, also demanded Chinese exclusion.[8] In 1892, the Victoria Trades and Labour Council circulated copies of a petition requesting Ottawa to raise the Chinese head tax.[9] Similar petitions were circulated by the Anti-Mongolian Society.[10] In response to these petitions, the Legislative Assembly passed a resolution in 1899 to ask the federal government to increase the head tax to at least $500;[11] in the same year, the assembly passed an act declaring it unlawful for any person to enter the province who failed to "write out and sign in the characters of some language of Europe."[12] Although the act was later disallowed, there was an increasing demand to restrict Chinese immigration. This prompted the Chinese Consolidated Benevolent Association of Victoria (CCBA) to send a circular to China in 1899 which related the hardship of the Chinese immigrants in Victoria and told the Chinese people in China not to come to Canada.[13] Evidently the circular had no effect, for within the first three months of the following year, 1,325 Chinese landed in Victoria.[14]

The fifty-dollar head tax, labour petitions, and Chinese circulars all failed to curb Chinese immigration, and in 1901 the federal government passed another Chinese Immigration Act, increasing the head tax to $100.[15] This act was replaced in 1903 by another which raised the tax to $500.[16] This virtually prohibitive tax had an immediate and noticeable effect: during the first half of 1904, no Chinese immigrants entered. Eight Chinese entered in 1905, paying the $500 head tax, and in the following year, twenty-two immigrants arrived. The Victoria Trades and Labour Council still regarded the $500 head tax as too low to be effective, and it sent a letter to Prime Minister Wilfrid Laurier suggesting total exclusion of all Chinese from Canada or an increase of the head tax from $500 to $1,000. The council argued that "the depreciation in the value of gold which is expressed in the increase in the price of the necessities and commodities of life, has reduced the potency of the $500 head tax on Chinese as a bar against the immigration."[17]

EXCLUSION ACT, 1923

The Chinese in Canada always feared that the federal government would restrict Chinese entry completely if the Manchu government made no attempt to stop or restrict emigration. In March 1903, the CCBA sent a letter directly to the Chinese ambassador in Britain, begging him to ask the viceroy of Guangdong Province to post a notice urging people not to come to Canada.[18] The Canadian government also wanted to solve the problem of Chinese immigration, as violence against Chinese labourers increased.[19]

In March 1909, W.L. Mackenzie King, then Deputy Minister of Labour, attended the International Opium Conference in Shanghai.[20] He went to Beijing to meet Liang Tun-yen, Chinese Minister for Foreign Affairs, and proposed to him that if China, like Japan and India, restricted emigration by issuing only a limited number of passports, Canada would remove the head tax on Chinese immigrants.[21] But no agreement was reached; the Manchu government refused to restrict emigration and the Canadian government continued its $500 head tax.

In spite of the heavy tax and strong discrimination, Canada was still a place of hope for many poor Chinese labourers. While they could earn about two dollars a month in China, their monthly income in Canada might be ten to twenty times more. If a labourer worked hard and did not gamble, he might save enough money to pay off his debts for the passage and head tax and be able to support his family in China. On the other hand, he would find it difficult to make any kind of living in his impoverished homeland. Since the mid-nineteenth century, China's economy had been deteriorating, affected by wars and revolutions such as the Taiping revolution, the wars against foreign powers, Dr. Sun's revolution, and warfare among the warlords.

Consequently a steadily increasing number of Chinese emigrated. In 1913, 7,078 Chinese left their homes to come to Canada.[22] The enormous influx of Chinese to Canada also worried Chinese community leaders. In 1913, the CCBA sent another circular to China, advising people not to come to Canada. It claimed that "innumerable overseas Chinese here are jobless and many are suffering from cold and hunger. They are so miserable that their anguish is beyond description."[23] The CCBA also wrote to the Chinese consul generals in Ottawa and Vancouver, imploring them to encourage the Chinese government to prohibit Chinese workers from emigrating to Canada. Lin Shih-yuan, the consul in Vancouver, suggested to the CCBA that various county associations in Victoria send a telegram to the civil magistrate of Guangdong Province, requesting him to persuade people not to come to Canada, at least temporarily. Such a telegram was sent to China on 17 April 1914, signed by eleven county association representatives.[24] Like previous circulars, the telegram carried no weight in China since its content was unofficial.

Canada experienced a general economic slump after the First World War. Factory workers were laid off; returned soldiers were unemployed. White men again blamed Chinese labourers for taking away their jobs.[25] Their resentment heightened as the number of Chinese immigrants soared—from 650 in 1918 to 4,066 in 1919.[26] Clearly the system of restricting immigration by means of the head tax was not working. The federal government was so alarmed at the postwar influx of Chinese that in June 1919 it announced that immigrants of any nationality who belonged to the labouring class were prohibited from landing in Canada at any of twenty ports of entry in British Columbia.[27] Obviously this measure was primarily aimed at Oriental labourers, since virtually no Europeans entered Canada through British Columbia.

A year before the 1921 federal election, a new section of the Dominion Franchise Bill decreed that "persons who by the laws of any province in Canada are disqualified from voting for a member of the Legislative Assembly of such province in respect of race, shall not be qualified to vote in such province of this act."[28] Thus Chinese people were not permitted to vote in the 1921 federal election. Chinese immigration again was used as bait for votes in the election; almost every candidate in British Columbia promised to restrict Chinese immigration.[29] On 8 May 1922, W.G. McQuarrie, an MP for New Westminster, introduced a resolution that the federal government take immediate action to exclude future immigration of Oriental aliens.[30] The resolution was unanimously supported by all thirteen MPs from British Columbia, and in April 1923 the House of Commons completed the final draft of a new Chinese Immigration Act, which the Chinese people called the "Forty-Three Harsh Regulations on Chinese Immigration," or "Exclusion Act." In spite of Chinese protest, this act was passed on 30 June 1923, going into effect the next day, Dominion Day.[31]

Under the new act, people of Chinese origin or descent were prohibited from coming into Canada except for those in exempted classes, such as consular officials, children born in Canada, merchants, and students.[32] The Minister of Immigration and Colonization defined "merchants" loosely, and "students" referred to people studying in a Canadian university, who were required not to work while they were studying.[33] The "merchants" and "students" were only allowed to enter Canada through Vancouver or Victoria.

Unlike previous immigration acts, the Chinese Immigration Act of 1923 was very effective in excluding the Chinese. Under the act, every Chinese in Canada had to register with the Ministry of Immigration and Colonization before he left the country. A person so registered was only permitted to leave Canada for two years; if he did not return to Canada within two years, he would not be allowed in. As a consequence, many Chinese people were effectively separated from their families in China.

IMMIGRATION AND POPULATION GROWTH, 1886–1940s

Before introduction of the 1885 Chinese Immigration Act the number of Chinese immigrants entering and leaving Canada was not recorded. As well, an undetermined number of Chinese left the province by being smuggled into the United States after the American government passed the Chinese Exclusion Act in 1882. The Americans found it difficult to patrol the northwestern frontier along the Washington Territory, separated from British Columbia by the narrow Strait of Juan de Fuca; a revenue officer at Port Townsend estimated that at least 1,000 Chinese had been smuggled across the strait during the first ten months of 1883.[34]

In 1885, the federal government began to record the annual entry and departure of Chinese immigrants by means of the head tax and departure fee, and the actual numbers recorded were quite revealing. Before 1904, the average annual arrival of Chinese immigrants exceeded the annual departure (Figure 8); after the head tax was raised to $500, the number of Chinese im-

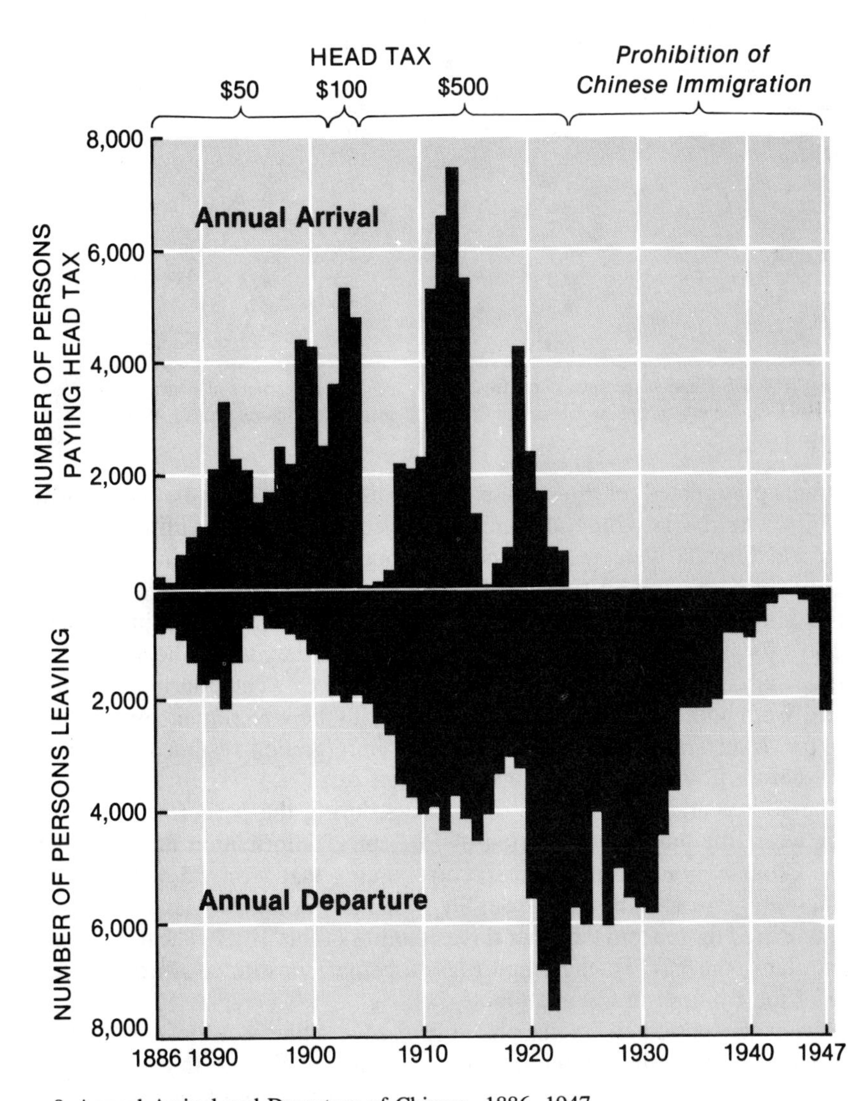

8 Annual Arrival and Departure of Chinese, 1886–1947

TABLE 10: Home county origins of Chinese in Canada, 1890s–1940s

County	% of total		
	hospital donors, 1892–1915	shipment of deceased, 1937	reburial of deceased, 1937–45
The Four Counties			
Taishan	34.0	43.2	36.9
Xinhui	18.4	19.1	13.3
Kaiping	15.7	16.7	14.4
Enping	4.6	11.8	7.5
Panyu	9.1	0.1	17.6
Zhongshan	6.7	2.7	4.5
Haoshan	2.2	2.5	0.2
Other	9.3	3.9	6.5
Total	100.0	100.0	100.0

Source: stubs of donation receipts, CCBA, 1892–1915; records of shipment of bones, CCBA, 1937; and reburial records of crates of bones in Chinese Cemetery of Victoria, CCBA, 1960

migrants plummeted for three years, then soared again. After the outbreak of the First World War, Chinese immigration declined as ships sailing between Asia and Canada were requisitioned as troop ships. Indeed, in 1916 only twenty Chinese came to Canada. In the meantime, the federal government was urging unemployed Chinese labourers to leave Canada, permitting them to leave for more than a year with a guarantee that they would be allowed to re-enter after the war without paying the head tax.[35] Thus during the First World War about 12,000 Chinese registered with the government and left the country. Many returned after the war, and immigration resumed for a few years before it was stopped by the Exclusion Act.

The source areas of Chinese immigrants during the period of restricted entry were like those in the period of free entry. Information derived from three Chinese non-census sources corroborates that over 75 per cent of Chinese immigrants still came from Siyi (the Four Counties); Taishan people outnumbered those from the other three counties (Table 10).[36] The three dominant clans, the Lis, Huangs, and Mas, together constituted nearly 30 per cent of the Chinese in Canada (Table 11).

During the period of exclusion, only twelve Chinese were admitted to Canada as immigrants; ten belonged to the exempted classes. In the same period, 61,213 Chinese registered for leave.[37] After reaching a record high of nearly 47,000 in 1931, the Chinese population in Canada began to decline for three main reasons. First, after the Exclusion Act was passed, not only did no new Chinese immigrants enter Canada, but many Chinese also started leaving the country (Table 12). Between 1921 and 1930, for example, a record

TABLE 11: Clan origins of Chinese in Canada, 1890s–1940s

	% of total		
Clan	hospital donors, 1892–1915	shipment record, 1937	reburial record, 1937–45
Li	10.5	12.3	9.2
Huang	10.4	10.4	10.6
Ma	7.5	8.5	9.5
Zhou	6.8	3.4	6.1
Chen	6.6	5.0	4.3
Lin	4.8	4.4	4.8
Liu	2.6	2.0	2.6
Zhang	2.5	2.4	1.8
Yu	2.4	1.6	1.7
Guan	2.2	2.2	1.8
Zheng	2.2	4.4	2.1
Liang	2.1	4.0	3.0
Wu	2.0	3.1	2.6
Xie	1.8	1.0	0.1
Other	35.6	17.3	39.8
Total	100.0	100.0	100.0

Source: stubs of donation receipts, CCBA, 1892–1915; records of shipment of bones, CCBA, 1937; and reburial records of crates of bones in Chinese Cemetery of Victoria, CCBA, 1960

TABLE 12: Number of Chinese entering and leaving Canada, 1886–1947

Period	Entry	Departure	Net gain or loss
1886–90	2,686	5,424	−2,738
1891–1900	26,345	10,429	15,916
1901–10	23,495	25,453	−1,958
1911–20	32,244	38,899	−6,655
1921–30	5,572	58,857	−53,285
1931–40	4	24,794	−24,790
1941–7	4	4,244	−4,240
Total	90,350	168,100	−77,750

Sources: compiled from annual reports of Superintendent of Immigration, Department of the Interior, 1886–1917; Annual Reports of Department of Immigration and Colonization, Dominion of Canada, 1918–30; and Report of the Department of Mines and Resources, Dominion of Canada, 1947, 245

TABLE 13: Chinese population growth in Canada, 1881–1986

	Chinese population			
Census year	No. of persons	Annual increase	% increase	% of Canada's population
1881	4,383	–	–	0.10
1891	9,129	4,746	108.3	0.19
1901	17,312	8,183	89.6	0.32
1911	27,831	10,519	60.8	0.39
1921	39,587	11,813	42.5	0.45
1931	46,519	6,932	17.5	0.45
1941	34,627	–11,892	–25.6	0.30
1951	32,528	–2,099	–6.1	0.23
1961	58,197	25,669	78.9	0.32
1971	118,815	60,618	104.2	0.55
1981	289,245	170,430	143.4	1.20
1986	414,040	124,975	43.1	1.65

Sources: Censuses of Canada, 1881–1981; Statistics Canada, *The Daily: Ethnic Origin*, 3 Dec. 1987

high of nearly 59,000 Chinese registered for leave. Most were middle-aged or older and had no desire to stay in Canada.

The second factor for the population decline was the unbalanced sex ratio. In 1921, for example, there were 37,163 Chinese men in Canada and 2,424 Chinese women. The ratio was about twenty-five males for every female. As well, there were very few child-bearing Chinese females. In British Columbia, for example, there were 685 births and 623 deaths between 1923 and 1925, giving an aggregate surplus of only sixty-two persons for the three years; the natural increase in population therefore was extremely insignificant.[38] In 1931, for example, only 12 per cent of the Chinese population in Canada were native-born.[39]

The last factor affecting the population was the economic depression of the 1930s, during which an undetermined number of Chinese died from starvation or malnutrition-related illnesses. To induce hungry and unemployed Chinese to return to China and to save the cost of supporting them, the Canadian government passed an Order-in-Council (PC 3173) on 29 December 1931 allowing registered Chinese to be absent from Canada for four years instead of two. Furthermore, those who were willing to return to China permanently would be repatriated at the expense of the government (the sixty-five-dollar one-way fare). This "lenient policy" towards the Chinese was supported by the public as a way to ease the unemployment situation in Canada and reduce the number of Chinese applying for relief.[40] As a result, several hundred Chinese returned to their homeland.[41] In addition, many Chinese returned to China without bothering to register out, since they had no intention of coming back to Canada. In 1934–5, for example, 889 Chinese did

TABLE 14: Distribution of Chinese in Canada, 1881–1986

	Distribution by percentage of national total					
Census year	British Columbia	Ontario	Prairie provinces	Quebec	Other provinces	Total
1881	99.2	0.5	0.1	0.2	0	100
1891	97.6	1.1	0.3	0.4	0.6	100
1901	86.0	4.2	2.8	6.0	1.0	100
1911	70.5	10.0	10.0	5.7	3.8	100
1921	59.4	14.2	19.0	5.9	1.5	100
1931	58.3	14.9	19.5	5.9	1.4	100
1941	53.8	17.1	19.9	6.9	2.3	100
1951	49.0	21.5	20.8	5.9	2.8	100
1961	41.6	26.0	21.5	8.2	2.7	100
1971	37.2	33.1	17.2	10.0	2.5	100
1981	33.5	41.0	17.5	6.7	1.3	100
1986	30.3	43.7	18.4	6.5	1.1	100

Sources: Censuses of Canada, 1881–1981; and Statistics Canada, *The Daily: Ethnic Origin,* 3 Dec. 1987

not register before their departure and 1,462 Chinese who had registered simply remained outside Canada beyond the time limit.[42]

POPULATION DISTRIBUTION, 1886–1940s

Throughout the six decades between 1881 and 1941, the Chinese never constituted more than 0.5 per cent of Canada's population (Table 13). However, their distribution in the country had undergone great changes since completion of the CPR. During the late 1880s and early 1890s, most new Chinese immigrants still chose to settle in British Columbia with relatives and friends. However, the growing hostility towards Chinese in the province induced many to move east to other provinces. As a result, Chinese communities appeared in Calgary, Moose Jaw, Regina, and other small railway towns. Some Chinese went further east to Ontario and Quebec where job opportunities were greater and the host society was less hostile. In 1901, only 4 per cent of Canada's Chinese people lived in Ontario and 3 per cent in the prairie provinces, but by 1921, the Chinese people in Ontario and the prairie provinces accounted for 14 per cent and 19 per cent of the national total respectively (Table 14). On the other hand, British Columbia's share of Canada's Chinese population had been declining. Before the turn of the century, British Columbia was home to over 85 per cent of the country's Chinese, but by 1921 the percentage had dropped to 59 per cent.

Similarly, the percentage of Chinese in British Columbia had dropped

TABLE 15: Chinese population growth in British Columbia, 1881–1986

	Population		
Census year	Chinese	Provincial total	Percentage of total
1881	4,350	49,459	8.80
1891	8,910	98,173	9.08
1901	14,885	178,657	8.33
1911	19,568	392,480	4.99
1921	23,533	524,582	4.49
1931	27,139	694,263	3.91
1941	18,619	817,861	2.28
1951	15,933	1,165,210	1.37
1961	24,227	1,629,082	1.49
1971	44,315	2,184,620	2.02
1981	96,915	2,713,615	3.60
1986	125,535	2,883,367	4.35

Sources: Censuses of Canada, 1881–1981; Statistics Canada, *The Daily: Ethnic Origin,* 3 Dec 1987

TABLE 16: Geographical shift of Chinese population in Canada, 1901, 1941

	Number of persons		Expected no.		
Province	1901	1941	of persons in 1941	Net shift	% of net shift
British Columbia	14,885	18,619	29,773	– 11,154	– 99.9
Ontario	732	6,143	1,464	+ 4,679	+41.9
Alberta	235	3,122	470	+ 2,652	+23.8
Saskatchewan	41	2,545	82	+ 2,463	+22.1
Quebec	1,037	2,378	2,074	+ 304	+ 2.7
Manitoba	206	1,248	412	+ 836	+ 7.5
Nova Scotia	106	372	212	+ 160	+ 1.4
New Brunswick	59	152	118	+ 34	+ 0.3
Prince Edward Is.	4	45	8	+ 37	+ 0.3
Yukon and NWT	7	3	14	– 11	– 0.1
Total	17,312	34,627	34,627	| 11,165 |	| 100 |

Sources: Censuses of Canada, 1901, 1941

steadily from 9 per cent in 1891 to only 4 per cent in 1921 (Table 15). As well, the distribution of Chinese within the province had already changed drastically. After the Exclusion Act of 1923, Chinese settlements in the interior mainland began to decline as people returned to China or moved to the more secure communities in Vancouver and Victoria.[43] By 1941, nearly 39 per cent of the 11,000 Chinese people in British Columbia resided in Vancouver; 16 per cent lived in Victoria. The remaining 45 per cent were widely scattered in small towns and cities in the province.

The geographical shift, a method of measuring regional shift in population

TABLE 17: Chinese population in metropolitan areas, 1941

Locality	City population	Metropolitan population	Percentage of Metro's total
Vancouver	7,174	7,880	91.0
Victoria	3,037	3,435	88.4
Toronto	2,326	2,559	91.0
Montreal	1,703	1,865	91.3
Winnipeg	719	762	94.4
Ottawa	272	289	94.1
Windsor	259	267	97.0
Hamilton	236	253	93.3
Halifax	127	146	87.0
Quebec City	130	134	97.0

Source: compiled from Census of Canada, 1941, Table 33, and Table 43, 508–17. This table includes only those cities with 100 Chinese residents or more.

over an intercensal period, was used to compare the distribution of Chinese people in Canada in 1901 with that in 1941; a positive difference signifies a shift of the population into the province and a negative difference indicates a shift out of the province.[44] The geographical shift of the Chinese population before and after the Great Depression was significant (Table 16). British Columbia accounted for 99 per cent; a loss of Chinese population in comparison with other provinces (Figure 9). Of all the provinces with a gain in Chinese population, Ontario had nearly 41 per cent of the inward shift, doubling the percentages of both Alberta and Saskatchewan.

The 1941 census reveals that half of Canada's 35,000 Chinese lived in ten metropolitan cities. The largest Chinese community was in Vancouver. Victoria, which had most of the country's Chinese residents in the 1880s, trailed behind Vancouver, but it had more than Toronto and Montreal (Table 17). In most metropolitan cities, over 90 per cent of the Chinese population was in the central city. In Metropolitan Vancouver, for example, 91 per cent of its 8,000 Chinese lived in Vancouver City in 1941 and the remaining 9 per cent resided in Richmond and Burnaby municipalities.[45] Similarly, 91 per cent of Metro Toronto's Chinese lived in the City of Toronto; the remaining 9 per cent resided in the adjoining rural subdivisions of York and Etobicoke.[46] The City of Montreal had 91 per cent of Montreal's Chinese residents, and Westmount, Verdun, and Outremont shared the remaining 9 per cent.

The Chinese were highly concentrated in large cities such as Vancouver, Toronto, and Montreal because they provided a larger market for Chinese laundries and restaurants and provided more opportunity to find jobs in stores, factories, and other trades and businesses than did small towns and villages (Figure 10). For example, Montreal had nearly 80 per cent of Quebec's Chinese population, and Winnipeg had 61 per cent of Manitoba's

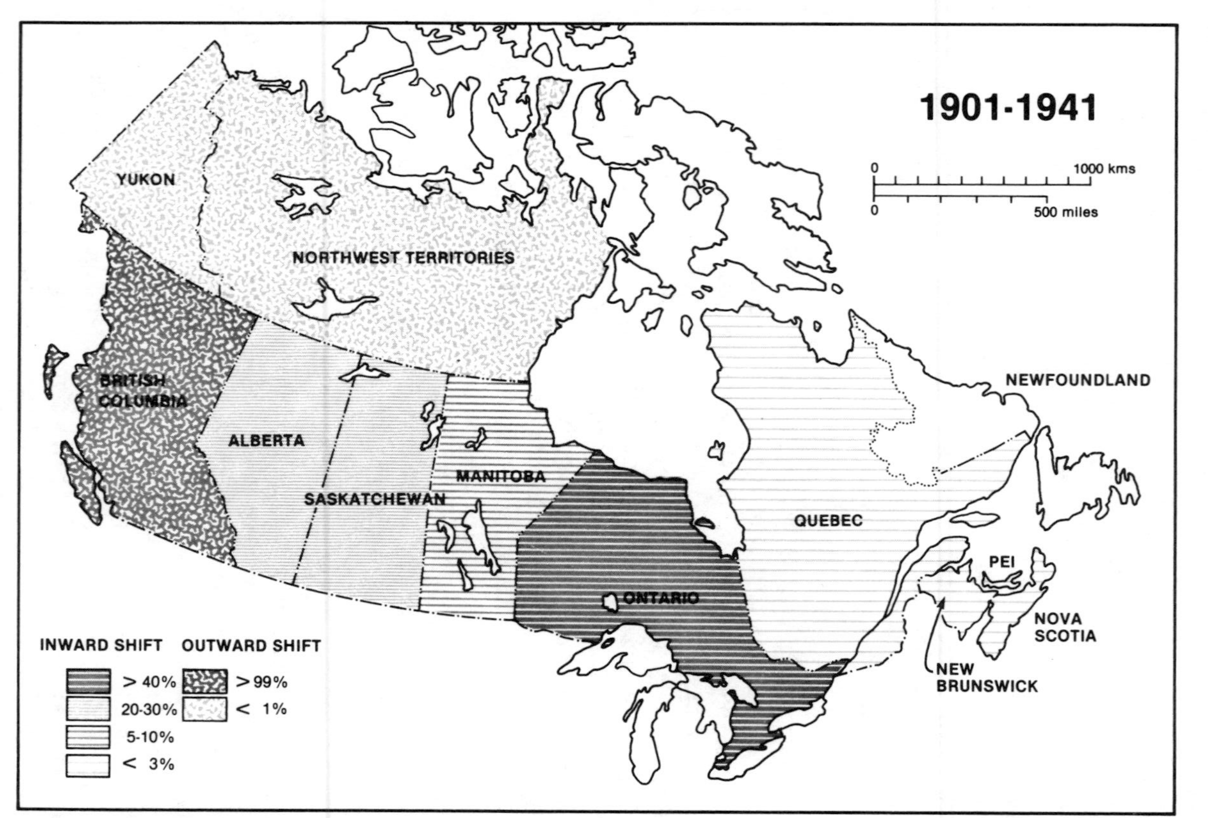

9 Geographical Shift of Chinese Population in Canada, 1901, 1941

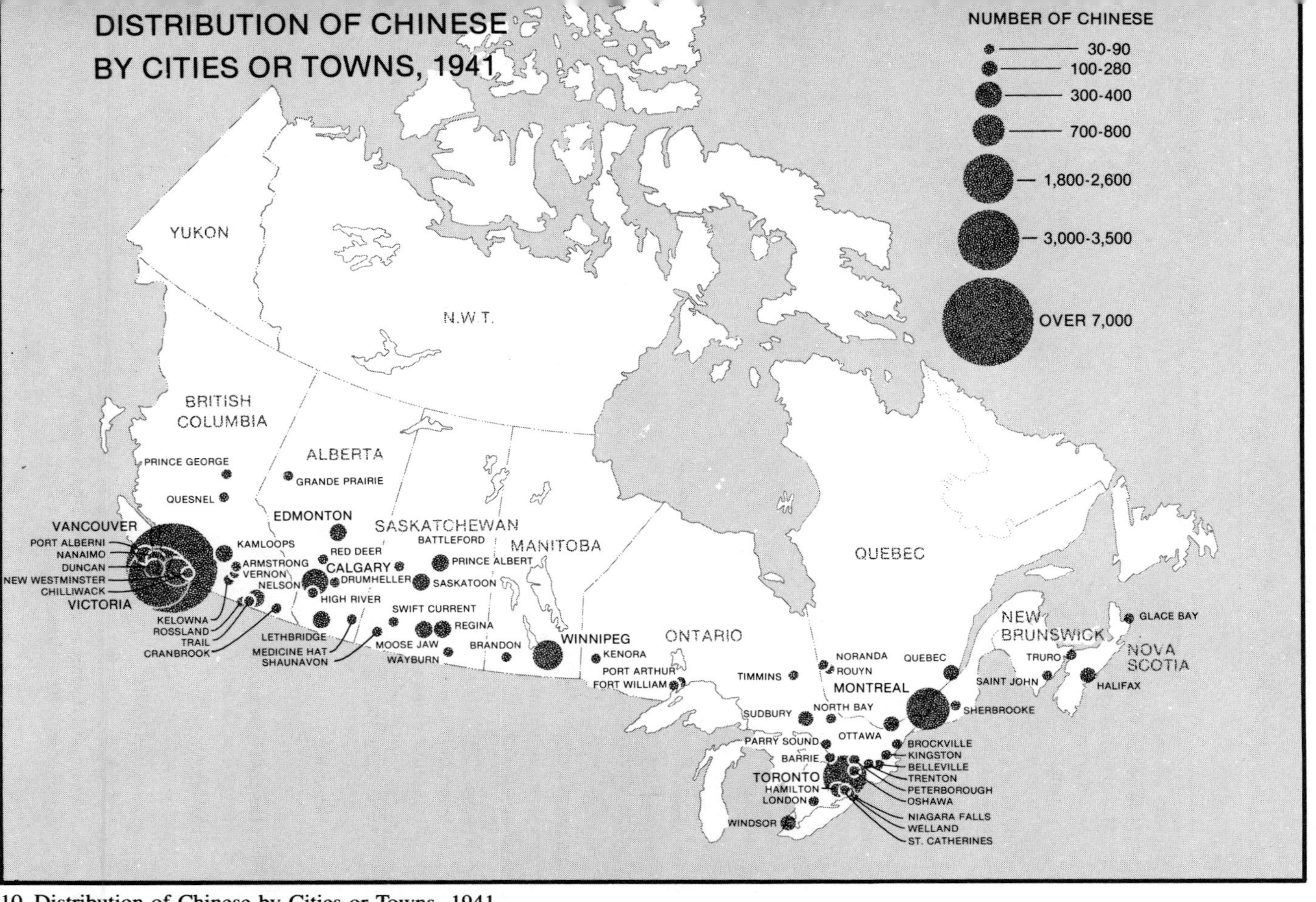

10 Distribution of Chinese by Cities or Towns, 1941

TABLE 18: Distribution of Chinese by cities, 1941

City	Number of Chinese	% of province's total Chinese population
BRITISH COLUMBIA		
Metropolitan Vancouver	7,880	42.3
Metropolitan Victoria	3,435	18.4
New Westminster	400	2.1
Nanaimo	298	1.6
Kamloops	281	1.5
Duncan City	165	0.9
Port Alberni	137	0.7
Vernon	112	0.6
Nelson	103	0.6
Other places	5,808	31.3
ALBERTA		
Calgary	799	25.6
Edmonton	384	12.3
Lethbridge	248	7.9
Other places	1,691	54.2
SASKATCHEWAN		
Moose Jaw	261	10.3
Regina	247	9.7
Saskatoon	206	8.1
Prince Albert	103	4.0
Other places	1,728	67.9
MANITOBA		
Metropolitan Winnipeg	762	61.0
Other places	486	38.9
ONTARIO		
Metropolitan Toronto	2,559	41.7
Metropolitan Ottawa	289	4.7
Metropolitan Windsor	267	4.3
Metropolitan Hamilton	253	4.1
Sudbury	109	1.8
Other places	2,666	43.4
QUEBEC		
Metropolitan Montreal	1,865	78.4
Metropolitan Quebec City	134	5.6
Other places	379	16.0
NOVA SCOTIA		
Metropolitan Halifax	147	40.0
Other places	225	60.0

Source: compiled from Census of Canada, 1941, Table 33 and Table 43, 508–17. This table includes only cities with 100 Chinese residents or more.

(Table 18). Vancouver, Toronto, and Halifax each had about 40 per cent of their province's total Chinese population. This drift towards large urban centres continued when Chinese immigration resumed after the Second World War.

5

Vicissitudes of Old Chinatowns

During the period of restricted entry, Chinatowns of varying sizes emerged in cities and towns across Canada. The largest Chinatowns in Canada were in Vancouver and Victoria. The six major Chinatowns on the prairies were in Calgary, Edmonton, Lethbridge, Saskatoon, Moose Jaw, and Winnipeg. In Ontario, half the province's Chinese people lived in Toronto's Chinatown and the other half in other cities such as Ottawa and Hamilton. In Quebec, Montreal's Chinatown, with 90 per cent of the province's Chinese population, overshadowed the small Chinatown in Quebec City. The Chinese population in the Atlantic Provinces was so small that no Chinatowns were established. During the period of exclusion, all Chinatowns in Canada were declining, and some became extinct.

TOWNSCAPES AND IMAGES

Canadian Chinatowns were not completely alike, and varied in age of development, size of population, and economic conditions. The variations tended to increase as one moved eastward, reflecting the general movement of the Chinese population from British Columbia to eastern Canada. Nevertheless, all Old Chinatowns went through the budding, blooming, and withering stages and had much in common. Given this sequence of evolution, the stage-development model was quite appropriate. For example, most Old Chinatowns had similar settings during the budding stage, namely on the fringes of downtown areas and physically and culturally separated from adjacent neighbourhoods. The types of business were remarkably constant from one Chinatown to another: Chinese restaurants, grocery stores, laundries, and the like. Institutional buildings, store signs, advertisements, merchandise, number of pedestrians, and other components of the townscape were firmly established and visibly Chinese. The strong visual image of an Old Chinatown blended with its mythical image as a mysterious and dangerous place with many narrow alleys and underground tunnels.

Consistencies in townscapes and images are typical in the evolution of Chinatowns. During the 1910s, they shared Canada's economic boom and expanded both vertically and horizontally. Many wooden structures were replaced by two- or three-storey brick buildings, and new buildings were constructed outside main streets as growth fanned out to the periphery. An important aspect of the Chinatown landscape was the pattern of streets and the arrangement of buildings along them. Larger Chinatowns usually had one or two main streets, such as Victoria's Fisgard and Cormorant streets and Vancouver's Pender and Keefer streets. Perpendicular to these main streets were numerous side streets and narrow alleys. Fan Tan Alley and Theatre Alley in Victoria and Shanghai Alley, Canton Alley, and Market Alley in Vancouver, for example, were well-known spots within the precincts of Chinatown. In Victoria, Fan Tan Alley once contained numerous gambling clubs and small restaurants, and Theatre Alley housed a Chinese theatre, opium dens, and brothels. City blocks inside Chinatown sometimes contained so many enclosed courtyards, alleys, and passageways that they became a "forbidden city" to outsiders; only a resident could find his way from place to place inside these labyrinthine blocks.

Most buildings in larger Chinatowns were built between the late 1880s and the 1920s and designed by Western architects, and included Western features such as bay windows. They were common in Chinatown structures because they increased the amount of interior floor space; their effectiveness is best illustrated by the Sam Kee Building in Vancouver (Plate 3). The building is a long, narrow, two-storey structure, averaging 1.8 metres (six feet) in width, but its bay windows increase the average width of its upper floor to 2.3 metres (7.5 feet).[1] A "cheater floor" is also a common Western architectural feature in Chinatown structures; it is a low-ceiled mezzanine so named because in the early days, tax assessments were based on the height of a building, which had an intermediate, untaxed storey, the "cheater floor" (Plate 4). But although buildings in Chinatown are constructed in Western styles, Chinese decorative details display the Chinese influence and create a unique townscape. The most obvious Chinese architectural features are upturned eaves and roof corners, extended eaves covering main balconies, and structural components such as tiled roofs, latticed windows, moon-shaped doors, and recessed balconies. In South China, recessed balconies are extremely common because they are open spaces for children to play in, and in which adults can worship during festivals and hang up clothes on rainy days. As most early Chinese immigrants came from South China, they probably asked Western architects to include recessed balconies in their institutional buildings.

Chinatown streetscapes are notable for the decorative motifs of dragons, phoenixes, and lions carved or painted on columns, walls, and shop signs. A wide variety of colours, particularly golden yellow, mandarin red, emerald green, and imperial gold are used to highlight decorative details, and numerous ornamented components such as pagodas, lanterns, chopsticks, and artistic Chinese characters enhance the Oriental nature of the structures. Most buildings are two- or three-storeyed, their strong vertical lines in direct harmony with vertical store signs and advertisements written in Chinese calligraphy. The movement of people is vertical: most tenement buildings are accessible via long narrow staircases. Ground floors are usually rented to businesses, and upper floors are reserved for institutional or residential use. In some ornate institutional buildings, the roofline has an ornamental parapet displaying in Chinese or English the association name and date of construction.

In the old days, Chinatowns were composed of many small shops such as hand laundries, tailor shops, small cafés, restaurants, and grocery stores. These stores were essential to a Chinatown's predominantly male society, where numerous bachelors needed someone to do their cooking, washing, and ironing. Those who had families tended to open their own small grocery stores, providing jobs and food for the entire family. Hand laundries were also set up outside Chinatown to cater to customers of the host society.

Although Chinatown was a place of comfort, security, and companionship to the Chinese, it was still perceived by the white community as a place of filth and sin, where gamblers, prostitutes, pimps, and other social outcasts congregated. Only undaunted missionaries, determined policemen, and inquisitive reporters would go there. The following excerpt describes a white woman's perception of Vancouver's Chinatown: "Secret tunnels. Opium dens. White slavery. Inscrutable celestials in pigtails, with knives hidden up their silk sleeves. Gorgeous slant-eyed beauties with bound feet. Shadowy, sinister doorways. Ancient love drugs. Medicines made from the tongues of wild serpents."[2]

CHINATOWNS IN GOLD-MINING TOWNS

At the beginning of the twentieth century, most Old Chinatowns in gold-mining or coal-mining towns in British Columbia were extinct, and the streetscapes of the few surviving Chinatowns had changed drastically. Few residents were left in Chinatowns, and many abandoned clapboard shacks or wooden cabins had deteriorated and collapsed. The opium dens, gambling halls, and brothels which once abounded had largely disappeared.

British Columbia was the only province where Chinatowns were established in gold-mining towns. After the turn of the century, Quesnelle Forks, Keithley, and other small mining towns were abandoned after their gold was exhausted, and their Chinatowns approached extinction. For example, by the late 1910s Chinese miners could make only twenty cents a day in the once-rich gravel of Quesnelle Forks; in 1922, they abandoned it.[3] Similarly, Keithley's Chinatown had about ten to twenty Chinese miners, but it was deserted in the 1920s as miners could not eke out a living.

Several instant Chinatowns emerged in the 1880s, but they did not survive more than a few years. For example, within a year an instant town of about 700 people, including 200 Chinese, sprang up in Granite Creek after gold was discovered there in 1885; the town had two restaurants, two saloons, one butcher shop, and seven stores.[4] But as soon as the returns from diggings diminished in the spring of 1886, the white miners abandoned the territory, followed a few years later by the Chinese miners. The instant Chinatown was soon gone completely.

In the early 1910s, Barkerville's Chinatown, once the largest in Canada, had entered the withering stage, having a Chinese population of about 150. A few Chinese prospectors operated placer mines in Begg's Gulch, Coulter Creek, and other Cariboo creeks.[5] Numerous Chinese miners worked in John Hopp's mines at Mosquito Creek, Burns Creek, and other places for two to three dollars a day, and some Chinese cut wood for less than two dollars a cord. Others worked as cooks or helpers in restaurants, earning about twenty dollars a month. During the 1910s there were still six Chinese stores in Barkerville's Chinatown, but by the 1920s only one store remained and the Chinese population had dwindled to thirty-five.[6] In the 1930s Barkerville's Chinatown continued to decline and was virtually empty by the 1940s.

CHINATOWNS IN COAL-MINING TOWNS

Coal mining, rather than gold mining, was the lifeblood of some small Chinatowns on Chinese Island. During the late 1880s and early 1890s, Nanaimo's Chinatown was blooming because it was the commercial hub, with small satellite Chinatowns in South Wellington, Wellington, Extension, Ladysmith, Bevan, and Cumberland (see Figure 12).

South Wellington, about nine kilometres south of Nanaimo, was a new town on the E & N Railway line. Robert Dunsmuir, owner of the E & N Railway Company, employed many Chinese labourers to work at the South Wellington mine, on the railway, and in local logging operations. By the 1890s, about a hundred Chinese lived in South Wellington. At the turn of the century, however, coal reserves in the region began to wane, and the mine closed in 1902. As a result, most Chinese moved out and the small Chinatown at South Wellington disappeared.[7] The stage-development model thus is not applicable to South Wellington's Chinatown, which was an instant Chinatown and completed the four phases of evolution in less than two decades.

Wellington's Chinatown, another instant Chinatown, boasted several hundred Chinese residents during the 1880s. In the late 1890s, the Wellington pits, owned by Wellington Collieries, were worked out, but the company discovered a rich deposit on the southern slope of Mount Benson, about eight kilometres south of Nanaimo. The company opened a new mine there and a new mining village emerged, called Extension. Oyster Harbour, eighteen kilometres away from Extension, was constructed as a port for its coal. As a result of a fire in the Wellington mines in 1899, the remaining buildings were moved to Oyster Harbour, which was renamed Ladysmith in 1900.[8] Wellington's Chinatown was finished after the Chinese miners were transferred to Extension and Ladysmith. In 1901, the Chinese living quarters at Extension and Ladysmith had a combined population of about 700.[9]

In September 1912, the "Big Strike" started in Cumberland, soon spreading to the Extension collieries and idling about 1,600 miners.[10] The Canadian Collieries Company immediately ordered the striking miners to remove their tools from the mines, evicted them from company houses, and negotiated a two-year agreement with Chinese miners. By 14 October, over 300 striking miners, including many Chinese, had returned to work in the Cumberland mines. To fight the company's strike-breaking measures, the United Mine Workers' Union started to intimidate working miners, instigating violence. Over a hundred special constables had to be despatched to Cumberland and Ladysmith to maintain order.[12]

In June 1913, the striking miners in Cumberland and Extension convinced a few hundred Nanaimo miners to join their strike.This led to a confrontation between the strikers and the workers, city police, and special constables. On 13 August, miners from Nanaimo and South Wellington marched to Ladysmith and Extension and burned, wrecked, or looted houses there.[13] Then the mob harried Chinatown, breaking into buildings and destroying property. The Chinese were forced to flee to the nearby woods. Similar riots also broke out in Nanaimo, where strikers smashed windows and bombed the Temperance Hotel, which housed imported strike-breakers. Order was restored only after troops from Victoria took control. The rioters were arrested and imprisoned, and the victims, including the Chinese in Ladysmith and Extension, were compensated by the federal government for their losses.

After the 1920s, the Chinatowns in Extension and Ladysmith continued to decline as coal production decreased. The Extension mines were finally closed in April 1931, and its Chinatown was deserted; in Ladysmith, by 1932, the population had sagged to 1,700 and its Chinatown was no longer in existence.[14]

In 1888, Dunsmuir's Union Colliery operated a small mine on the northern slope of a steep-walled, east-west valley at Union, ninety-seven kilometres north of Nanaimo.[15] The main road, Dunsmuir Avenue, ran through the mining village above a railway line on the valley bottom (Figure 11). A small lane branched down Dunsmuir Avenue to the southern slope of the valley, where the Chinese lived. When the townsite of Union was planned in 1893, its population had reached about 3,000, but since the town was hemmed in by steep slopes it was ill adapted for expansion.[16] Therefore James Dunsmuir, eldest son of Robert Dunsmuir, chose a new townsite east of Union called Cumberland.[17] In 1897, the town of Cumberland was incorporated. However, Chinatown, Japtown, Coontown (for Black people) and other ethnic settlements remained in the mining village of Union. Another small Chinatown had also been established at Bevan, a small mining community near Number Seven Mile south of Puntledge River.[18]

In 1906, Dunsmuir's Wellington Collieries Company employed about 500 Chinese at the Union mines. In many cases, Dunsmuir paid the head tax and passage fare for the miner, in return deducting seventy-five cents a day from the daily salary of $1.25. He also provided Chinese miners with company housing in Chinatown, free water supply, a regulated amount of free light and oil, and the land on which the house stood.[19]

In 1910, Wellington Collieries Company, which owned the mines in Cumberland and Extension, was sold to Canadian Collieries (Dunsmuir) Ltd. The new company continued to operate, with as many as 540 Chinese miners in 1920. All lived in Cumberland's Chinatown, which had become a self-sufficient community of several hundred people.[20] Cumberland's Chinatown contained scores of one- or two-room wooden shacks, each with a tiny vegetable garden (Plate 5). There were also many boarding houses. At its peak, Cumberland's Chinatown had no less than eighty business establishments, including twenty-four grocery stores, four restaurants, five drugstores, two 400-seat theatres, a temple, and eighteen gambling houses.[21] The Lum Yung Club, the largest wooden structure in the Chinatown, was a popular gambling establishment. The Dart Coon Club, another organization of the Hongmen Society, and the Chinese Nationalist League also had established branches in Cumberland's Chinatown.[22] During the early 1920s, two explosions shattered the Union mines. As a result, the mine was closed and many Chinese labourers left.[23] After half the town was destroyed by fire in 1936, nearly all the Chinese left, leaving Cumberland's Chinatown virtually extinct.[24]

Nanaimo's Chinatown was the last remaining Chinatown in the coal district of Vancouver Island. In 1901, the city had a Chinese population of about 600, most of whom lived in Chinatown. In 1905, the Western Fuel Company of San Francisco, which owned the Chinatown land, was sold to the Canadian Western Fuel Company (CWF), which started selling some of the property. Mah Bing Kee and Ching Chung Yung, two wealthy merchants in Nanaimo, together bought forty acres of company land on which Chinatown was located.[25] Soon after their purchase, they started to increase rents. As a

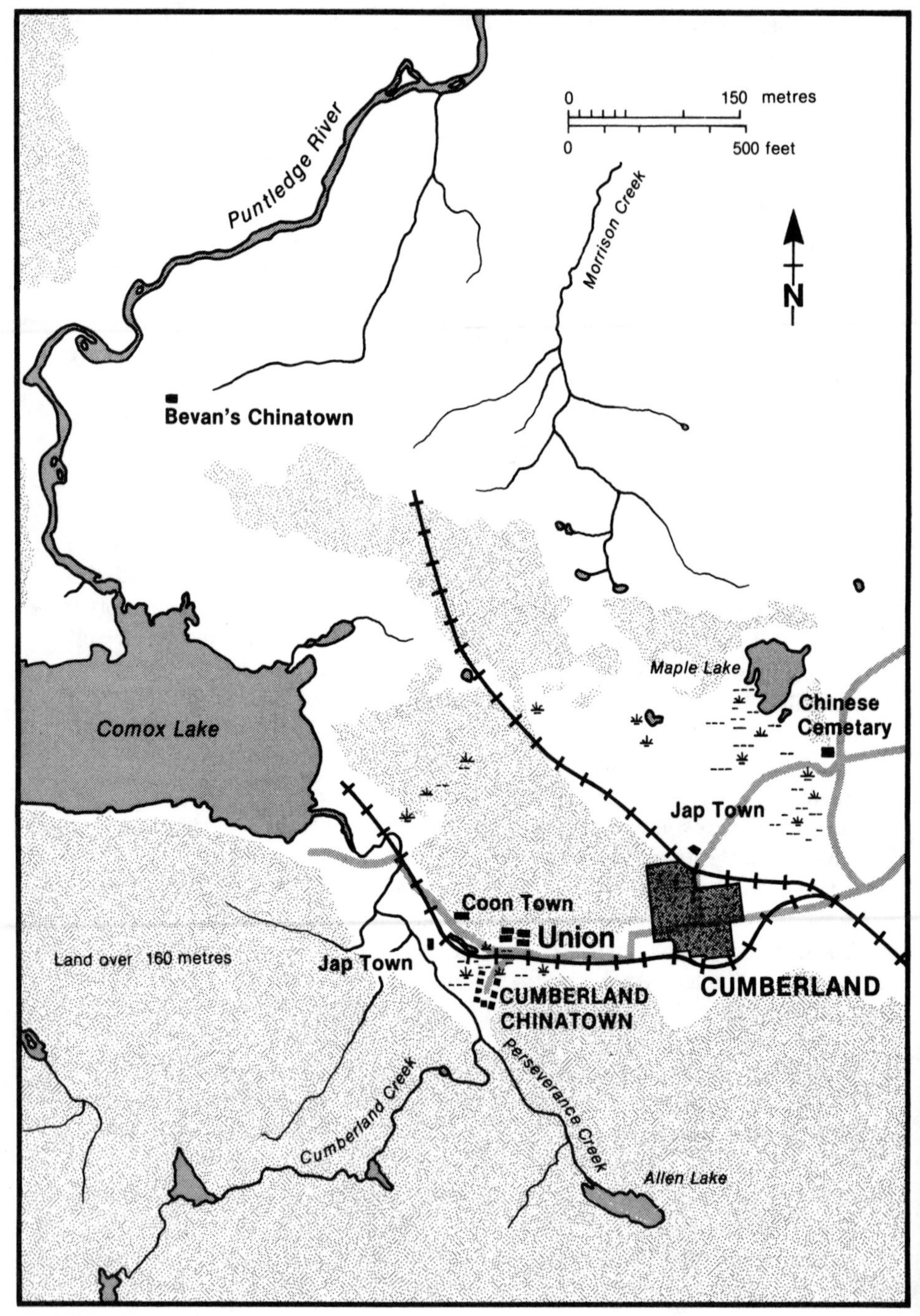

11 Cumberland's Chinatown, 1900s

result, the Chinatown merchants and residents decided to pool their resources and buy another piece of CWF land, relocating their homes and shops. Thus a third Chinatown was born in Nanaimo.

The exodus from the second Chinatown was led by the owners of the Yick Chong, Lung Kee, Hope Lee, Wing Fung, Wing Tai, San Chong, and Wong Sun Wo companies. Together with other merchants and Chinese workers, these men formed a new company known as the Lianyitang (United Benefit Association). They registered it with the government as the Land Yick Land Company Ltd; by selling 2,171 shares at five dollars per share, the company generated a starting capital of $10,855 with which to purchase some wooded land on a bluff. The property, covering an area of 11.25 acres, was about 250 metres north of the second Chinatown (Figure 12). Straddling the city boundary, the site was flanked on the left by steep cliffs and on the right by the railway track. After the trees were cleared and the land levelled, the property was divided into twenty-three lots, allocation of which was determined by drawing lots. Merchants could lease the lots from the company at a monthly rent of five dollars, but they had to erect their own buildings. As workers could not afford to build their own houses, the company erected residential buildings for them, charging a monthly rent of ninety cents each. Unemployed workers had free accommodation. As well, the company maintained the right to run all gambling clubs in the new Chinatown, an important source of income for the company.

Some buildings in the second Chinatown were built by the merchants themselves. When they began to move their buildings to the new sites, the owner of the second Chinatown, Mah Bing Kee, took them to court. However, the court ruled that the buildings constructed by individual merchants were chattel property, not part of Bing Kee's freehold, and that the merchants could relocate their buildings.[26] The second Chinatown then completed its life-cycle in 1908 according to the stage-development model after its structures were removed or demolished and the site was subdivided.

By the summer of that year, an array of wooden structures on both sides of the unpaved Pine Street constituted the beginning of the third Chinatown.[27] Its streetscape was like that of an old western town. On both sides of the main street, Pine Street, were dozens of plain frame buildings with false fronts, overhanging balconies, and wooden sidewalks (Plate 6). The buildings were quite large, usually two stories. Pine Street ran along the city's boundary and its northern side was within the city of Nanaimo, its southern side belonging to the provincial district of Harewood; consequently, neither the city nor the provincial government felt responsible for services such as schools, water, or street maintenance in Chinatown. Eventually the Chinese had to pay the city part of Chinatown's infrastructure cost and to pay taxes to both governments. During the 1910s, Chinatown was blooming; it expanded northward along

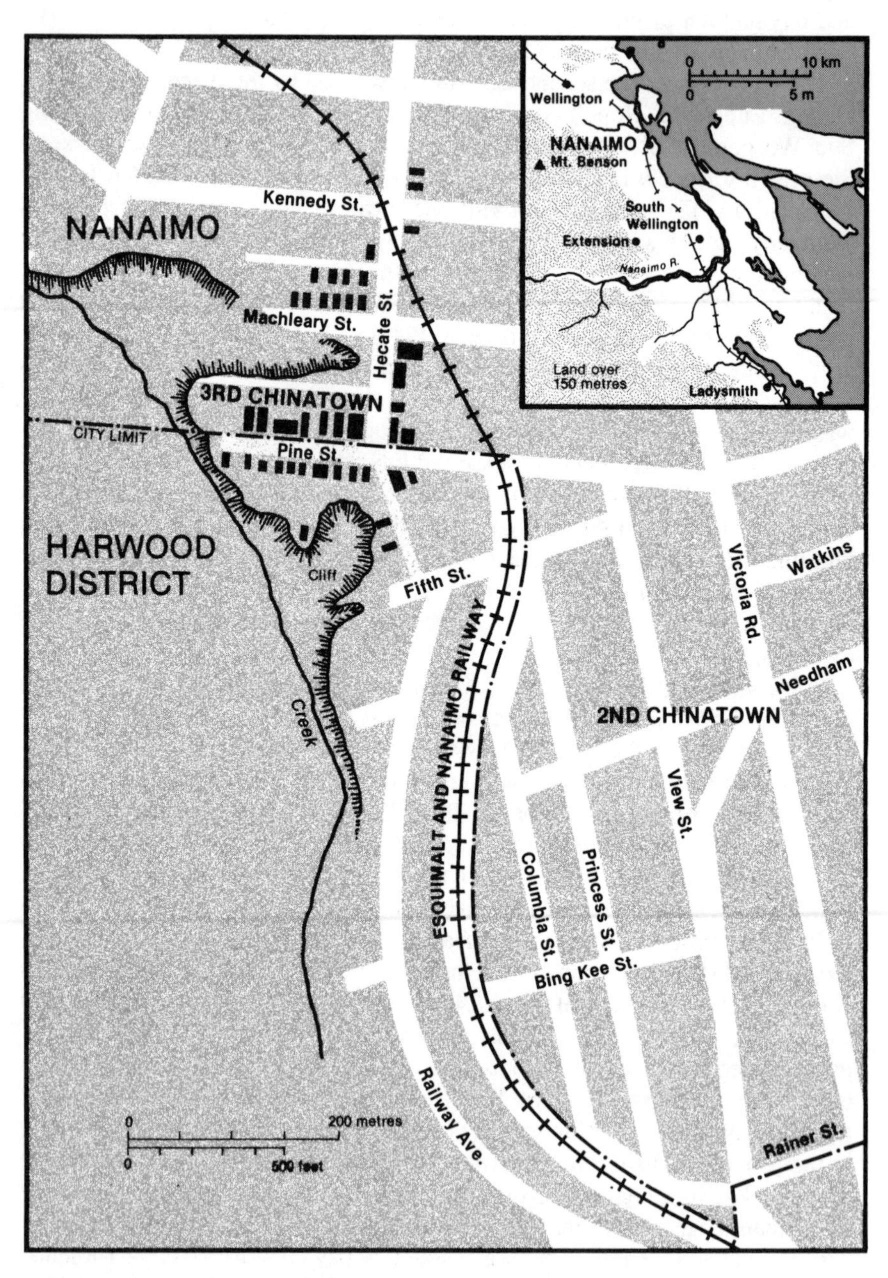

12 Nanaimo's Chinatown, 1910s

Hecate Street and westward along Machleary Street, thus becoming U-shaped, with a deep gully between Pine Street and Machleary Street.

The coal-mining industry in Nanaimo began to decline in the early 1920s as a result of the depletion of coal deposits and the rising competition from oil in the United States and elsewhere in Canada.[28] Nanaimo's Chinatown entered the withering stage; towards the end of the 1920s hundreds of Chinese miners were laid off and several Chinatown merchants went bankrupt. With diminishing rental incomes, the Land Yick Land Company also faced bankruptcy. To save Chinatown, Chinese merchants set up a non-profit company known as the Wah Hing Land Company and appealed to Chinese across Canada to buy company shares. Eventually 4,000 shares of stock were sold, and, in 1929, the Wah Hing Company had enough capital to purchase all the properties owned by the Land Yick Land Company.[29] As Wally Chang, an old-timer from Nanaimo, remarked, "the Chinese are the only race that deliberately buys stock in a company which they know will never pay any dividends."[30] However, although Chinese merchants succeeded in keeping their properties from going into receivership, they could not stop Chinatown's depopulation. By 1941, the Chinese population in Nanaimo had dropped to 298, about two-thirds of what it was in 1921.[31]

OTHER MAJOR CHINATOWNS IN BRITISH COLUMBIA

New Westminster's Chinatown was the largest on the mainland of British Columbia before Vancouver's developed. Although New Westminster was levelled by fire in 1898, reconstruction began immediately and a new Chinatown emerged.[32] In addition to new structures for business and residents, two new institutions were established: the Chinese Methodist Church and the Chinese Reform Society.[33] During the housing boom of the 1910s, modern blocks of reinforced concrete structures were built in Chinatown. In 1913, for example, Lee Din, a successful and wealthy Chinese merchant, constructed a concrete, seven-storey apartment building, and two other Chinese merchants built a large frame building on Carnarvon Street between McInnes and Tenth streets.[34] During the blooming stage, the commercial centre of Chinatown had gradually shifted northward to Columbia and Carnarvon streets, and several blocks away was the Chinese cemetery at Eighth Avenue.[35]

By the 1920s, most Chinese in New Westminster were concentrated along Columbia and McInnes streets (Figure 13). In 1921, for example, seventeen of thirty-eight Chinese business concerns in Chinatown were concentrated on Columbia Street between Eighth and McInnis streets.[36] After the Exclusion Act of 1923, Chinatown began to decline, and by the 1930s only a few Chinese businesses and residents remained, mainly along McInnes between

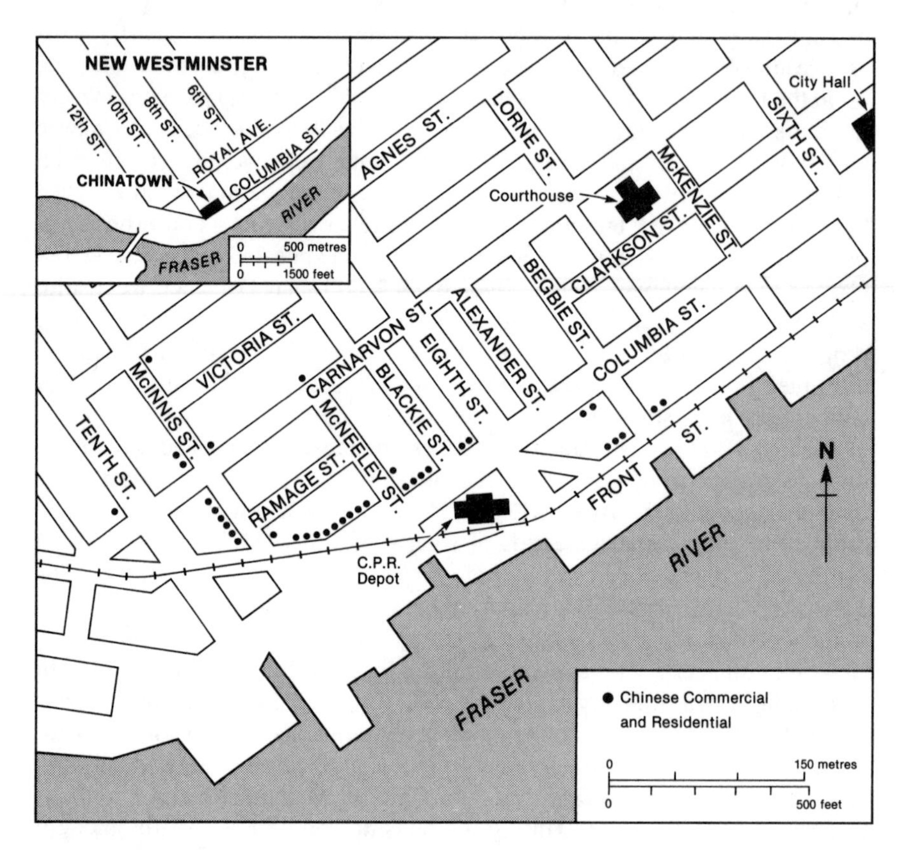

13 New Westminster's Chinatown, 1921

Columbia and Agnes streets. The city's Chinese population dropped from about 750 in 1921 to 600 in 1931, mainly because many Chinese residents had taken better jobs in Vancouver. In 1938, only two hand laundries, two grocery stores, and a barber shop remained on McInnes Street, and there were few residents left on Agnes Street.[37] Most Chinese tenement buildings had been demolished. New Westminster's Chinatown reached the stage of extinction in the 1940s when the 400 Chinese in New Westminster were scattered in different parts of the city.[38]

Kamloops's Chinatown was another important one in the 1880s. After the railway was completed, many Chinese labourers remained in Kamloops, working on farms, ranches, mines, or canneries, or as cooks or domestic servants in the white community. In 1892, Chinatown had a population of about

100 people, including six merchants.[39] The following year, a large part of Chinatown was destroyed by fire, but it was rebuilt immediately.[40] It was divided into northern and southern sections by the CPR track running along the middle of Main Street (Victoria Street) (see Plate 2). New structures such as the Chinese Methodist Mission Hall, the Chinese Freemasons Building, and the Chinese Nationalist League Building were built on either side of the street. When the CPR track was moved north of Main Street in 1914, the northern half of Chinatown was levelled. And in 1919 a bylaw was passed which restricted all Chinese hand laundries to Main Street west of Nicola Road, in order to protect white-operated laundries.[41] The bylaw effectively limited the Chinese community to the western edge of the city. After the Exclusion Act of 1923, both the population and economy of Kamloops's Chinatown began to decline; by 1927, most of the city's 300 Chinese residents did not live in Chinatown, and most of the city's forty-one Chinese businesses were in the white community.[42]

Vancouver's Chinatown developed much later than many other Chinatowns on the mainland, but its rapid growth soon overshadowed all other Chinatowns in the province. On 2 April 1886, Vancouver was incorporated on the former site of Granville Village. Within weeks of incorporation, Vancouver had 8,000 business establishments and a population of about 2,000.[43] Scores of Chinese were among these early settlers. A local newspaper reported in June 1886 that Chinese passengers came by stage and boat into Vancouver, where there were "two nests for these pests... one of which was located in Water Street."[44] Most Chinese were railway labourers and some were laundrymen. In the summer of 1887, several hundred Chinese were laying track to extend the CPR from Port Moody to Vancouver.[45] On 22 May 1887, Vancouver welcomed the first train from the Atlantic to the Pacific, becoming the western terminus of the transcontinental railway.[46]

Vancouver was virtually burnt to the ground by a great fire on 13 June 1886, but the city was rebuilt immediately.[47] The government needed labourers to clear the land for settlement, so it leased sixty hectares (or 160 acres) of forested land on Westminster Avenue (Main Street), on the northern shore of False Creek, to the Chinese, rent-free for ten years, on the condition that they clear and cultivate the land.[48] A small Chinese community thus grew up on the northern tidal flat of False Creek on the outskirts of the city (Plate 7). By the end of 1886, a tiny Chinatown, with a population of about ninety, had emerged on Dupont Street (East Pender Street) west of Westminster Avenue.[49]

Despite its rapid growth, Vancouver was still a frontier town on the densely wooded side of Burrard Inlet. A group of Chinese labourers were brought from Victoria by John McDougall, a contractor, to clear the forested Brighouse Estate on the western edge of Vancouver. The importation of

Chinese labourers immediately raised the wrath of the local white people. On 8 January 1887 a large group of white men rushed to the estate, pulled down the labourers' tents, drove them to the dock, and ordered them to return to Victoria.[50] Not long after this incident, R.D. Pitt, the local master workman of the Vancouver Knights of Labour, chaired a public meeting at which some 300 people signed a pledge to do all in their power to discourage the location of Chinese within the city limits.[51] The Vancouver Vigilance Committee was formed. It sent a threatening notice to the Chinese living on the north shore of False Creek, which said: "Warning all Chinamen to move with all their chattels from within the corporation of the city of Vancouver on or before January 15, 1887, failing which all Chinamen found in the city after the above date shall be forcibly expelled therefrom."[52]

Throughout February, the committee held a series of meetings and adopted various measures. At one meeting, it was decided that cards be distributed to businessmen to display in their windows.[53] These read:

VANCOUVER ANTI-CHINESE PLEDGE

> To appreciate freedom, we must prohibit slave labor. The undersigned pledges himself not to deal directly or indirectly with Chinese or any person who encourages them by trade or otherwise.[54]

On 24 February, 200 members of the Anti-Chinese League assembled on the CPR wharf to await the arrival of a steamer from Victoria which, according to rumour, carried a hundred Chinese labourers.[55] However, only twenty-four Chinese labourers disembarked. The police stood by on the wharf while the agitators gave the Chinese a couple of derisive cheers and then dispersed.

That afternoon, a man walked through the streets carrying a placard which read "The Chinese Have Come! Mass Meeting in the City Hall Tonight!"[56] Those at the meeting were provoked by the speeches and began to surge to the Brighouse Estate to storm the camps. The Chinese labourers were kicked and knocked about, and their bedding and clothing were thrown into a fire.[57] During the attack, some Chinese managed to escape into the bush, but others were forced to run into the icy waters of Coal Harbour, only to emerge half-frozen after the mob left. There was an unconfirmed report that four Chinese were tied together by their pigtails and thrown into the harbour. The following day, the Chinese labourers left their camps and took refuge in Chinatown. But the Chinatown residents themselves received a notice to leave, and they hastily barred the doors and windows of their houses, packed their belongings onto wagons, and departed for New Westminster.[58] The Chinese labourers from the Brighouse Estate joined the exodus; no Chinese felt safe in Vancouver.

The unsettled situation forced the provincial government to send thirty-five special provincial constables from Victoria to Vancouver to restore order.[59] Under the protection of the "Victoria specials," the Chinese residents who had fled to New Westminster gradually returned to their Chinatown on Dupont Street, and the Chinese labourers who had been recruited from Victoria went back to work on the Brighouse Estate. Ironically, John Robson, a key spokesman for the anti-Chinese movement, encouraged the provincial government to send the specials to "protect" the Chinese labourers, and "to show to the nations of the earth that we were Britons and not in the name alone." His sense of British justice towards the Chinese was in fact generated by his own interest; he owned lots in the estate and wanted to have the land cleared as cheaply and quickly as possible.[60]

Vancouver's Chinatown was still in the budding stage in the 1880s but it was growing rapidly. By 1889, there were twenty-nine Chinese business concerns in Vancouver.[61] Except for three laundries on Water Street and one on West Pender Street, all the businesses were located in Chinatown (Figure 14). They included ten merchandise and grocery stores, seven laundries, two opium importers, two labour contractors, two tailors, one butcher, and one boot- and shoemaker. Chinatown was situated in the city's red light district, where Hart's Opera House, brothels, saloons, gambling dens, and other entertainments were available.

Vancouver's Chinatown entered the blooming stage in the 1890s; it had more than 1,000 Chinese residents and overtook New Westminster's as the second largest Chinatown after Victoria. The Chinese people began to call Vancouver *Erbu* (the Second Port). In order to avoid confusion, some people called Vancouver *Xianshui Erbu* (the Second Port on Brackish Water) and New Westminster *Danshui Erbu* (the Second Port on Fresh Water). Gradually, Vancouver became known as *Xianshuibu* (Brackish Water Port), and New Westminster as *Erbu,* which two terms are still in use.

Vancouver's Chinatown displayed the characteristic features of the development model. Its population was predominantly male labourers and most of them lived in long, narrow, wooden tenement buildings or small cabins. Like other Chinatowns, it was extremely crowded. Most of its sleeping quarters were much smaller than required by the city's bylaw, which stipulated an area 10.9 cu. m (384 cu. ft.) as the minimum space for a living room.[62] According to a city inspector, many shacks in the back alleys were crowded almost to suffocation.[63] Since Chinatown was situated on a tidal flat, it was smelly, particularly when the community saw Chinatown as "an incubation of leprosy, smallpox, cholera, and other diseases," and suggested that the only way to clean up its stinking Dupont Street was to run the Chinese out of the city.[64]

Several organizations in Vancouver's Chinatown provided residents with

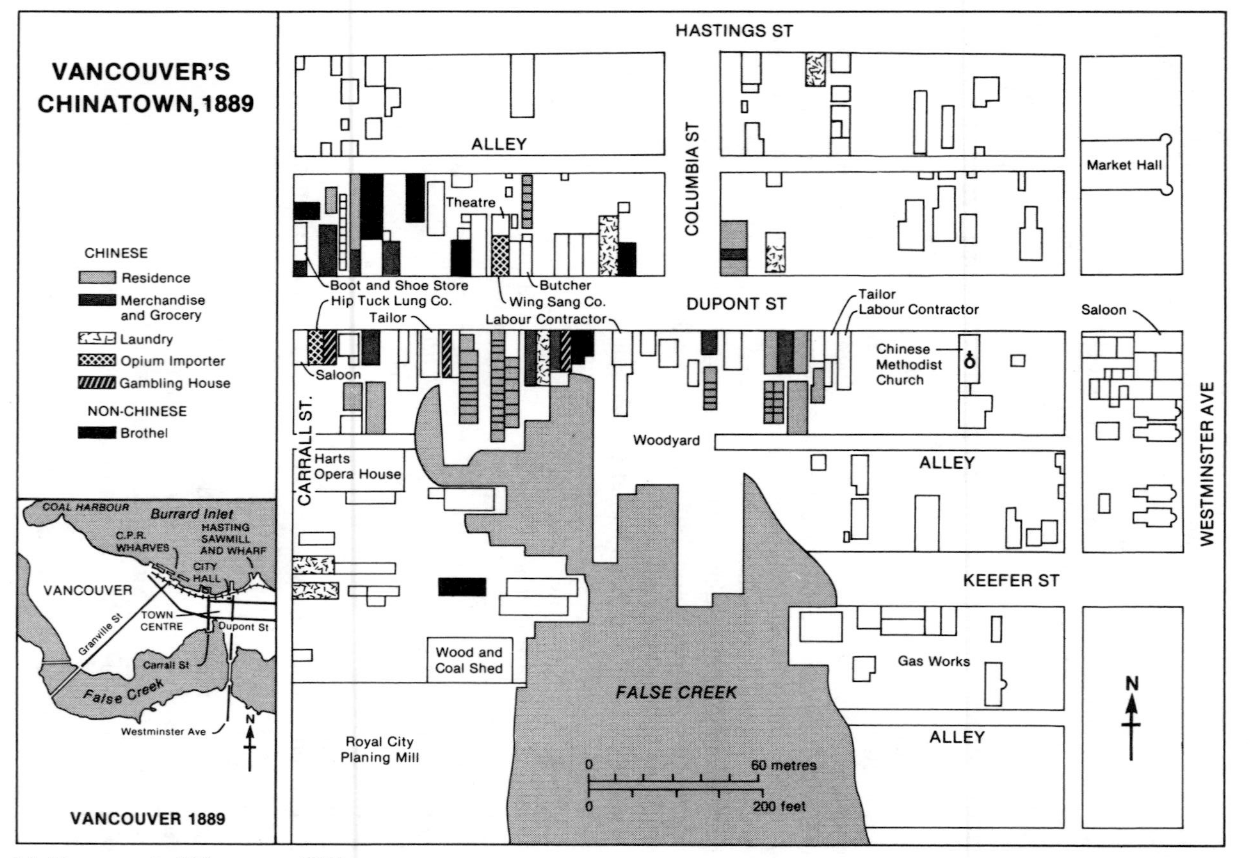

14 Vancouver's Chinatown, 1889

educational and social activities. The Methodist Mission was probably the first important institution. As early as 1888, Mrs. M. Monk, daughter of the Reverend Ebenezer Monk, established the first mission school on Hastings Street, conducting classes for the Chinese.[65] The following year, the mission erected the Chinese Methodist Church on Dupont Street, with Chan Sing Kai as pastor. The church was then the largest structure in Chinatown, and consisted of a chapel, schoolroom, reading room, missionary residence, and lodging rooms for Christian Chinese.[66] Another important organization in Chinatown was Chee Kung Tong, which was officially opened on Dupont Street on 9 September 1892.[67] During the 1890s other facilities such as a Chinese hospital and a Chinese theatre were built.[68]

With the advantages of its deep harbour and location as the western terminus of the CPR, Vancouver gradually replaced Victoria as the major seaport on Canada's Pacific coast. Vessels plying between Asia and Canada bypassed the island city to call at Vancouver. As a result, new Chinese immigrants landed at Vancouver first and the city's Chinese population rose to 2,840 in 1901, only 138 less than Victoria's.[69] After 1901, Vancouver's Chinese population outnumbered Victoria's and its Chinatown replaced Victoria's as a major target for racial hostility in British Columbia.

During the early 1900s, an increasing number of Chinese business concerns were established on Carrall Street between Hastings Street and the shore of False Creek (Plate 8). As the demand for commercial and residential premises increased, rows of tenements were built, constructed close to one another in order to maximize land use. Several new alleys were created as accesses to these tenement buildings. In 1902, a new alley from the rear of City Market to Carrall Street was named Market Alley, to which Chinese businesses and residents moved.[70] Some white businessmen on Hastings Street feared that if the Chinese continued to buy lots on the southern side of Hastings Street, Chinatown might be expanded into their territory. Because of their complaints the city health inspector, Robert Marrion, made use of the fire and sanitary regulations to condemn Chinatown and restrict its expansion to the north. On 18 June 1902, for example, he took "a whole line of chinks" to court for operating rooming houses without a city licence.[71]

As the northern expansion of Chinatown was impossible because of the opposition of local residents and businessmen, Chinatown grew westward along the northern bank of False Creek. In 1904, the Wing Sang Company's principals formed a syndicate and raised $50,000 to develop the Canton Alley tenements.[72] That same year, Dupont Street had twenty-eight, Carrall Street had thirty, and Columbia had five Chinese businesses.[73] The following year, Shanghai Alley was established. By 1905, the twenty-seven Chinese business concerns on Canton Alley included eleven merchandise and grocery stores, five tailor shops, three restaurants, three barber shops, and other businesses. On the other hand, the Wah Yick June was the only Chinese company on Shanghai Alley.[74] After 1904, Dupont Street became known as East Pender Street and its streetscape became dominated by three-storey buildings such as the Sun Ah Hotel, the Lee Association Building, the Chinese Freemasons building, and the Chinese Benevolent Association Building.[75] By the end of the 1900s, Chinatown covered about four city blocks bounded by Canton Alley on the west, Hastings Street on the north, Keefer Street on the south, and Westminster Avenue (named Main Street after 1910) on the east.

Reacting to the growing Chinese population, the Asian Exclusion League, first formed in San Francisco, established a branch in Vancouver on 5 August 1907.[76] Its membership was made up of Americans living in the city, members of the Knights of Labour, and an odd blend of jobless labourers and prominent business, religious, and military leaders.[77] On Saturday, 7 September 1907, the Vancouver League organized an anti-Oriental parade which included hundreds of American agitators from Bellingham and Seattle. The parade marched from Cambie Street to City Hall, where a number of inflammatory speeches against Orientals were delivered.[78] By evening the crowd had increased to several thousand, and A.E. Fowler, secretary of the Seattle Asian Exclusion League, spoke to the crowd. He told how, in Bellingham on 5 September, 500 white men attacked the Sikhs and Hindus, dragging them from their beds and driving them out of the city. Fowler hinted that Vancouver could achieve what Bellingham had done. Incited by his speech, the mob marched to Chinatown, where they beat up dozens of Chinese, wrecked stores, and smashed windows. The Chinese barricaded their doors but did not resist. The mob then surged to Japanese Town, where the Japanese put up a fight. The whites did not disperse until midnight, and on Sunday morning they tried to raid Japanese Town again. The Japanese, armed with sticks and guns, beat off the rioters, who then assaulted Chinatown again. By that time, however, police reinforcements had driven the rioters off. More trouble was expected; some Chinese purchased firearms and others went on an unofficial strike to protest the assault. Order was not restored for several days.

When news of the Vancouver riots reached Ottawa, Governor-General Earl Grey was furious and requested a report on the riots. He later appointed W.L. Mackenzie King as a commissioner to go to British Columbia and investigate the losses sustained by the Chinese. After his inquiry, King recommended compensation for them totalling $26,900: $3,185 for actual damage, $20,236 for resultant damage, $2,569 for losses incurred by the Chinese Board of Trade, and $1,000 for legal expenses.[79] The compensation was later paid by the federal government.

Despite these attacks, Vancouver's Chinese population of 3,559 outstripped Victoria's of 3,458 in 1911 and continued to rise rapidly to 6,484 in

1921, and to 13,011 in 1931.[80] It was during the 1910s and 1920s that most of the buildings standing today in Vancouver's Chinatown were constructed. Chinese business concerns and residents were concentrated on Pender Street from Canton Alley to Gore Street (Figure 15). In its prime, Chinatown was virtually self-contained, with two Chinese theatres, six schools, a hospital, a library, and a large number of clan, county, and other associations. The headquarters of some associations, such as Lung Kong Kung Shaw and Shon Yee Association, were removed from Victoria to Vancouver because there were more members in the latter. The townscape of Vancouver's Chinatown was dominated by both Chinese institutional buildings and other structures such as City Hall, the Vancouver Public Library, bank buildings, and white-operated hotels.

After the 1923 Exclusion Act, Vancouver's Chinatown entered the withering stage; it's Chinese population began to decline steadily until it reached 7,174 in 1941, only 55 per cent of the level ten years before. The death rate among the Chinese in Vancouver was nearly sixteen per thousand, whereas the birth rate was only nine per thousand.[81] This was partly because of the increasing number of elderly men and because of the unbalanced sex ratio: there were more than eleven males per female in 1931, improving to five males per female in 1941.

While Chinatown had always been criticized as a filthy, overcrowded slum, the government made no effort to improve it. Instead it merely condemned buildings in Chinatown and evicted tenants. For example, in August 1944, 300 Chinese tenants were removed from several tenement buildings on Shanghai Alley after the structures were declared unsanitary.[82] A health officer claimed that the rate of death by tuberculosis in Chinatown was six or seven times greater than that in other parts of Vancouver. The assertion was probably exaggerated, as a public health nurse reported in May 1945 that only twenty-six of the 160 deaths from tuberculosis in Vancouver were Chinese—only 16 per cent.[83]

Before the 1940s it was not easy for Chinese to move into better residential areas. For example, after a young, educated Chinese couple bought a home in a new block in West Point Grey, a delegation representing eighty-three local residents went to city council to complain that their property values had dropped 20 per cent after the Chinese couple had bought the house.[84] It urged the zoning committee to prohibit the sale, to prevent Orientals from owning or occupying homes in their area, and to pass legislation restricting Orientals to certain areas of the city. In response, city council set up a committee to recommend bylaw changes which would restrict Orientals to owning and occupying homes in stipulated localities. No recommendations were made by the committee, but the incident revealed the strength of prejudice against the Chinese. After 1945, some Chinese began to move into the former Japanese

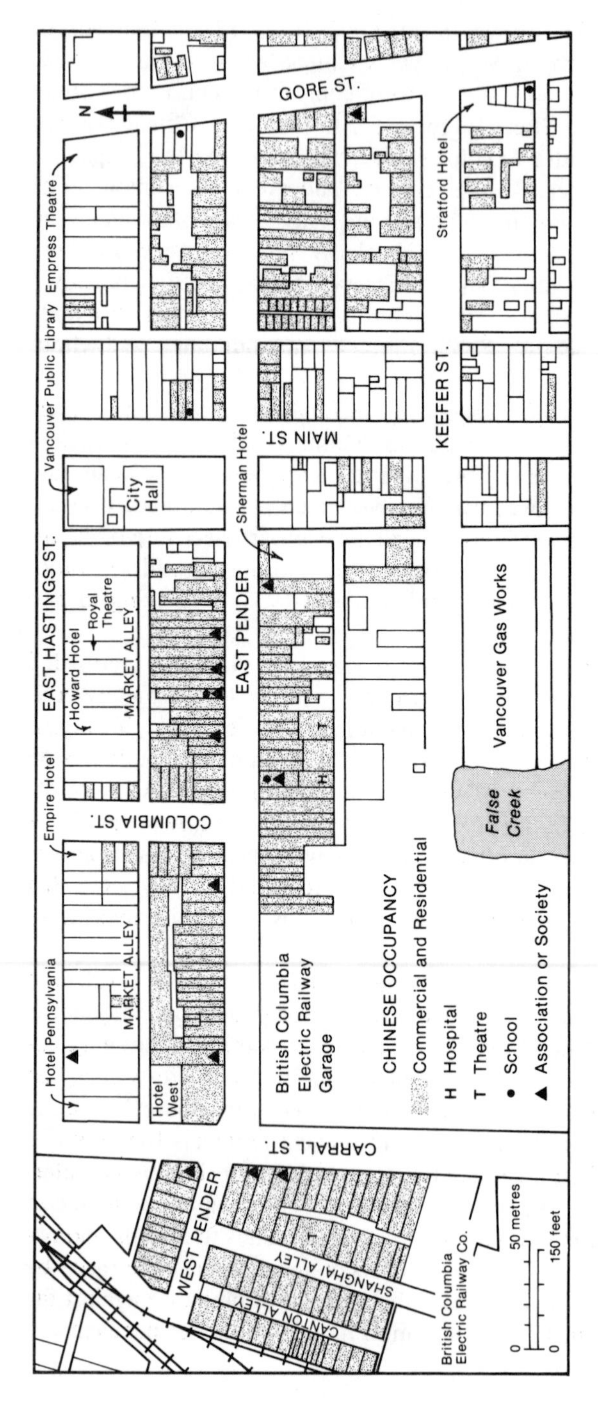

15 Vancouver's Chinatown, 1927

Town at Powell Street, but most still preferred living in their crowded Chinatown.[85]

MAJOR CHINATOWNS ON THE PRAIRIES

After the 1880s, the CPR brought prosperity to Calgary, Moose Jaw, Regina, and other towns.[86] Scores of more adventurous Chinese immigrants came to these railway towns in the hope that the racial antagonism on the prairies might be less severe and job opportunities greater. Towards the turn of the century, Chinatowns emerged in a few growing prairie towns.

Although these Chinatowns had much in common with the stage-development model, they had some dissimilarities during the budding stage because of different economic conditions. Unlike the Chinatowns in British Columbia, the Chinatowns on the prairies were not associated with gold-mining, coal-mining, or railway construction during the process of inception. Their pioneer Chinese were mostly laundrymen. They came to the small prairie towns to open hand laundries, partly because there was a great demand for this service and partly because the work required neither training nor much capital. They tried to establish their businesses far from one another in order to avoid competition, and to be closer to their white customers. As a result, their presence in the city was less noticeable, the white community had no objection to them, and a Chinatown usually was formed much later.

In the 1870s, Fort Calgary, at the confluence of the Bow and Elbow rivers, was a small village with a population of less than one hundred. It was incorporated in 1882 and began to grow after the arrival of the CPR line in August 1883.[87] A small town centre was developed along Centre Street, north of the railway station (Figure 16). A few Chinese laundrymen came to Calgary and started business on Atlantic Avenue (now 9th Avenue SW) near the station; by 1888, several Chinese hand laundries were in operation.[88] A few Chinese residents were concentrated on the eastern fringe of the town centre, which became the site of a small Chinatown.

The existence of this Chinatown did not draw the attention of the host society until the outbreak of smallpox in 1892. In June, a Chinese resident in a laundry contracted smallpox, and the building and all its contents were burned by civic authorities; as well, all its occupants were quarantined in a shack outside town.[89] Nine Chinese fell ill, and three later died. Blame was put upon the Chinese. When four of the ill Chinese were released from quarantine on 2 August 1892, a mob of over 300 men smashed the doors and windows of the Chinese laundries, trying to drive the Chinese out of town. An article in the *Calgary Herald* commented that Calgarians, as law-abiding citizens, should not condone mob violence, but rather should implement new laws to make Chinese residence illegal or boycott them by not giving them

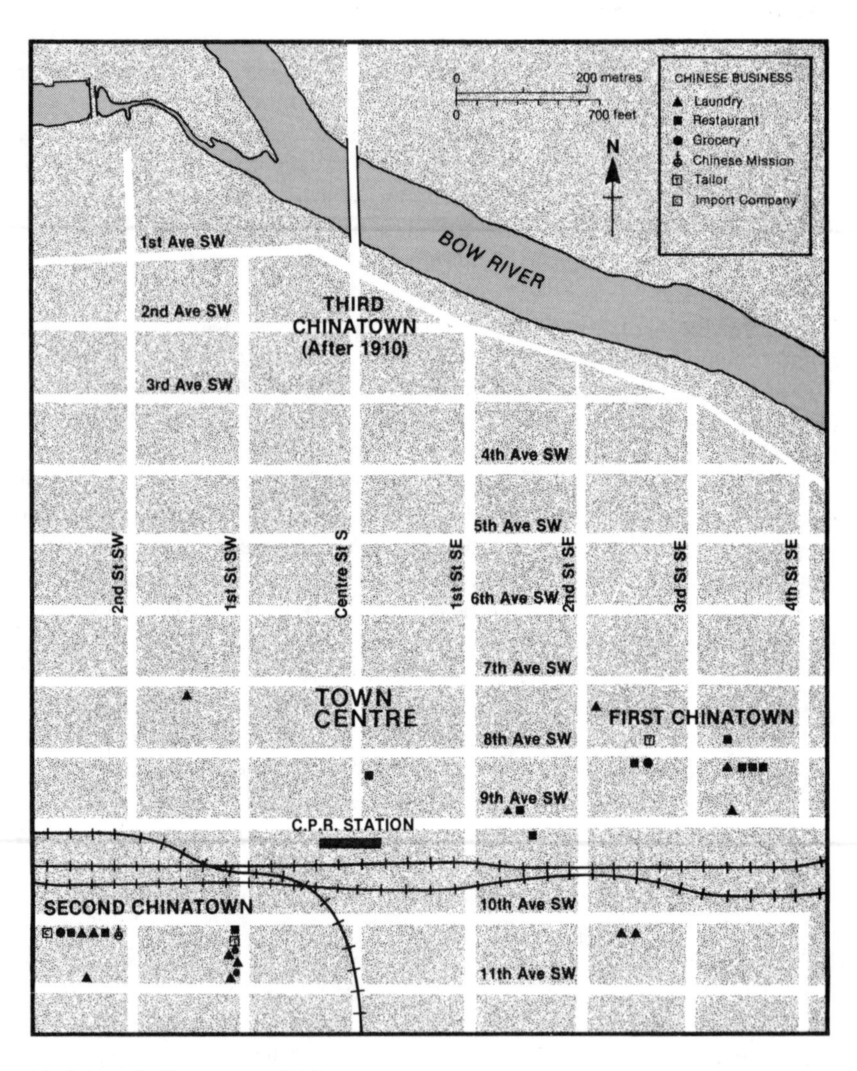

16 Calgary's Chinatown, 1900s

work.[90] For three weeks, the North West Mounted Police had to protect Chinatown and its residents against further attack by the local whites. This hostility discouraged many Chinese from coming to the city. By 1901, for example, there were only sixty-three Chinese in Calgary, only 1.5 per cent of the city's population.[91] Calgary's Chinatown was still in the budding stage.

In 1901, Dr. J.C. Herdman, the minister of Knox Presbyterian Church, tried in vain to get a room for mission work in Chinatown. Local people did not support the idea of introducing Christianity to unassimilated Chinese heathens, but eventually Mayor Thomas Underwood, a devout Baptist as well as a prosperous real estate developer, offered to let a small wooden building on First Street SW near Smith Avenue (Tenth Avenue SW) at a reduced rate.[92] Rev. Herdman also rented nearby properties to Chinese businessmen. As a result, another small Chinatown began to emerge around the mission. Thus, by 1909, Calgary had two small budding Chinatowns separated by the railway tracks. The first consisted of eight restaurants, a grocery store, a tailor shop, and several laundries on Eighth and Ninth Avenues East between Centre and Fourth streets.[93] The second Chinatown established around the Chinese Mission included three restaurants, six laundries, three grocery stores, a tailor shop, the Chinese Mission, and a row of Chinese dwellings on Tenth Avenue West between Centre and Third streets. Both Chinatowns were still in the budding stage at the beginning of the twentieth century.

These Chinatowns in Calgary followed one of the evolutionary paths of the model: they entered the stage of extinction without passing through the blooming and withering stages. In 1910, the Canadian Northern Railway announced its proposed route into the city and proposed construction of a hotel depot near the second Chinatown site.[94] Soon after this announcement, the value of the properties in the two Chinatowns soared; the landlords immediately expelled the Chinese and sold their properties. Because of this expulsion, several wealthy Chinese merchants decided to buy their own property elsewhere for a Chinatown. In September 1910 therefore they bought a parcel of land at the intersection of Second Avenue and Centre Street, where they established a new Chinatown near the Centre Street Bridge over the Bow River. The host community considered the area a "cheap dumping ground" because it was near the dwellings of low income families. Thus by the early 1910s the first two Chinatowns were gone and the third Chinatown in Calgary had been begun.

Calgary's third Chinatown had a very short budding stage, and boomed throughout the late 1910s and 1920s. It had several associations such as Chee Kung Tong, the Chinese Nationalist League, the Chinese Public School, the Shuo Yuan Association, and the Mah Association.[95] The Chinese YMCA (later known as Calgary Chinese Mission) was organized in Chinatown as well. By the late 1910s, Calgary's Chinatown had over seventy Chinese Christians, who accounted for 64 per cent of all Chinese Christians in Alberta.[96] Its Chinese population had grown from 485 in 1911 to its peak of 1,054 in 1931. Meanwhile an increasing number of Chinese laundries and grocery stores were set up outside Chinatown to cater to the white clientele. Their proprietors tended to live behind, above, or close to their stores. Although these people did not live in Chinatown, they went there regularly to buy Chinese goods or meet friends. Like other Chinatowns, Calgary's Chinatown began to decline after the Immigration Act of 1923 and the "Dirty Thirties." By 1941, its Chinese population had dropped to about 800, of which less than half lived in Chinatown.[97]

Outside Calgary, small Chinatowns emerged in Lethbridge, Edmonton, Medicine Hat, and Red Deer, of which the Chinatowns in Lethbridge and Edmonton were the most significant. When Lethbridge was incorporated in 1891, there were only three Chinese laundries.[98] The local white landlords were reluctant to lease their properties to Chinese if the land was attractive to the white community. As a result, the Chinese could only rent wooden shacks on the edge of town. A few Chinese business firms were established on Ford Street (Second Avenue South) on the northern edge of a coulee, and a Chinatown began to emerge (Figure 17). In 1904, a few hundred Chinese workers went to work on the Knight Sugar Company's sugar beet farm in Raymond, a small town southeast of Lethbridge.[99] Some of these workers later moved to Lethbridge and opened laundries or ran market-gardens. By 1909, the Lethbridge city directory included 102 Chinese, eighty-one living in Chinatown. It also listed five Chinese laundries, two Chinese restaurants, and two Chinese stores.[100] Most of the Chinese businesses were in Chinatown, which was bounded by Galt, Ford, Smith, and Baroness streets. In 1911, some white laundrymen complained that Chinese laundries, such as Sing Mah and Lee on Fifth Street South, were too close to the town centre. Accordingly, city council passed bylaw no. 83, under which Chinese laundries were permitted only inside the "restricted area," bounded on the north by First Avenue, on the east by Fourth Street, on the south by Sixth Avenue, and on the west by the coulees.[101] Bylaw no. 83 took effect on 1 January 1911; although it was rescinded five years later, the attitude represented did not die then.[102]

In 1915, a branch of the Chinese National League was established, the first Chinese institution in Lethbridge's Chinatown. Seven years later, the Chee Kung Tong also established a lodge in Chinatown.[103] The Chinese population in Lethbridge remained very small, increasing from 170 in 1921 to 230 in 1931 and to its record high of 248 only in 1941. Most of the Chinese in Lethbridge were surnamed Leong and had come from Kaiping County. The Liang Zhongxiaotang and Kaiping Huiguan were the largest clan and county associations in Chinatown.

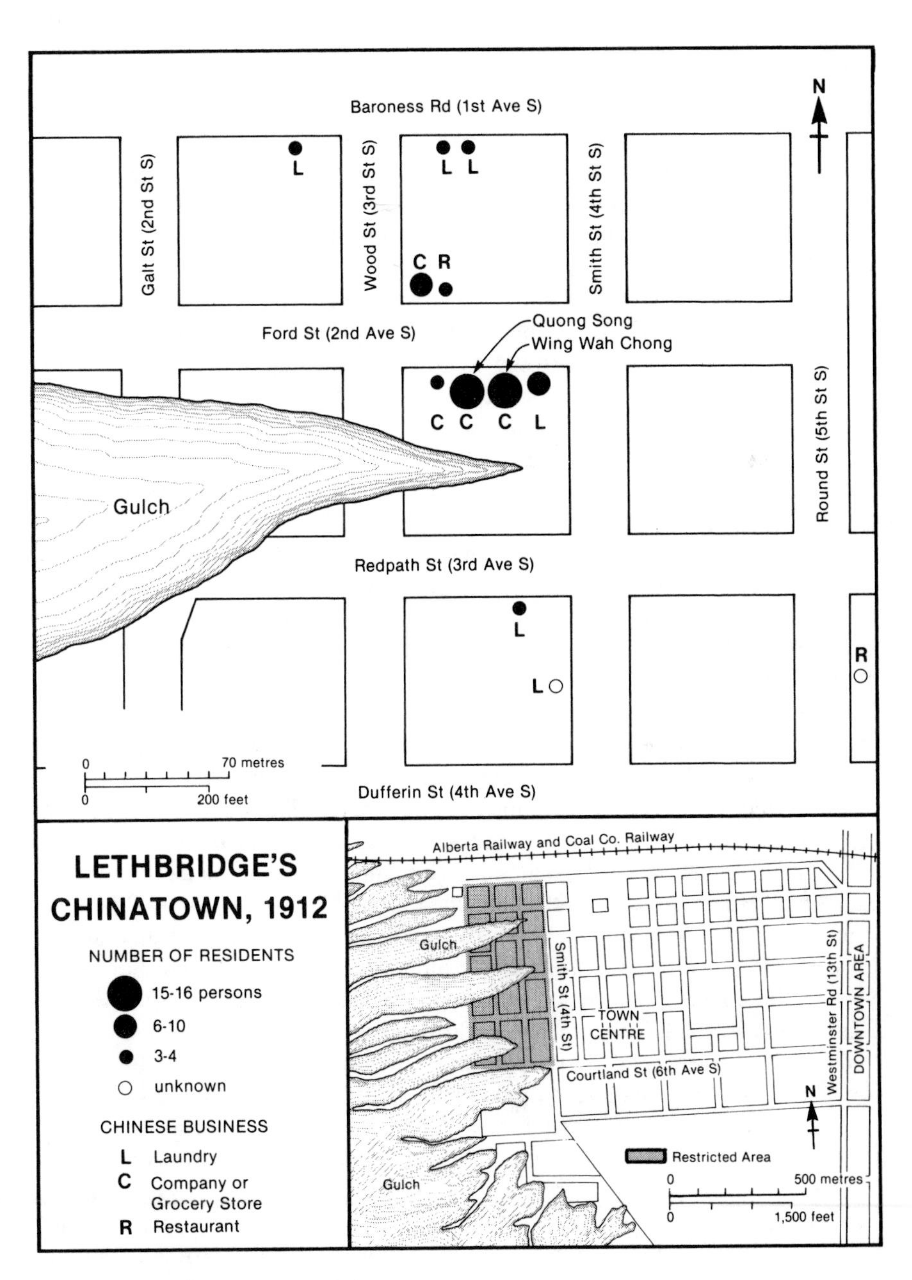

17 Lethbridge's Chinatown, 1912

Unlike in Calgary, the CPR had no impact on the growth of Fort Edmonton, established by the Hudson's Bay Company on the northern bank of the North Saskatchewan River. Principally a fur-trading post, permanent settlement outside Fort Edmonton did not begin until the Reverend George McDougall, a Methodist missionary, built his church outside the palisades in 1871. Even then, settlement was slow to develop.[104] In 1891, a branch of the CPR reached Strathcona, opposite Edmonton on the south side of the river. When Edmonton was incorporated in 1892, with a population of 700, Chung Gee and his brother Chung Yan were probably the only Chinese residents. They had come from Calgary, probably to escape smallpox, and arrived in Edmonton in July 1892, opening the town's first laundry.[105] During the Klondike gold rush of 1898, hundreds of gold-seekers and their families swarmed into Edmonton en route to the Klondike goldfields. Many died along the trail and many others turned back, settling in Edmonton.

Edmonton was incorporated as a city in 1904, a year before Alberta became a province. It was chosen as the provincial capital, and its city limit was extended when it merged with Strathcona in 1912. By the early 1910s, a small Chinatown had been established on Rice Street (101A Avenue) between Fraser Street (98th Street) and Namayo Street (97th Street) on the eastern fringe of downtown Edmonton (Figure 18). Its Chinese population had grown from 130 in 1911 to a record high of over 500 in 1921. Chinatown had also expanded eastward from 97th Street as far as 95th Street; it consisted of many stores as well as a few clan associations such as the Mah Association, Gee Association, Wong Society, and Lee Association.[106] Missionary work had also been carried on for some time by the Methodist and Presbyterian churches, notably by Westminster Church, which organized Sunday study classes among the Chinese.[107] In the early 1920s, the Chinese Freemasons and the Chinese Nationalist League were also formed in Chinatown. However, the Chinese Benevolent Association (Zhonghua Kungsuo, later known as Zhonghua Huiguan) was not established until 1932; the Chinese United Church was built the same year.[108] However, the prosperity of Edmonton's Chinese did not last long; after the Immigration Act of 1923 and the Dirty Thirties the Chinese population in Edmonton declined steadily, from 467 in 1931 to 384 in 1941. Its Chinatown was depopulated and became a skid row in the Boyle Street area, where there were already many cheap hotels, rooming houses, shabby theatres, taverns, dance halls, and second-hand stores.[109]

Saskatchewan became a province in 1905, its growth also closely related to the CPR. The city of Regina, for example, developed after it was chosen as a point where the railway would cross Wascana Creek.[110] Moose Jaw, sixty-five kilometres (forty miles) west of Regina, became a booming railway town virtually overnight when in 1882 it was chosen as the divisional point of the CPR. Soon a few Chinese hand laundries were opened in the town.[111] By

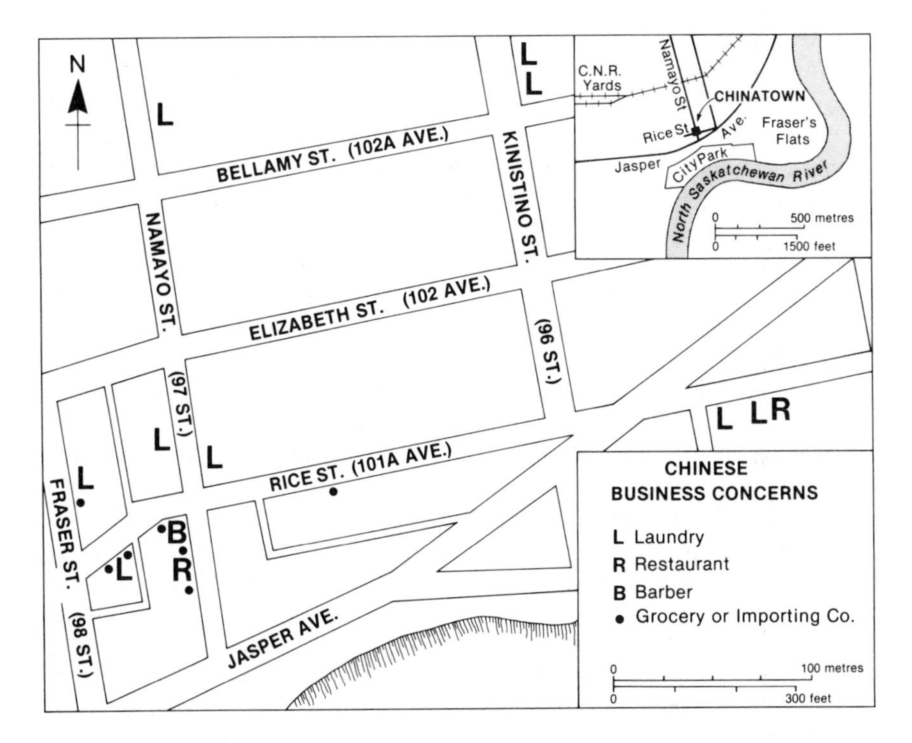

18 Edmonton's Chinatown, 1911

the 1910s, a small Chinatown had emerged on River Street with a population of about 150 and a few restaurants and stores.[112] However, a Chinese old-timer recalled that around 1913 Moose Jaw had about 450 Chinese men and two women, thirty-five to thirty-eight Chinese laundries, and three Chinese restaurants.[113] Because Moose Jaw had the largest Chinese population in Saskatchewan, the Reverend Dr. Salton of the Zion Methodist Church in 1911 organized the Moose Jaw Chinese Mission and Yip Sam was appointed evangelist in charge.[114] He not only laboured for the Chinese in Moose Jaw, but also carried out missionary work in other places such as Saskatoon.

The small Chinese community in Moose Jaw was little known outside Saskatchewan until the employment of white females by Chinese businessmen became an issue. In 1912, a Chinese restaurant owner in Moose Jaw was arrested by the police after his employee, a white waitress, lodged an assault complaint against him. The case was widely publicized by local newspapers and resulted in passage of an act by the Saskatchewan Legislature that all

Japanese, Chinese, or other Oriental people were prohibited from employing any white woman or girl; any Oriental who violated the act would be fined $100.[115] The act came into force on 1 May 1912, but the words "Japanese or other Oriental persons" were deleted after the Japanese complained. Two years later, the legislatures of British Columbia and Ontario passed a similar act. Throughout the late 1910s, Chinese communities across Canada united to fight these acts. In Saskatchewan, the act was not repealed until 1918, to be replaced by the Female Employment Act, which required Chinese businessmen to obtain a special licence from the municipality in order to hire a white female.[116] Like other Chinese communities, the introduction of steam laundry and the 1930s Depression put many Chinese laundries out of business, and the Chinese population in Moose Jaw dropped steadily from 320 in 1921 to 260 in 1941. Its small Chinatown was extinct by the 1940s.[117]

The Chinese population in Saskatoon was merely 228 persons in 1921, increasing to 308 in 1931.[118] A tiny Chinatown consisting of a few stores was established on 19th Street between First and Third avenues.[119] It did not survive long, and by the late 1930s no longer existed.

Regina did not have a Chinatown, partly because of the small Chinese population and partly because of the mutual agreement made by early Chinese settlers that they would avoid competition by not setting up businesses close to each other. In 1907, for example, there were four Chinese laundries, two Chinese restaurants, and one Chinese grocery store in Regina, scattered throughout the city's downtown area.[120] The Chinese population in Regina was only eighty-nine in 1911. By 1914, the number of Chinese laundries had increased to twenty-nine, but the number of Chinese grocery stores had only increased to two, and there were still only two Chinese restaurants.[121] These were not confined to one particular street or locality. After the 1920s the Chinese hand laundry business declined steadily, and by 1940, only eight laundries remained in the city.[122] In 1941, Regina had a Chinese population of only 250.

In Manitoba, Winnipeg's population included Chinese residents even before the arrival of the CPR. In November 1877, the *Winnipeg Free Press* reported that three "Heathen Chinese" had arrived from the United States by stage coach.[123] But it was not until the completion of the CPR that Chinese came to Winnipeg in larger numbers. Most who came were surnamed Lee and had originated from Chenshan Village in Heshan County; by 1886 they had opened eight Chinese laundries and tried to prevent non-Heshan people from settling.[124] They waylaid non-Heshan people at the railway station, beat them up, and forced them to continue heading east.[125] As a result, an undetermined number of Chinese immigrants whose original destination was Winnipeg ended up in Fort William and other eastern cities.[126] On a few occasions, the Chinese people in Fort William returned in groups to Winnipeg to assault the Heshan people there. After the Lee Association of Vancouver was informed of the trouble, it sent representatives to Winnipeg to persuade the Heshan people to stop harassing other Chinese coming to the city. For several years, however, the Heshan people in Winnipeg tried to monopolize the laundry business, cutting prices to provide unfair competition for other non-Heshan laundrymen.[127] On the other hand, the Heshan people agreed not to set up their laundries or stores too close to each other; as a result, no Chinatown developed in Winnipeg for many years.[128] It was not until 1909 that a few stores concentrated at the intersection of King Street and Alexander Avenue, forming the centre of a Chinatown.[129] In 1901, there were only 109 Chinese residents in Winnipeg, but by 1911 the Chinese population was five times that.[130] Chinatown was still small, consisting of three or four stores on King Street, and the Chinese laundries were widely spaced (Figure 19).

Winnipeg's Chinatown entered the blooming stage in the 1910s, during which time associations such as the Chee Kung Tong and the Kuomintang were founded.[131] The Chinese Christian Association, established on Logan Avenue, organized services and English classes for Chinese residents and played an important role in assimilating them in the host society.[132] Winnipeg's Chinatown reached its peak in the 1920s when the umbrella organization, the Chinese Benevolent Association, and various clan associations such as Gee How Oak Tin Association were formed.[133] In 1921, Chinatown covered six city blocks bounded by Princess and Main streets, and Logan and Rupert avenues, with King Street as its business street. About one-third of the 800 Chinese in Winnipeg worked in the city's 300 laundries.[134] The remaining two-thirds worked as cooks, domestic servants, or labourers.

Winnipeg's Chinatown entered the withering stage after the 1920s. The number of Chinese laundries dropped from over 300 in the 1920s to 125 in 1938, mainly because of mechanization of the laundry business.[135] During the economic depression of the 1930s, many Chinese restaurants and grocery stores went out of business because of the lack of Chinese patrons. Some Chinese landowners could not afford to pay taxes, lost their property, and had to move into rented premises. It was difficult to rent a place outside Chinatown; rental agents for the James Street properties on the southern border of Chinatown, for example, warned their white tenants that their leases would be endangered should they sublet to Chinese.[136]

ONTARIO'S MAJOR CHINATOWNS

Chinese immigrants from China and British Columbia began to come to Ontario by rail after completion of the CPR. However, an undetermined number of immigrants had also arrived from the United States, particularly from New York City, Buffalo, Detroit, and other Atlantic and Midwestern Amer-

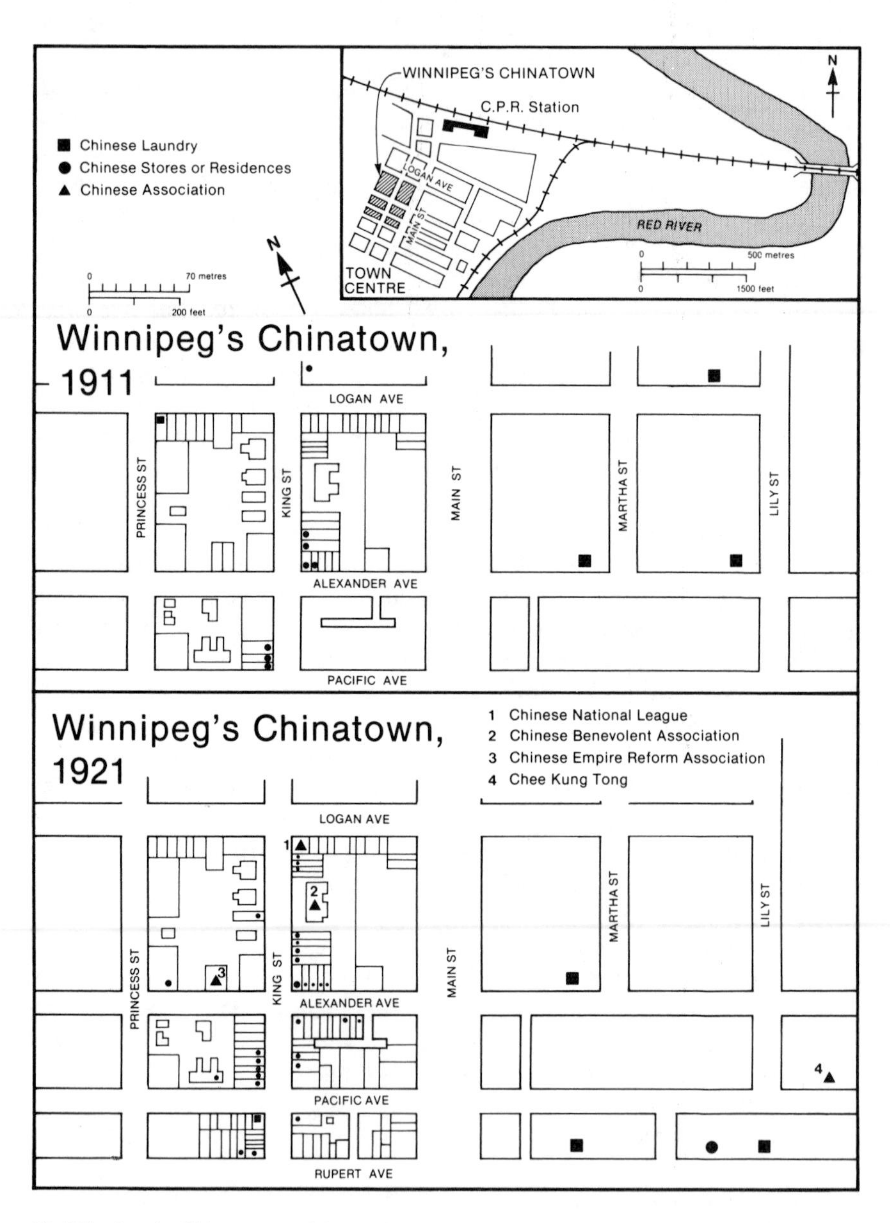

19 Winnipeg's Chinatown, 1911, 1921

ican cities.[137] Chinese communities were established in many cities and towns in Ontario, but many were so small and scattered that no Chinatowns were formed. The province's major Chinatowns were in Toronto, Ottawa, and Hamilton.

As early as 1878, a Chinese came to Toronto to open a laundry at No. 9 Adelaide Street East, but by 1881 there were only ten Chinese residents and four Chinese laundries in the city.[138] After the CPR was completed, a steady stream of Chinese immigrants came to Toronto by rail; by 1894 the city's Chinese population had increased to about fifty.[139] By 1900, the city's 200 Chinese residents and 95 Chinese businesses were widely scattered, although a few laundries and stores were concentrated on Church and Yonge streets, and along Queen Street East and West.[140] Chinese associations were not yet in existence. However, missionaries had established five Chinese Sunday schools at the Young Men's Christian Association, Metropolitan Church, and other churches; for many years, these schools were the focal points of Chinese activities in Toronto.

By 1910, two clusters of Chinese business establishments emerged.[141] Seven business concerns were set up on Queen Street East (Figure 20), and the Chinese Empire Reform Association also established a local office there. (This association is discussed in detail in Chapter 9.) Another group of nine Chinese business concerns or lodgings was located on York Street; on the same street was the Chee Kung Tong Lodge. These two Chinatowns were still in the budding stage in the early 1910s. However, after the downfall of the Manchu government, the Chinese Empire Reform Association became defunct; Chinese businesses near the association moved out and the budding Chinatown on Queen Street East was dead. On the other hand, the budding Chinatown on York Street was growing as the Jewish community there moved to the suburb north of Queen Street and the Chinese people moved in.[142] The increasing number of Chinese residents and businesses in the York Street area prompted the editor of Toronto's *Saturday Night* to warn the white public of the Chinese influence on the city and advocate the policy of "keeping the Chinese on the move" so no Chinese quarter could develop in Toronto.[143] *Jack Canuck,* another local labour paper, supported this policy, since the development of a Chinatown would have "dangerous consequences" for the city.[144] However, these policies had the opposite effect, inducing the Chinese to locate their businesses and residences closer to one another for mutual protection. By 1911, Toronto had a Chinese population of about one thousand, and the Chinatown on York and Elizabeth streets began to boom. Throughout the 1910s, Toronto's Chinatown expanded rapidly northward along Elizabeth Street to Dundas Street West; after the 1920s, Toronto's Chinatown was the third largest Chinatown in Canada after Vancouver's and Victoria's, a position it held until the end of the Second World

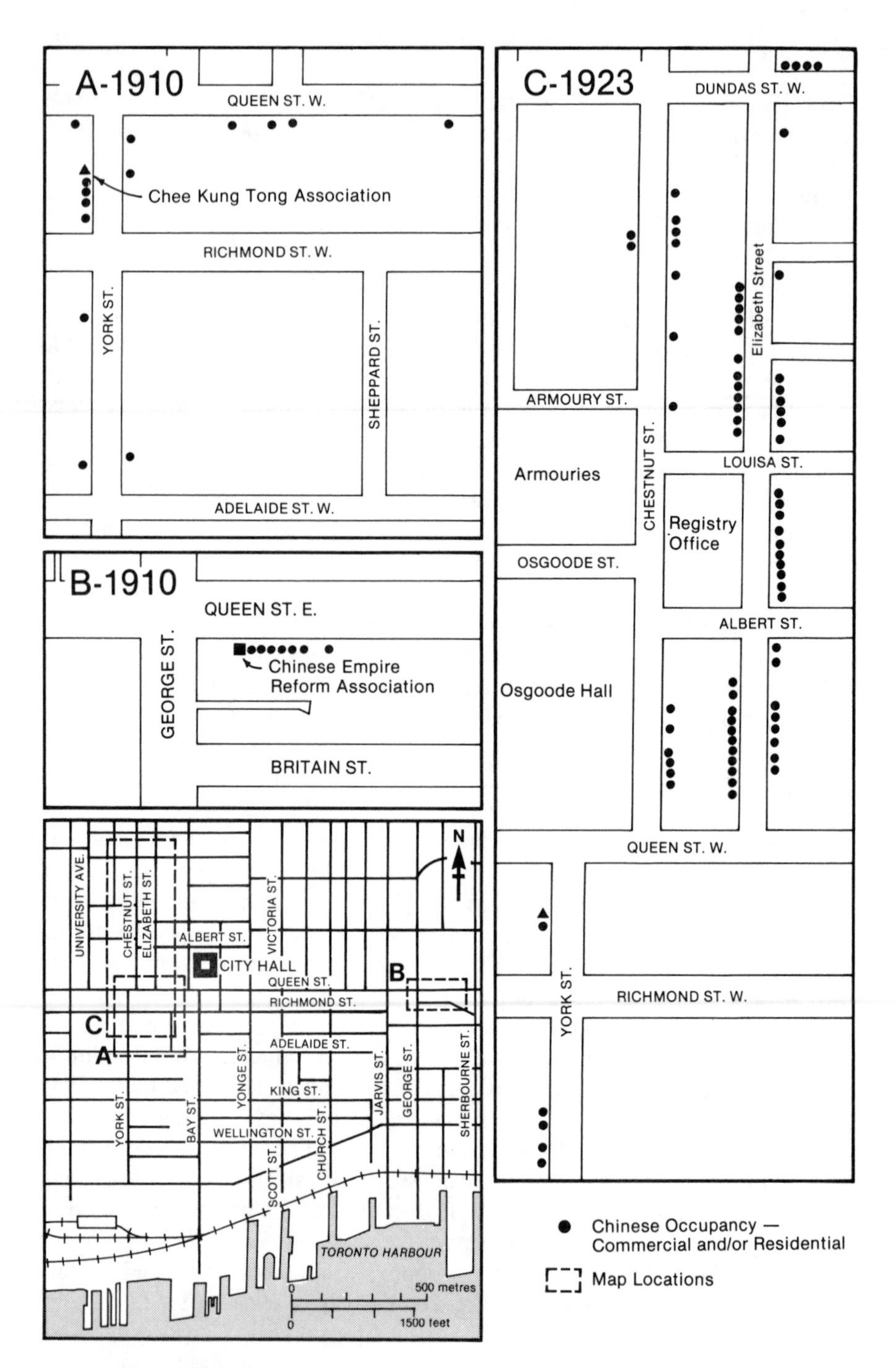

 Toronto's Chinatown, 1910, 1923

War (Plate 9). In addition to a great variety of small businesses, Toronto's Chinatown had many clan and county associations, Chinese churches, schools, theatres, and opera houses.[145]

Outside Toronto, the only significant Chinatowns were found in Ottawa and Hamilton. In 1911, Ottawa had a Chinese population of about 170 and did not have a Chinatown.[146] The Chinese community grew slowly until it had a population of about 300 in 1931.[147] That same year a small Chinatown, consisting of three grocery stores, two laundries, two recreation clubs, and one gift shop, developed on Albert Street between Kent and O'Connor streets.[148] By 1941 Ottawa's Chinatown had not expanded territorially, although its business establishments had increased to twelve: four restaurants, three laundries, two grocery stores, and three other shops. In addition, it had four organizations: the Oriental Club of Ottawa, Dai Lou Club, Moo Chung Chinese Club, and the Chinese Nationalist League.[149] An anti-Japanese organization, formed in Chinatown after the Japanese invasion of China, was reorganized as the Chinese Benevolent Association during the late 1940s.

Unlike Ottawa's Chinatown, the Chinatown in Hamilton was established much earlier. During the 1910s, a small, budding Chinatown formed along King William Street between John and Hughson streets, helped indirectly by several city bylaws. One bylaw prohibited the Chinese from building or using buildings as laundries, stores, or factories in areas close to City Hall and the central business district.[150] Another bylaw required Chinese laundrymen to renew their licences every year and stated that renewal would be denied if any resident objected. Under these restrictive bylaws, Chinese laundrymen could not set up their business anywhere at all but were confined in or near their tiny Chinatown. In 1913, for example, fifteen Chinese laundrymen applied to relocate their laundries to other places, in order to serve the non-Chinese clientele, but the city refused them all after local residents objected.[151] Under another bylaw, Chinese laundrymen were prohibited from gambling in their premises; they had to go to clubs or associations in Chinatown which helped sustain the Chinese clubs.[152]

After the 1930s, Hamilton's Chinatown began to decline, partly because the hand laundry business was being phased out and partly because of depopulation. Many Chinese residents in Hamilton had moved to Toronto, where job opportunities were greater. By 1941, fewer than 200 Chinese remained in Hamilton, and towards the end of the 1940s, Hamilton's Chinatown was virtually extinct.

QUEBEC'S MAJOR CHINATOWNS

Completion of the CPR also had resulted in the migration of Chinese immigrants from the west coast to Quebec, although an undetermined number of

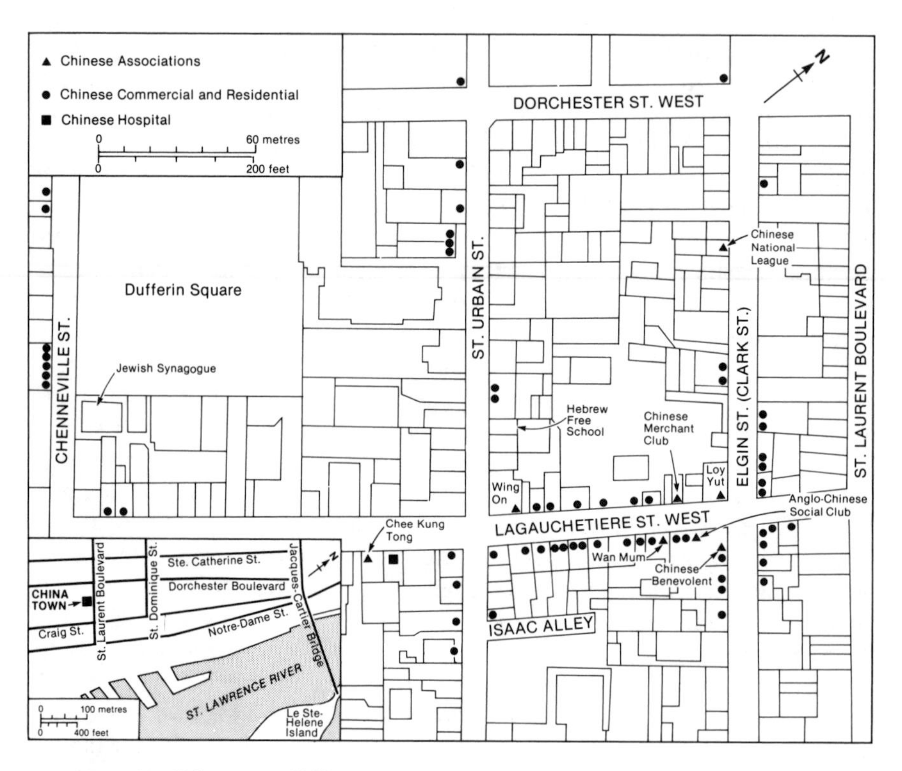

21 Montreal's Chinatown, 1921

them had come by train from the United States.[153] Nearly all the Chinese arrivals settled in Montreal; only a few went to Quebec City. In 1911, for example, Quebec City's Chinatown, the second largest Chinatown in the province, had only sixty-eight residents, compared with nearly 1,200 residents in Montreal. Quebec City's Chinatown, situated around Côte d'Abraham, remained very small and its population reached a peak of only 140 in 1941.[154] Its growth had been eclipsed by the development of Montreal's Chinatown.

During the 1890s, Montreal's Chinatown began to emerge on Laguachetière Street West between St. Urbain and St. Charles Borromée streets (Clark Street) in Dufferin District. At that time, Dufferin District, on the northern edge of Montreal's business area, was a rundown residential area with warehouses, machine shops, and some light industries.[155] Many property owners in the district partitioned their premises into rooms and took in lodgers.

The large number of male residents working in the district's factories demanded laundry, café, restaurant, and other services, which were provided by Chinese immigrants. During the 1910s, Montreal's Chinatown, consisting of about twenty Chinese business concerns, had expanded from the intersection of Lagauchetière Street West and St. Urbain Street.[156] In October 1918, an epidemic of Spanish influenza among the Chinese, and the lack of a Chinese hospital in Chinatown, prompted some Chinese community leaders to rent two houses on Clark Street as a temporary hospital. They later bought a building at 112 Lagauchetière Street West in 1920 for the Montreal Chinese Hospital. Montreal's Chinatown was thriving through the late 1910s, and by 1921, it had a Chinese population of 1,735 and became the fourth largest in Canada. It was bounded by Dorchester, Clark, Vitre, and Chenneville streets (Figure 21). A large number of Chinese businesses and institutions such as the Chee Kung Tong, Chinese Hospital, and Chinese Methodist Church were concentrated on Lagauchetière Street West, the commercial spine of Montreal's Chinatown.[157]

COMMUNITIES IN ATLANTIC PROVINCES

Chinese communities in the three Atlantic Provinces of New Brunswick, Nova Scotia, and Prince Edward Island were very small; in 1921 the total Chinese population in the three provinces was less than 600. Except for Saint John and Halifax, which had a Chinese population of about 140 each, Chinese communities in other towns had only ten to twenty people. No Chinatowns had been formed because most of the Chinese were laundrymen and were scattered in the towns. Newfoundland, which entered Confederation in 1949, had about eighty Chinese residents in 1922, mostly in St. John's.[158] Again, no Chinatown had been established in the city because the Chinese laundrymen did not confine themselves to one section of a street.

6

Postwar Arrival of New Immigrants

Repeal of the Chinese Exclusion Act marked the beginning of the period of selective entry, during which regulations against Chinese immigration were rescinded one by one. However, there were still restrictions, as seen in Prime Minister Mackenzie King's 1947 statement that "the people of Canada do not wish, as a result of mass immigration, to make a fundamental alteration in the character of our population. Large scale immigration from the Orient would change that fundamental composition of the Canadian population."[1] Therefore Chinese immigrants were still discriminated against until the federal government changed its immigration policy in 1967 and for the first time treated Chinese immigrants as it did immigrants of other nationalities. After 1967, different types of Chinese immigrants entered Canada, playing important roles in the transformation of Canadian Chinatowns; the distribution of Chinese Canadians by provinces and by cities also changed drastically.

REPEAL OF CHINESE EXCLUSION ACT

After the Sino-Japanese War broke out in 1937, Canadians in general were sympathetic to China and supportive of local Chinese people in their fund-raising campaigns to send relief goods to their war-torn country. China became an ally of Canada and the United States soon after the outbreak of the war in the Pacific in 1941, and about 400 Chinese Canadians from British Columbia joined the Canadian forces. The war initiated a change in the American government's attitude towards Chinese migration to the United States: the Chinese Exclusion Act was abolished on 17 December 1943, and an annual quota of 105 Chinese immigrants was permitted.[2] However, a similar change of immigration policy did not occur in Canada. After the war was over, the Chinese people in Canada started lobbying Parliament to repeal the Chinese Exclusion Act while Chinese veterans demanded equal treatment.[3] In Europe, the revelation of Nazi atrocities shocked the world and triggered a United Nations human rights movement.

All these developments inside and outside Canada prompted the Canadian

TABLE 19: Chinese immigration to Canada, 1946–65

Period	Number of persons	
	Total	Annual average
1946–50	2,654	531
1951–55	11,524	2,305
1956–60	10,407	2,081
1961–5	11,785	2,357

Sources: Canada, Department of Citizenship and Immigration, Immigration Statistics, 1946–65

government to reconsider its own policies concerning the Chinese. Eventually, the Chinese Exclusion Act was repealed on 14 May 1947.[4] The control of Chinese immigration was then regulated by Order-in-Council PC 2115, by which a Chinese-Canadian citizen living in Canada could bring his wife or unmarried children under eighteen years of age to Canada on an immigrant basis.[5] However, very few Chinese were Canadian citizens; under Order-in-Council PC 1378 certificates of naturalization were issued only to Chinese subjects who could prove that they had renounced their Chinese allegiance.[6] Many Chinese could not produce that evidence or were unwilling to repudiate their loyalty to China; consequently, very few could be naturalized. In 1941, for example, of 34,627 Chinese in Canada, only 2,055 Chinese were naturalized, or 6 per cent of the Chinese in the country.[7] Therefore few Chinese were qualified to bring their families to Canada. In the first few years after the Exclusion Act was repealed, an average of about 500 Chinese immigrants were admitted each year to Canada (Table 19).

On the other hand, preferential treatment was given to European immigrants in order to foster growth of a white population in Canada. For example, under the Close Relatives Plan, European Canadians could bring to Canada their parents, sons, daughters, brothers, or sisters together with their spouses and other close relatives. Under the Group Movement Plan, European immigrants, rather than being nominated individually by Canadian residents, were selected in accordance with the recognized manpower needs by Canadian Immigration Labour Teams travelling in Europe.[8] Chinese and other Asians did not have the same right of entry into Canada as Europeans chosen under these plans. Of the 282,164 immigrant arrivals in 1957, for example, 36 per cent were British, 11 per cent Hungarian, 10 per cent Italian, 9 per cent German, and 4 per cent Dutch. Only 0.6 per cent were Chinese, 0.3 Indian, and 0.1 Japanese.[9]

IMMIGRATION REGULATIONS, 1957–66

Throughout the late 1950s and early 1960s, regulations on Chinese immigration were relaxed. In 1957, Order-in-Council PC 2115 was rescinded, and

Chinese residents were permitted to apply for their families to come to Canada before they acquired Canadian citizenship. In following years, the admissible classes to Canada were widened by a series of Orders-in-Council. For example, the age of unmarried children of Chinese Canadians was raised from eighteen to twenty-one and then to twenty-five on compassionate grounds. Later, fathers over the age of sixty-five and mothers over sixty were permitted to enter.[10]

Many Chinese immigrants posed as members of other Chinese-Canadian families or formed fictitious families in order to come to Canada; about 11,000 Chinese entered the country illegally during the 1950s.[11] In the spring of 1960, the Royal Canadian Mounted Police invited Hong Kong policemen to come to Canada to help gather evidence against Chinese immigration racketeers, raiding more than thirty Chinese homes and businesses in Vancouver, Edmonton, Calgary, Regina, Brandon, Winnipeg, Sarnia, Peterborough, Toronto, and Montreal.[12] In June 1960, the Canadian government introduced the Chinese Adjustment Statement Programme, under which all Chinese immigrants who entered Canada illegally before 1 September 1964 and came forward to have their cases reviewed by the Minister of Citizenship and Immigration would be permitted to remain in Canada if they were considered of good character. Under this program, over 12,000 Chinese were allowed to stay in Canada.[13]

In 1962, Prime Minister John Diefenbaker announced that Canada would accept one hundred refugee families who had fled from Communist China to Hong Kong and that the federal government would finance their trip and pay the costs of their settling in Canada.[14] The same year, the Canadian government began a more relaxed policy of Chinese immigration, removing the emphasis on the country of origin as a major criterion for admission to Canada.[15] Four years later, the Department of Citizenship and Immigration was replaced by the Department of Manpower and Immigration, reflecting the increased demand for manpower in the country. In British Columbia, for example, the desperate need for labour led to a demand for Chinese workers just as had happened a century before. In January 1965, the British Columbia Federation of Agriculture presented a brief to the provincial government, suggesting the importation of Chinese workers to ease the labour shortage.[16] The brief stated that fruit and vegetables were rotting on the vine; there were insufficient pickers, and many farmers went out of business simply because they had no workers to pick their crops. Chinese immigrants would be the solution to the farmers' problem: they would be paid standard Canadian wages to pick crops during the summer and prepare seeds and other chores in greenhouses in the winter.[17]

From 1956 to 1965, about 22,000 Chinese immigrants entered Canada, 66 per cent from Hong Kong (Table 20). Since the People's Republic of China

TABLE 20: Country of last permanent residence of Chinese immigrants, 1956–65

Last residence	Number of persons	% of total
Hong Kong	14,648	66.0
Taiwan	4,686	21.6
Other Asian countries	717	3.2
Britain	400	1.8
United States	344	1.6
Other countries	1,298	5.8
Total	22,193	100.0

Sources: Canada, Department of Citizenship and Immigration, Immigration Statistics, 1956–65

did not have a direct diplomatic relationship with Canada, Chinese emigrants having relatives in Canada first obtained the Chinese government's permission to go to Hong Kong, where they could apply for entry to Canada. Thus, many Chinese listed as immigrants from Hong Kong originally had come from China, mostly from the traditional source areas: the Siyi, Sanyi, or Zhongshan counties.

NEW TYPES OF IMMIGRANTS AFTER 1967

Canada usually follows a "swinging door" policy of immigration. When it needs labour, its doors are wide open for immigrants. When there is a shortage of jobs, its doors are shut. In 1967, having considered Canadian economic and demographic needs, the federal government adopted an immigration policy based on non-discrimination and universality.[18] The new immigration regulations gave people from all parts of the world an equal opportunity to qualify for admission. For the first time in Canadian history, prospective Chinese immigrants were treated exactly the same as immigrants of other nationalities and were selected for admission according to education, training, skills, and other criteria linked to the economic and manpower requirements. Race was no longer a criterion for immigration, and after 1967 the government ceased to identify the ethnic origins of immigrants in immigration statistics, listing instead the country of their last permanent residence. In statistics, an "immigrant from Hong Kong" merely designated that an individual was living in Hong Kong before he or she migrated to Canada. The immigrant might be Chinese, Indian, English, or another race. Therefore, the exact number of Chinese immigrants coming to Canada after 1967 is not known. However, most Chinese immigrants traditionally came from Hong Kong, Taiwan, and China although there was an undetermined number of Chinese immigrants from southeast Asian countries such as Vietnam and Malaysia. Between 1972 and 1978, nearly 67,000 landed immigrants came

TABLE 21: Landed immigrants from Hong Kong, Taiwan, China, and Vietnam, 1972–84

	Number of landed immigrants			
Year	Hong Kong	Taiwan	China	Vietnam
1972	6,297	859	25	–
1973	14,662	1,372	60	418
1974	12,704	1,382	379	373
1975	11,132	1,131	903	2,269
1976	10,725	1,178	833	2,269
1977	6,371	899	798	243
1978	4,740	637	644	659
1979	5,966	707	2,058	19,859
1980	6,309	827	4,936	25,541
1981	6,451	834	6,551	8,251
1982	6,542	560	3,572	5,935
1983	6,710	570	2,217	6,451
1984	7,696	421	2,214	10,950

Sources: Canada, Department of Manpower and Immigration, Immigration Statistics, 1972–7; and Department of Employment and Immigration, Immigration Statistics, 1978–84

from Hong Kong, 7,000 from Taiwan and 4,000 from China (Table 21).

Most Chinese immigrants entering Canada after 1967 were different from their nineteenth-century predecessors with respect to their origin, wealth, education, occupation, aspirations, and motives. Before the Exclusion Act of 1923, Chinese immigrants to Canada were mostly poor, unemployed rural people. Driven by hunger and poverty, they came to Canada to make a living. Without any idea of urban culture or Western society, they landed in Canada as complete strangers. Unable to speak the language, and unprepared for the Western way of life, they maintained contacts mainly with their fellow countrymen. They did not try to understand the Canadian culture nor did they intend to integrate into Canadian society; they planned to return to China as soon as they could save enough money to support their families and themselves for the rest of their lives.

On the other hand, post-1967 Chinese immigrants came from many lands and cultures: Hong Kong, Taiwan, China, Southeast Asia, Britain, the United States, and other places. Although it is not known how many and where these Chinese immigrants have come from, it is commonly believed that most were former residents of Hong Kong, Taiwan, and China. Unlike the Chinese pioneers, many of them were mainly independent immigrants and were admitted according to their educational background, occupational skill, knowledge of English, and personal qualities. In 1981, for example, 7 per cent of the landed immigrants from Hong Kong and Taiwan listed their intended occupations as professionals such as doctors, engineers, and architects (Table 22). Other intended occupations included clerical workers,

TABLE 22: Intended occupations of landed immigrants from Hong Kong, Taiwan, and China, 1981

	Hong Kong		Taiwan		China	
Occupation	No. of persons	% of total	No. of persons	% of total	No. of persons	% of total
Professionals	497	7.7	62	7.4	177	2.7
Clerical	400	6.2	43	5.2	158	2.4
Entrepreneurs	53	0.8	65	7.8	29	0.4
Farming	9	0.1	4	0.5	953	14.5
Machine technicians	221	3.4	25	2.9	430	6.6
Teachers, etc.	58	0.9	18	2.2	76	1.2
Other	969	15.0	65	7.8	990	15.1
Non-Workers (retired, spouse, children, etc.)	4,244	65.8	552	66.2	3,737	57.1
Total	6,451	100.0	834	100.0	6,550	100.0

Sources: Canada, Department of Employment and Immigration. Immigration Statistics, 1981

teachers, and machine technicians. A considerable number of Chinese immigrants from Hong Kong were entrepreneurs of moderate or great wealth who set up businesses and created employment in Canada; their businesses ranged from restaurants to banking and their influence is seen today in all walks of life.

Most post-1967 Chinese immigrants came to Canada because of educational and economic opportunities for themselves as well as for their children. For example, many Hong Kong Chinese worried that they might not have the same freedom, economy, and life-style after Hong Kong was returned to China in 1997, and they emigrated to Canada with their families, intending to remain permanently. Within a short time, some of these immigrants have already distinguished themselves by their self-reliance, industriousness, and ability to integrate easily into the business, professional, and social fabric of Canada. They have improved the quality of life in Canada for themselves and their children. Some investors and professional people, however, returned to Hong Kong and continued with their businesses there after they had settled their wives and children in Canada. They returned to Canada once every six months in order to maintain their landed immigrant status. Like the Chinese immigrants of the nineteenth century, they have become "sojourners"; the difference is that they are sojourning in Hong Kong instead of in Canada. When they have earned enough money in Hong Kong, or when the economic or political situation in Hong Kong deteriorates, they will return to their Canadian home.

Another significant impact of the 1967 Immigration Act was a drastic change in the racial distribution of immigrants. In the 1950s, nearly 90 per cent of new arrivals were from Europe, while Asian immigrants constituted

TABLE 23: Country of last permanent residence of immigrants to Canada, 1951–7, 1968–73

	Period			
	1951–7		1968–73	
Last residence	No. of persons	% of total	No. of persons	% of total
Europe	1,103,539	89.1	459,881	49.9
North and Central America	76,277	6.2	216,840	23.5
Asia	28,465	2.3	154,864	16.8
South America	10,182	0.8	32,827	3.6
Australasia	11,571	0.9	21,345	2.3
Africa	7,265	0.6	30,820	3.4
Oceania	1,652	0.1	4,747	0.5
Total	1,238,951	100.0	921,324	100.0

Source: Ottawa, Department of Manpower and Immigration, highlights from *Green Paper on Immigration and Population* (1975), 35–9

an insignificant 2 per cent. By the late 1960s and early 1970s, however, the proportion of European immigrants had slipped to 50 per cent, while immigrants from Asia had risen to nearly 17 per cent (Table 23).

In 1973, a provision in the Immigration Act permitted permanent residency to any person who had been in Canada on or before 30 November 1972, including visitors and students.[19] As a result, many Chinese students from Hong Kong and other places became landed immigrants. Before 1973, very few Chinese immigrants came directly from China. After Prime Minister Pierre Elliott Trudeau and Chinese Premier Zhou Enlai reached an agreement whereby some Chinese were permitted to join their families in Canada, the number of immigrants from China immediately soared from sixty in 1973 to 379 in 1974 and 903 in the following year (see Table 21).

IMMIGRATION POLICIES AFTER 1978

In September 1973, the federal government began to review its immigration policy. A collection of briefs and letters submitted by national organizations and a series of discussion documents were compiled and published in January 1975 as the Green Paper on Immigration and Population.[20] According to the Green Paper, significant increases in immigration from Asian and Caribbean countries had the potential to cause social tension in large cities such as Toronto, Vancouver, and Montreal. Therefore, close attention needed to be paid to the racial distribution of immigrants.

The Green Paper's proposals were highly criticized by both non-white Canadians and by some people from the white community. For example, the Vancouver Chinese community immediately formed the Immigration Action

Committee, and submitted in September 1975 a report to the members of the Special Joint Committee of Parliament on Immigration, expressing disagreement with the recommendations of the Green Paper.[21] After many discussions, the federal government finally passed a new Immigration Act in 1976, effective in April 1978.[22] It listed the three principal objectives of Canada's immigration policy: to stimulate economic growth and encourage social and cultural development; to encourage family reunion; and to alleviate the plight of refugees through humanitarian programs. The intake of immigrants was based on a policy of non-discrimination. The Minister of State for Immigration, in consultation with the provincial governments, would set the immigration levels, which were initially limited to about 140,000 persons per year.

The significant effect of the new Immigration Act was the admission of a large number of "boat people." After the fall of Saigon in 1975, hundreds of Vietnamese people fled from the country, many entering Canada as landed immigrants. Their ethnic origins were unknown, but many were Chinese. Before 1978, China had opened its door to overseas Chinese from Vietnam, but after China closed its border, the Vietnamese government gave the remaining Chinese the choice of either living in undeveloped "new economic zones" or accepting "assisted departures" from Vietnam by boat, for which they had to pay seven ounces of gold per person. Initially, most of the boat people were Chinese; later an increasing number of Vietnamese joined the exodus. Meanwhile, thousands of refugees entered Thailand from Laos and Kampuchea. Canada was one of the countries to offer permanent resettlement for these Indochinese refugees. In July 1979, Canada offered to provide homes for up to 50,000 Indochinese refugees over a period of two years, through a partnership arrangement between the federal government and private sponsors.[23] Thus, from Vietnam alone nearly 46,000 landed immigrants entered Canada during 1979 and 1980. In 1983, the government-assisted refugee intake was set at 12,000, of which 3,000 were from Indochina.[24] An undetermined number of them were Chinese.

In 1978, the Chinese government relaxed its control of emigration and the number of immigrants from China more than tripled. Since they had been admitted on the basis of family reunion, most new immigrants from China were rural people from traditional source areas in Guangdong Province, where their sponsors had come from. They were mainly farmers or farm labourers. In 1981, for example, nearly 15 per cent of the immigrants from China intended to take up farming as their occupation (see Table 22). Like the Chinese immigrants during the nineteenth century, many recent immigrants have not been exposed to Western culture and so face the problem of a language barrier. Unlike professional people from Hong Kong and Taiwan, many of these new immigrants from China, together with Chinese Vietnamese refugees, still prefer living inside or near Chinatowns and rely on them as stepping stones into Canadian society.

In the early 1980s, Canada again had to close its door to immigrants because the unemployment rate had increased more than 10 per cent, and immigrants were perceived to be taking jobs from Canadians. Therefore immigration levels were reduced from between 130,000 and 140,000 for 1982 to a record low of 85,000 and 90,000 for 1985.[25] However, according to Statistics Canada, the birth rate in Canada had dipped to an average of only 1.66 children for every woman of childbearing age, down from 3.85 three decades ago. Unless more immigrants were accepted, the population decline in Canada eventually would affect the country's economy. Towards the end of 1985, Walter McLean, Minister of State for Immigration, announced that the immigration door, which had been virtually closed since 1982, would be open again in 1986.[26] To prevent a decline in Canadian population by the turn of the century, the minister set the immigration levels at between 105,000 and 115,000 for 1986.[27] In the same year, a new "investor" immigrant category was introduced, under which an investor and his family would be admitted as immigrants if the investor had a successful track record in business in his native land and a net worth of at least half a million dollars, and invested $250,000 in an approved business for three years in Canada. This program proved to be very successful because it provided Canada with risk capital and international experience and helped create jobs in the country.

This program has attracted many wealthy Hong Kong entrepreneurs and investors to Canada as they do not have confidence in the future of Hong Kong. In 1986 about eighty investors applied to come to Canada under the "investor" category; more than one-third were from Hong Kong.[28] The immigration levels for 1987 was raised to 125,000, and the levels for 1988 were increased further to 135,000.[29] Gerry Weiner, Minister of State (Immigration), said in April 1987 that many Canadians had misconceptions about the arrival of immigrants, such as believing they would take most existing jobs in Canada. Instead, studies have suggested that new Canadians created jobs, earned 10 per cent more than the national average income during the first ten months of their arrival, and tended to pay more taxes than many older Canadians. Many Hong Kong entrepreneur immigrants have helped Canadian industries get back on their feet. For example, Cee-Dee Log Buildings Inc., a Calgary-based cedar home manufacturer, had been near financial collapse for five years and was back into the black in 1987 only after Hong Kong investors made an investment of $1.8 million in the company.[30] The same year, Li Ka Shing, a Hong Kong billionaire, together with his son, Victor Li, purchased 52 per cent of Calgary-based Husky Oil Ltd.[31] In April 1988, Li's company, Concord Pacific Developments Ltd., purchased the 85-hectare site of Expo 86 on the north shore of False Creek for $320 million. In the same month, Barbara McDougall, Minister of Employment and Immigration, announced a revised immigrant investor program, under which qualified investors are eligible if: they have a net worth of $500,000 and make an investment of $150,000, locked in for three years, in a province which, the previous year, has received fewer than 3 per cent of Canada's business immigrants; they have a net worth of $500,000 and make an investment of $250,000 locked in for three years; or they have a net worth of $700,000 and make a minimum investment of $500,000 locked in for five years.[32] This program will attract wealthy immigrants to Canada whose investments will help the Canadian small business community access a broader range of financing so it can grow and create jobs in Canada.

New immigration rules, relaxed to help reunite families, went into effect on 8 July 1988. For example, never-married sons and daughters of immigrants, regardless of age, will be allowed to accompany parents when they come to Canada, or to join parents already in Canada. Previously only children under twenty-one years of age were admitted. Married sons and daughters, or brothers and sisters, of immigrants already in Canada will be credited with five extra points towards the 70-point total they need to qualify as legal immigrants. In 1987, Canada admitted 146,994 immigrants of whom 51,712 were in the family class, and 11,800 in the assisted-relative class. It is expected that the relaxed immigration rules will enable about 5,000 additional family members to get into Canada each year.

The flight of capital and people from Hong Kong to Canada and other countries such as the United States and Australia increased after China launched a campaign against "bourgeois liberalization" in the fall of 1986. The same year, the Chinese government told the British government that it objected to the introduction of direct elections for members of the Legislative Council, the lawmaking body in Hong Kong. Many Hong Kong people regarded this objection as an extension of China's "anti-bourgeois liberalization." That year more than 8,000 heads of families in Hong Kong obtained permits to emigrate to Canada and brought with them another 10,000 dependents from Hong Kong. In the financial year 1985–6 alone, nearly U.S. $900 million was siphoned out of Hong Kong by 25,500 emigrants to Canada, Australia, and the United States.[33]

Hong Kong people became more nervous about the future of Hong Kong after Li Hou, deputy director of China's Hong Kong and Macau Affairs Office, in June 1987 challenged the right of the Hong Kong government to propose any political change, such as creation of a representative government before 1997.[34] Li Hou's challenge dealt another blow to local confidence and raised suspicion that China would interfere in Hong Kong's internal affairs and that the "one country-two systems" concept was unworkable. In 1987, another 23,000 Hong Kong residents emigrated to Canada. Both the Hong Kong and Chinese governments expressed concern at the exodus of capital,

TABLE 24: Destination of Chinese immigrants, 1956–60

Destination	Number of persons	% of total
British Columbia	3,676	35.3
Ontario	3,194	30.7
Quebec	1,137	10.9
Alberta	1,068	10.3
Saskatchewan	662	6.4
Manitoba	306	2.9
Other provinces	258	2.5
Unspecified*	106	1.0
Total	10,407	100.0

Sources: Canada, Department of Citizenship and Immigration, Immigration Statistics, 1956–60.
*Destination of Chinese immigrants from the United States is unspecified.

professionals, and entrepreneurs to Canada and other countries, but they conceded that there was little they could do to stop the outflow.

DEMOGRAPHIC GROWTH AND DISTRIBUTION

The Chinese population in Canada had increased rapidly since the 1970s. In 1981, it stood at 289,245, a 143.4 per cent increase over 1971 and nearly 400 per cent over 1961. The distribution of Chinese across the country had also changed drastically, mainly because of the new immigrants' settlement choices. During the 1950s, most Chinese immigrants preferred British Columbia as their Canadian destination. In the second half of the 1950s, for example, 35 per cent of Chinese immigrants went to British Columbia, 30 per cent to Ontario, only 11 per cent to Quebec, and 10 per cent to Alberta (Table 24). However, an increasing number of new Chinese immigrants began to pour into Ontario and Alberta throughout the 1960s, mainly because of job and investment opportunities there. By the 1970s, Ontario overtook British Columbia as the leading destination for Chinese immigrants. Between 1971 and 1980, of 100,000 landed immigrants from Hong Kong, Taiwan, and China, 43 per cent settled in Ontario and 30 per cent in British Columbia, and 12 per cent in Alberta (Table 25). Under a 1975 federal-provincial agreement, Quebec immigration officers played a dynamic role in recruiting and counselling immigrants destined for Quebec. Since most Vietnamese refugees were French-speaking, a large number settled in Quebec. For example, 3,072 of 3,538 immigrants from Vietnam came to Quebec in 1975 and 1976.[35] In 1984, the Quebec government set up an office in Hong Kong to induce Chinese entrepreneurs to invest in Quebec. This service proved effective; in April 1985, for example, 87 applicants from Hong Kong were ap-

TABLE 25: Destination of immigrants from Hong Kong, Taiwan, and China, 1971–80

	Number of immigrants				
Destination	Hong Kong	Taiwan	China	Total	% of total
Ontario	36,244	5,207	3,422	44,873	43.0
British Columbia	24,502	2,493	4,529	31,524	30.2
Alberta	11,042	609	1,059	12,710	12.2
Quebec	5,852	790	887	7,529	7.2
Manitoba	2,873	260	285	3,418	3.3
Saskatchewan	1,921	143	297	2,361	2.3
Nova Scotia	583	80	56	719	0.7
New Brunswick	382	105	80	567	0.5
Newfoundland	335	50	48	433	0.4
Prince Edward Is.	53	9	6	68	0.1
Yukon and NWT	125	7	15	148	0.1
Total	83,913	9,753	10,684	104,350	100.0

Sources: Canada, Department of Employment and Immigration, Immigration Statistics, 1971–80

proved, of which thirty-three immigrants would invest in Quebec.[36]

A comparison of the geographical shift of the Chinese population in Canada between 1951 and 1981 reveals that the greatest gain was in Ontario and the great loss in British Columbia (Table 26). Between these two years, there was a nearly 800 per cent gain in the country's Chinese population, but three provinces and one territory had a proportional gain: Ontario had over 85 per cent of the inward shift, distantly followed by Alberta, Quebec, and Northwest Territories (Figure 22). The Chinese population in other provinces declined; British Columbia accounted for nearly 70 per cent of the outward shift. In 1981, 41 per cent of Canada's Chinese people lived in Ontario, 34 per cent in British Columbia, 13 per cent in Alberta, 7 per cent in Quebec, and the remaining 5 per cent in other provinces. During the first half of the 1980s there was a great increase in the Chinese population in Canada. Between 1981 and 1986, there was an increase of 124,795 persons; nearly 95 per cent were in Ontario, British Columbia, Alberta, and Quebec (Table 27). In terms of the percentage of increase, the highest percentages were in Newfoundland, Alberta, Ontario, and Manitoba. This reflected the trend of the rapid Chinese population growth during the first half of the 1980s.

A great majority of Chinese immigrants tended to make their homes in metropolitan cities because of greater opportunities for education, employment, and investment. The greatest percentage increases were also found in cities in Ontario and Alberta. For example, the 1981 Chinese population in Toronto, Calgary, Edmonton, Oshawa, and Kitchener was triple the 1971 population (Table 28). In 1981, Metropolitan Toronto had a Chinese community of nearly 90,000, closely followed by Metropolitan Vancouver's popula-

TABLE 26: Geographical shift of Chinese population in Canada, 1951, 1981

Province	Number of persons 1951	Number of persons 1981	Expected no. of persons in 1981	Net shift	% of net shift
Ontario	6,997	118,640	62,219	+56,421	+86.8
British Columbia	15,933	96,915	141,679	– 44,764	– 68.9
Alberta	3,451	36,770	30,687	+ 6,083	+ 9.4
Quebec	1,904	19,260	16,931	+ 2,329	+ 3.6
Manitoba	1,175	7,065	10,448	– 3,383	– 5.2
Saskatchewan	2,144	6,965	19,065	– 12,100	– 18.6
Nova Scotia	516	1,540	4,588	– 3,048	– 4.7
New Brunswick	146	875	1,298	– 423	– 0.7
Newfoundland	186	630	1,654	– 1,024	– 1.6
Yukon	37	220	329	– 109	– 0.1
Northwest Territories	4	200	36	+ 164	+ 0.2
Prince Edward Is.	35	165	311	– 146	– 0.2
Total	32,528	289,245	289,245	| 64,997 |	| 100 |

Sources: Censuses of Canada, 1951, 1981

TABLE 27: Chinese population in Canada, 1981, 1986

Province	Number of persons 1981	Number of persons 1986	Increase in no. of persons	% increase
Ontario	118,640	180,960	62,320	52.5
British Columbia	96,915	125,535	28,620	29.5
Alberta	36,770	56,760	19,990	54.4
Quebec	19,260	26,755	7,495	38.9
Manitoba	7,065	10,740	3,675	52.0
Saskatchewan	6,965	8,620	1,655	23.8
Nova Scotia	1,540	2,015	475	30.8
New Brunswick	875	1,015	140	16.0
Newfoundland	630	1,035	405	64.3
Yukon & NWT	420	420	0	0
Prince Edward Is.	165	185	20	12.1
Total	289,245	414,040	124,795	43.1

Sources: Census of Canada, 1981, and Statistics Canada, *The Daily: Ethnic Origin,* 3 Dec. 1987.

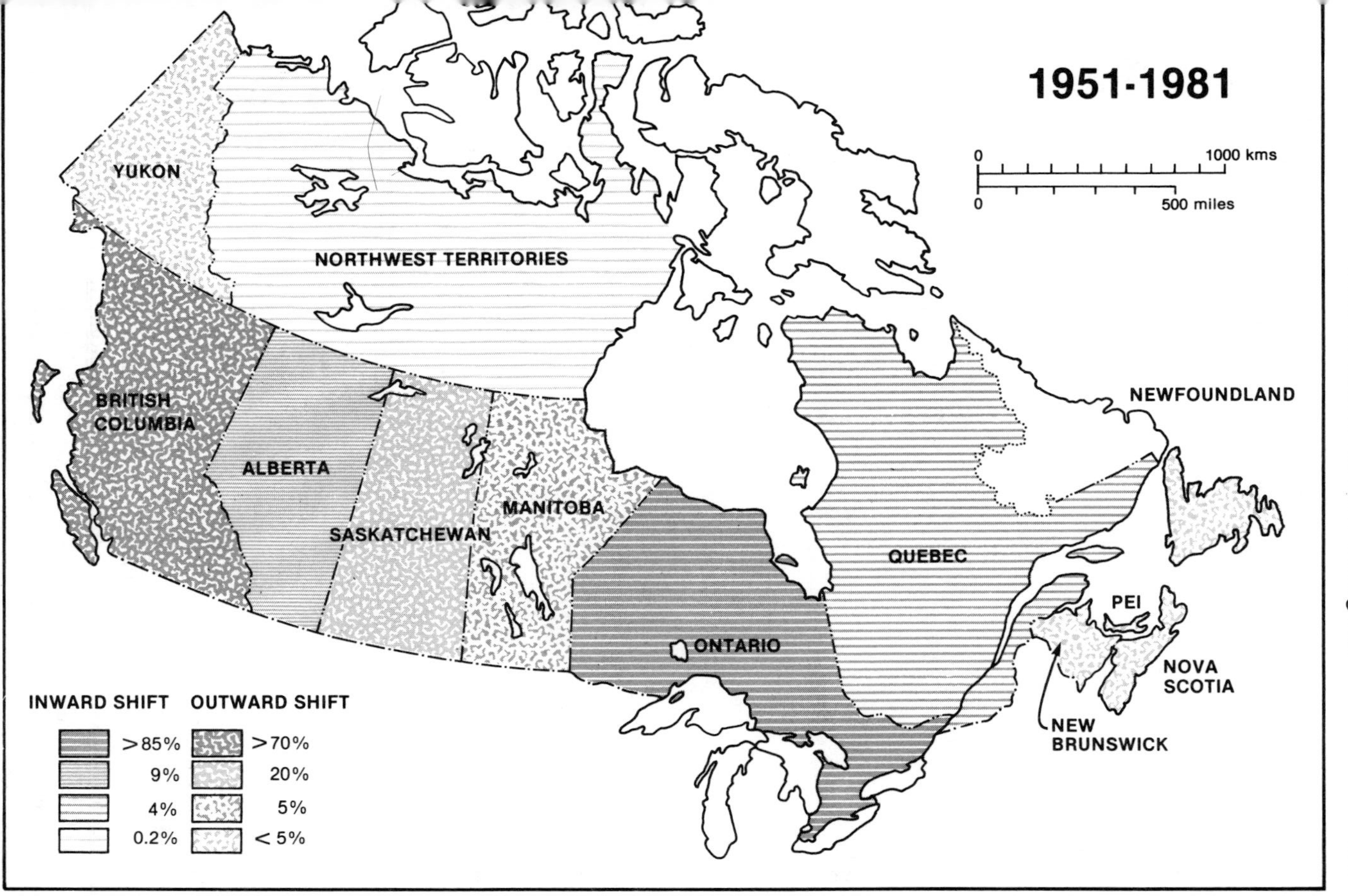

22 Geographical Shift of Chinese Population in Canada, 1951, 1981

TABLE 28: Chinese population increases in census metropolitan areas, 1971–81

	Chinese population		
Metropolitan area	1971	1981	% of increase
Toronto	26,285	89,590	241
Calgary	4,630	15,545	236
Edmonton	5,110	16,300	219
Oshawa	235	720	206
Kitchener	565	1,710	203
Ottawa-Hull	3,060	8,205	168
Hamilton	1,380	3,405	147
Saskatoon	975	2,410	147
Winnipeg	2,535	6,195	144
London	820	1,960	139
Vancouver	36,405	83,845	130
Windsor	1,070	2,325	117
St. Catharines-Niagara	495	980	98
Saint John	155	290	87
Victoria	3,290	5,825	77
Halifax	600	1,000	67
Montreal	10,655	17,200	61
Sudbury	345	545	58
Regina	1,275	1,835	44
Thunder Bay	305	440	44
Quebec	525	715	36
St. John's	275	360	31

Source: Census of Canada, 1981

ation of 84,000 (Figure 23). Lagging behind these two big cities are Montreal, Edmonton, and Calgary, each with a population of 16,000 to 17,000 in 1981. During the first half of the 1980s, the largest percentage increases of Chinese population were in Oshawa, St. John's, Calgary, Toronto, Edmonton, and Winnipeg, mainly because of the arrival of new immigrants from Hong Kong and other places (Table 29). According to the 1986 intercensus on the Chinese population in Canada, Toronto with 143,000 ranked first, followed by Vancouver (109,000), Calgary (26,000), and Edmonton (24,600); Montreal (24,200) had dropped to fifth position. However, many Chinese community leaders felt that these census data underestimated the Chinese population in their cities. For example, they claimed that Toronto has a Chinese community of about 250,000, Vancouver between 150,000 and 200,000, and Calgary, Edmonton, and Montreal about 60,000 to 80,000 each.

The intra-city distribution of the Chinese population had also undergone drastic changes. Population increases occurred in the suburbs rather than the

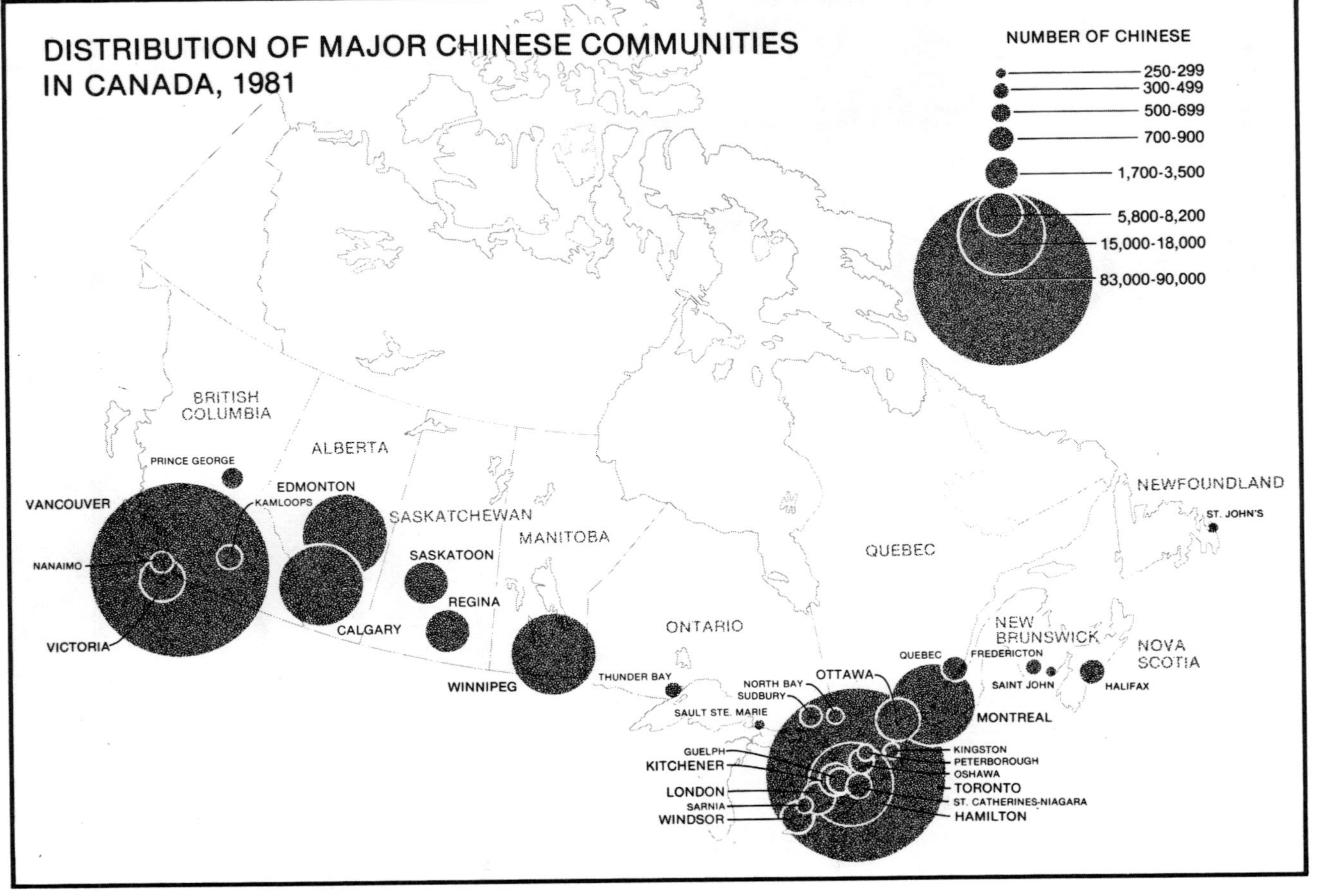

23 Distribution of Major Chinese Communities in Canada, 1981

TABLE 29: Chinese population in census metropolitan areas, 1981, 1986

	No. of persons			
Metropolitan area	1981	1986	Increase in no. of persons	% increase
Toronto	89,590	143,235	53,645	59.9
Vancouver	83,845	109,370	25,525	30.4
Calgary	15,545	26,175	10,630	68.4
Edmonton	16,300	24,560	8,260	50.7
Montreal	17,200	24,185	6,985	40.6
Ottawa-Hull	8,205	11,575	3,370	41.1
Winnipeg	6,195	9,300	3,105	50.1
Victoria	5,825	6,295	470	8.1
Hamilton	3,405	4,665	1,260	37.0
Saskatoon	2,410	3,245	835	48.8
Regina	1,835	2,595	760	41.4
Windsor	2,325	2,390	65	2.8
Kitchener	1,710	2,545	835	48.8
London	1,960	2,430	470	24.0
St. Catharines-Niagara	980	1,740	760	77.6
Oshawa	720	1,400	680	94.4
Halifax	1,000	1,250	250	25.0
Quebec City	715	895	180	25.2
Thunder Bay	440	755	315	71.6
St. John's	360	690	330	91.7
Sudbury	545	505	–40	–7.3

Source: Census of Canada, 1981; and Statistics Canada, *Summary Tabulation of Ethnic and Aboriginal Origins* (1986), Table 2

centres of cities, mainly because many Chinese people of moderate means had moved out of Chinatown to better residential districts in the suburbs. For example, there was a less than 100 per cent increase of Chinese people in the City of Toronto and City of Vancouver between 1971 and 1981 (Table 30). But at the same time the Scarborough Borough of Metro Toronto had an increase of nearly 900 per cent and the City of North York a nearly 500 per cent increase; Vancouver's Richmond municipality had a 600 per cent increase during the same period. This rapid growth of Chinese population led to the rise of New Chinatowns in the suburbs as well as the revitalization of the economy of some Old Chinatowns.

TABLE 30: Chinese population increase in Toronto and Vancouver metropolitan areas, 1971–81

Locality	1971	1981	% increase
METROPOLITAN TORONTO	26,285	89,590	241
Toronto City	17,755	32,390	82
Scarborough Borough	1,810	17,900	899
North York City	3,055	17,545	474
Mississauga City	n.a.	5,345	—
East York Borough	890	4,245	377
Etobicoke	955	4,075	327
Markham Town	135	2,990	2,115
York Borough	595	1,550	161
Brampton City	n.a.	1,300	—
Other areas	n.a.	2,250	—
METROPOLITAN VANCOUVER	36,405	83,845	130
Vancouver City	30,640	59,620	95
Richmond Municipality	935	6,675	614
Burnaby Municipality	2,040	8,380	311
Delta Municipality	230	1,125	389
Surrey Municipality	545	1,895	248
Coquitlam Municipality	380	1,415	272
North Vancouver Municipality	470	1,385	195
Port Coquitlam City	90	565	528
West Vancouver Municipality	255	665	161
New Westminster City	250	410	64
Other areas	570	1,710	200

Sources: Census of Canada, 1971, 1981

7

Postwar Chinatowns

During the period of selective entry, there were four types of Chinatowns: surviving Old Chinatowns; New Chinatowns, which emerged after the Second World War; Replaced Chinatowns, which were built to replace the demolished Old Chinatowns; and Reconstructed Historic Chinatowns, which were rebuilt from extinct Chinatowns as live museums. During the 1950s and 1960s, some Old Chinatowns could not survive depopulation and became extinct. Other Chinatowns which were still in the withering stage were partially demolished in the course of slum clearance or urban renewal programs. Many city governments perceived an Old Chinatown as a blighted inner-city neighbourhood and considered redevelopment to be the only solution. This meant that all the old buildings in Chinatown would be demolished to provide sites for new construction projects, which usually do not conform to the characteristic land uses of Chinatowns. In Toronto and Montreal, for example, large sections of Old Chinatowns were levelled; in Vancouver, several city blocks in the Strathcona District, the residential annex of Chinatown, were obliterated for redevelopment. Most city governments had never thought to preserve Old Chinatowns, and the Chinese themselves had no concerted views or opinions about their future. Some people wanted to save Old Chinatowns from extinction; others disdained them and wanted them gone; many others were uninterested. It was not until the late 1960s and early 1970s that some concerned Chinese leaders mobilized Chinese communities to fight for survival of their Chinatowns. City governments also began to realize that Chinatowns could be restored to take advantage of their history, preserved through the retention, repair, conservation, and improvement of historic buildings. New types of Chinese immigrants, particularly investors and entrepreneurs, helped revitalize Old Chinatowns and establish new suburban Chinatowns. In some Canadian cities, however, such as Regina, London, and Halifax, the Chinese population has increased rapidly in recent years but is so widely scattered that no Chinatown has been formed. Although the identity of these Chinese communities is not expressed in territorial terms, their presence is manifested by their cultural activities.

CHANGING ATTITUDES, IMAGES, AND TOWNSCAPES

Since the Second World War, social attitudes and practices in the Chinese community and the host society have been changing rapidly. Younger generations and new Chinese immigrants have integrated themselves well into Canadian society and have participated in community-wide public events. Some Chinese have also participated in local politics and have been successful in politics (see Appendix 1). On the other hand, the host society also began to take pride in its ethnic diversity and no longer regarded the Chinese people as aliens. Action has also been taken to recognize the contributions and / or suffering of the Chinese community in Canada that has resulted from previous discrimination and prejudice.

For example, on 25 September 1982, the Historic Sites and Monuments Board of Canada installed a bronze plaque at Yale, British Columbia, to commemorate the Chinese construction workers on the Canadian Pacific Railway.[1] The dedication, written in English, French, and Chinese, briefly describes the history of the railway construction, during which "hundreds of Chinese died from accidents or illness, for the work was dangerous and living conditions poor."[2] On 21 June 1986 four Canadian labour unions—the Canadian Association of Industrial, Mechanical and Allied Workers; the Canadian Association of Smelter and Allied Workers; the Pulp, Paper and Woodworkers of Canada; and the Sudbury Mine Miners and Smelter Workers' Union, Local 598—erected a cairn at the Chinese and Japanese Cemetery and dedicated it to the Chinese and Japanese miners killed in the coal mines of Cumberland.[3] In the same year, a bronze plaque listing the names of about 400 Chinese-Canadian veterans was erected in Vancouver's Chinese Cultural Centre to commemorate their sacrifices and efforts during the Second World War. In 1987, all three political parties—the Progressive Conservative, the Liberal, and the New Democratic—supported the introduction of an all-party parliamentary resolution to recognize the injustice and discrimination of the head tax and the Chinese Exclusion Act.[4] Since the 1970s many Chinese Canadians have been recognized nationally and locally for their achievements and contributions. For example, thirteen Chinese Canadians have been appointed to the Order of Canada, the country's highest honour (see Appendix 2), and eight Chinese Canadians have been named Honorary Citizens by the City of Victoria (see Appendix 3).

As the host society's attitudes towards its Chinese population change, so its images of Chinatowns have also changed. In the 1950s and 1960s Chinatown was no longer regarded as godforsaken; instead, it was perceived as an aging residential and commercial inner city neighbourhood where poor, elderly Chinese males lived. It was the place where a non-Chinese person could buy typical Chinese products unavailable outside Chinatown and could taste genuine but inexpensive Chinese food. However, despite their changing images of Chinatown, many Canadians still regarded Chinatown as a mysterious and exotic neighbourhood where gambling clubs proliferated. Some people even believed that Chinatown still had opium dens with subterranean passages.

Although the Chinatowns across Canada varied in their degree of decline during the 1950s and early 1960s, they had much in common and conformed closely to the withering stage of the development model. Late nineteenth-century buildings were fast disappearing, demolished to provide sites for new office buildings or parking lots. However, the marks of age were still visible and the townscapes were deplorable: narrow streets, filthy alleys, garbage-choked back yards, old brick two- or three-storey buildings, and overhead wiring. Most institutional buildings, once focal points of activities in Chinatown, were functionally obsolescent and physically decaying. In Victoria's and Winnipeg's Chinatowns, for example, some tenement buildings did not meet new fire and building regulations for dwellings. Since the cost of renovation would not be covered by the rents, the owners opted to leave the upper floors vacant and upgrade only the ground floors for commercial use. In other Chinatowns, such as those in Calgary and Edmonton, their "tong houses" and cheap hotel buildings were dilapidated and crowded with poor elderly people or families of low income. Drunkards, prostitutes, and drug addicts were commonly seen in Chinatowns and nearby neighbourhoods, particularly in the late evenings. In short, Old Chinatowns had all the ingredients of a skid row. Their environment was so deplorable during the 1950s and early 1960s that nearly every city with a Chinatown regarded it as an eyesore and wished to eliminate it.

DECLINE AND EXTINCTION, 1950s–1960s

There are several major reasons for the decline and eventual extinction of many old Chinatowns during the 1950s and early 1960s. Fire was one destructive force. Nanaimo's Chinatown, for example, continued to decline after the Second World War. By 1955, there were only four restaurants, two butcher shops, one grocery store, one drug store, and a barber serving a population of about 250.[5] Except for a few young families and newcomers, most residents were elderly pensioners who lived in derelict buildings, paying low rent to the Wah Hing Company, which owned most of the properties in Chinatown. During the week, Nanaimo's Chinatown was very quiet and appeared deserted. Only a few elderly men dozed or chatted on wooden benches on the sidewalks. After dark, however, Chinatown came alive as local people came to drink and dine. On weekends, Chinatown might be crowded with Chinese tourists from Victoria or other cities on the island. Nanaimo's Chinatown retained the characteristic of a frontier town: its frame buildings were lined with wooden sidewalks and tall verandahs stretched from structure to structure along its full length. Several wooden rooming houses, originally constructed for bachelors, were derelict and empty. And like so many Chinatowns in the days of the gold rushes, Nanaimo's Chinatown was burnt to the ground on 30 September 1960.[6] About seventy elderly pensioners lost their homes and had to be temporarily housed in the Nanaimo Military Camp. After the fire, the Wah Hing Company donated its land to the new Cathay Senior Citizen Housing Society for the construction of two buildings for the pensioners.[7] Chinatown has never been rebuilt in Nanaimo. Today its Chinese community is dispersed throughout the city and integrated into the host society.

Depopulation was another factor in the extinction of Chinatowns such as those in Barkerville, Cumberland, Hamilton, and other cities. Cumberland's Chinatown, for example, experienced a slow death caused by the gradual loss of its residents. By the late 1950s, time had changed it into a dusty shack town with only twenty elderly residents.[8] Most of them were former miners, over seventy years of age, and they paid only a dollar a month to Canadian Collieries for their ramshackle cabins.[9] As the shanties continued to deteriorate, these people were permitted to remain free of charge. When a shack was no longer inhabitable, its occupants moved to another one. By 1962, only twenty out of about sixty structures in Cumberland's Chinatown remained.[10] Towards the end of the 1960s, the entire town had been dug up by souvenir hunters in search of Chinese relics. Only four men, each over eighty years old, remained. Huang Geng, affectionately called "Jumbo" by the local people because of his stoutness, was the only one who was still alert (Plate 10). By the mid-1970s, three of the remaining residents had died. When Jumbo, the last resident, moved to a care home in Vancouver in the late 1970s, Cumberland's Chinatown was left as a quaint relic of the past.

A decline in Chinatown's Chinese population was also noticeable in large metropolitan cities. After the war, Chinatown was no longer regarded as a sanctuary, since its protection was unnecessary. The second generation of Canadian-born Chinese was better educated and better off economically. With diminishing discrimination, these people had moved out of Chinatown to better residential neighbourhoods. Consistent with the assimilation theory, the importance of Chinatowns was declining as the Chinese no longer segregated themselves. The third and later generations and postwar immigrants did not regard Chinatown as their home at all, since they had not been brought up there. By the late 1950s, most remaining residents of Chinatowns were elderly single males who could not afford better accommodations. In Winnipeg's Chinatown in 1951, for example, 86 per cent of its 700 residents in 1951 were old bachelors.[11] Twenty years later, its Chinatown population had decreased to 540 persons, representing only 20 per cent of the Chinese population in Metropolitan Winnipeg. Similarly, Montreal's Chinatown once had a residential population of over a thousand, but by the 1970s only 300 residents remained, mostly elderly men or widows.[12] Most of the Chinese had moved out to other parts of the city such as St. Laurent, LaSalle, Brossard, and Outremont.

Socioeconomic activities in Chinatown were much reduced, partly because of depopulation and partly because of the disappearance of the bachelor generation. Before the war, traditional Chinese associations had organized social activities in Chinatown and provided services essential to its bachelor community. As this society was gradually replaced by one composed of family units, the associations began to lose their importance and meaning. They could not compete with other social clubs, such as golf clubs and Lions clubs, which no longer excluded Chinese and which were more appealing to the younger generations. Many Chinese youths did not bother to join the clan or county associations in Chinatown; some even rejected all affiliation with these traditional societies. Only members of the older generation still regarded Chinatown's traditional associations as their second home, where they would meet their old friends, read Chinese newspapers, celebrate various Chinese festivals, and maintain Chinese customs. Many of these associations became extinct after the old members died.

The demise of Chinatown's traditional associations was also caused by disunity in Chinese communities across Canada. In each major Chinatown, there was an umbrella organization known as the Chinese Benevolent Association (CBA). Like the CCBA in Victoria, each CBA had functioned as a spokesman for a Chinese community, but after establishment of the People's Republic of China in 1949, some CBA's in Canada's Chinatowns refused to recognize the new government. Many directors of the CBA's were strongly affiliated with the Nationalist League and were more interested in enlisting the support of overseas Chinese to overthrow the communist government than in promoting the welfare of the local Chinese community. Accordingly, some Chinese organizations withdrew from the CBA's because of their collaboration with the Kuomintang in Taiwan. As a consequence, the CBA's gradually lost control of Chinese communities.

The economy of Chinatown continued to decline because of the reduction of its trade and commerce. As an increasing number of Chinese people lived outside Chinatown, new Chinese businesses were set up in other parts of the city, drawing away many former patrons. Furthermore, many of the younger generation were so assimilated to the host society that they did not bother going to Chinatown; it had little to offer them. The economy of Chinatown declined further when its small stores and cafés had to close after their original proprietors retired or died; there were no successors to operate them. The younger generation of educated children did not want to work long hours like their parents and refused to run the family business. Many small Chinatowns such as Ottawa's Old Chinatown on Albert Street disappeared as their last stores ceased to operate.[13]

Landlords themselves were also responsible for the deterioration and eventual destruction of Chinatowns. They were reluctant to repair or renovate old, dilapidated buildings because the rents were so low that when the buildings failed to meet new fire and building regulations, they would leave them vacant or demolish them to make room for parking lots. As a result, derelict houses were commonly interspersed with vacant lots throughout many Old Chinatowns. In Winnipeg's Chinatown, for example, after the city council passed a bylaw in 1973 which required that all rooming houses and apartment hotels in the downtown area be upgraded to meet new regulations, many Chinatown landlords felt that the cost of renovating their tenement buildings would not be met by the low rents. Accordingly, they evicted their tenants and left the rooms vacant.

Land speculation also played an important role in the decline and destruction of Chinatowns. In some cities, such as Calgary and Montreal, Chinatowns were located in the path of major downtown developments. Rising land prices enticed speculative developers to purchase derelict tenement buildings at low prices; waiting for development opportunities, the owners allowed these old buildings to deteriorate beyond repair, demolishing them when they were condemned. In Calgary's Chinatown, for example, about three-quarters of its residential properties were owned by investors.[14] These absentee landlords had no interest in maintaining the decaying homes; they sold them as soon as an attractive offer was made.

Slum clearance or urban renewal projects were the major causes of destruction of many Old Chinatowns. Physical deterioration of the buildings made them vulnerable to demolition in the course of downtown revitalization programs. Duncan's Chinatown, for example, was levelled during the early 1970s to provide parking space for the new provincial court house (Plate 11). Similarly, Kamloops's Chinatown was described in 1946 as an extreme fire hazard, unhealthy, and dilapidated.[15] Much of it was levelled during construction of the new Overlander Bridge in 1961. Kamloops's Chinatown was completely levelled in 1979 after the city government appropriated all its premises and demolished them to provide for expansion of Victoria Street.[16]

Throughout Canada, the destruction of Chinatowns had been accepted with equanimity in small towns and cities, but in large cities Chinese communities were stronger and more vocal. They organized campaigns to fight against destruction of their Chinatowns, sometimes successfully postponing development projects. For example, when the "Downtown Master Plan" threatened the survival of Calgary's Chinatown in 1969, the Chinese community organized the first "National Conference on Urban Renewal as It Affects Chinatowns," which attempted to unite Chinese communities across Canada in a concerted examination of the urban crisis threatening Chinatowns.[17] A study of the major Old Chinatowns across Canada will reveal the significance of community involvement in planning, which had changed the government's policy from one of obliterating Chinatowns to one of rehabilitation, restoration, and preservation.

VANCOUVER'S CHINATOWN

Vancouver's Old Chinatown is probably the only Chinatown in Canada which has not been drastically reduced by postwar urban renewal programs, although its survival has been threatened several times. Canton Alley, one of its oldest streets, was the only section destroyed after the tenement buildings on both sides of it were demolished in 1949. The same year, Dr. Leonard Marsh proposed the first major redevelopment scheme for the Strathcona district (Figure 24). Marsh's survey of the district revealed that it contained about 7,500 people: 44 per cent Europeans, 33 per cent Canadians, 11 per cent Chinese, and 12 per cent other national groups. Sixty per cent of their properties were structurally poor, out of date, or markedly dilapidated.[18] His report called for the acquisition and clearance of dilapidated houses in the area and their replacement with three types of rental housing: apartments, row-housing, and small suites. Throughout the 1950s, an increasing number of Chinese people moved into Strathcona District partly because its house prices and rents were lower than those in other downtown areas and partly because it was adjacent to Chinatown. In 1957, the city's Housing Research Committee conducted a survey of the East End which revealed that nearly half the residents in Strathcona District were of Chinese origin.[19] Thus, the district was perceived to be a residential section of Old Chinatown whereas Old Chinatown itself, on the western side of Gore Avenue, was characterized by many two- or three-storied buildings with a mixture of commercial, residential, and institutional uses, both horizontally and vertically.

In December 1957, the Housing Research Committee submitted a report to Council on the twenty-year Redevelopment Plan of the East End Area, which covered an area of about 160 city blocks in three districts, with a total population of 15,147.[20] The cost of acquisition and clearance was estimated at $75 million; the federal government would pay half the cost and the provincial and city governments would split the other half. The report proposed to demolish nearly all old buildings in the area and replace them with a carefully zoned complex of private dwellings, light industries, and government-subsidized public housing projects. The first stage of the redevelopment plan involved slum clearance in District Two, Strathcona District, where 37 per cent of its 274 structures suffered from inadequate original construction and

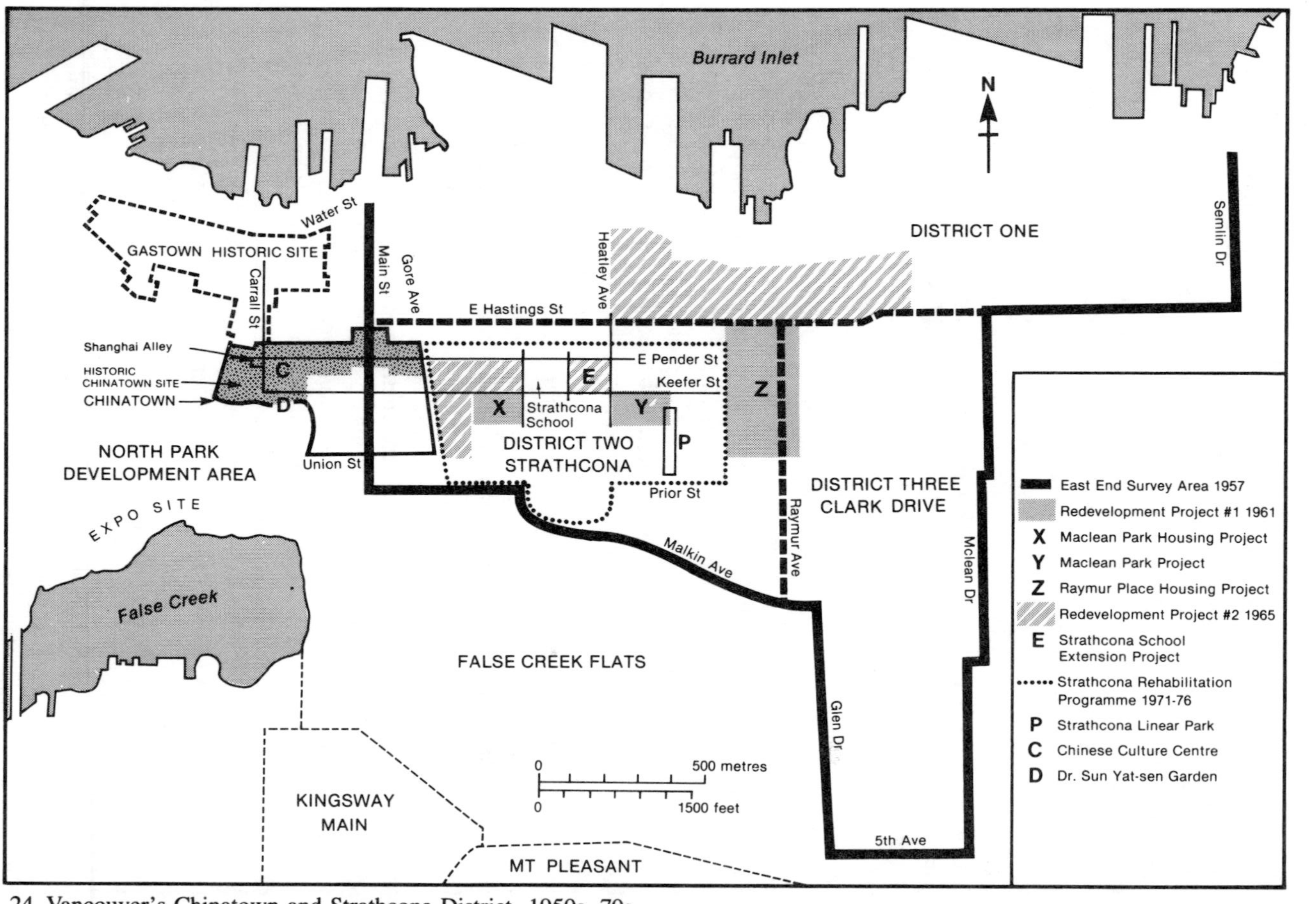

24 Vancouver's Chinatown and Strathcona District, 1950s–70s

nearly 24 per cent had poor or very poor interior conditions.[21] In January 1958, City Council approved the redevelopment plan, declaring Strathcona District a redevelopment area. The plan's objectives were to demolish all old houses in the area and replace them with townhouses and high-rise buildings which would eventually house between 7,000 and 8,000 persons, permitting open space as well as high population density.[22] No regular public works maintenance in Strathcona District would be undertaken by the city in the next ten years because all dwellings had been programmed for demolition. Meanwhile, all property values in the district were frozen and no private redevelopment or improvement permits were issued to property owners. In October 1960, a delegation of Chinese community leaders, headed by lawyer C.C. Locke, met City Council to seek a delay in the slum clearance in Strathcona, since its 4,500 residents, mostly Chinese, would be displaced and were reluctant to live in public apartment buildings.[23] Their request was rejected; the city firmly believed that clearance was the only way to improve the physical environment of the slum area.

In the fall of 1960, in preparation for rehousing about 1,200 residents who would be displaced by slum clearance, the government began to construct the 150-unit Maclean Park Housing Project at Maclean Park in Strathcona and another 234-unit public housing project at Skeen Terrace, about ten kilometres southeast of the first one.[24] As these two projects became available for occupancy in the spring of 1961, Redevelopment Project No. 1 was launched. The city forced owners to sell their properties at a negotiated price, which ranged from $6,000 to $8,000; they expropriated the house if an owner refused to sell. Most homeowners complained that the price offered was insufficient to purchase houses elsewhere in the city. However, within a year, over six city blocks in Strathcona had been appropriated and their structures, including some historic Chinese ''tong houses'' such as the Hing Mee Society house, were demolished to provide sites for the extension of the Maclean Park Housing Project, the replacement of Maclean Park, and construction of the 376-unit Raymur Place Housing Project.[25] About 300 Chinese residents were forced to move, some of whom were very bitter about the clearance, arguing that the social impact on the community had not been considered.[26] For example, twelve old men had been living together in the Hing Mee Society house for thirty-seven years, paying a monthly rent of eight dollars.[27] They wanted to remain together, but were forced to separate.

To appease opposition, Mayor Tom Alsbury set up a Redevelopment Consultative Committee in March 1961 to deal with complaints and problems caused by the slum clearance.[28] Since the Chinese were not used to confronting authority, few complained to the committee. According to its chairman, the displaced residents were either resigned to progress or happy about it.[29] This declaration reflected the committee's ignorance about Strathcona residents' reaction.

In December 1962, Council announced Redevelopment Project No. 2, which would involve the acquisition and clearance of nearly thirty acres of land and the displacement of about 2,300 people.[30] The project included five city blocks in Strathcona and would displace 770 persons, mostly Chinese, in order to provide sites for public and private housing, and for the expansion of Strathcona School.[31] Various Chinese associations immediately protested against the project, which was criticized as being "unwise, too ambitious and without regard for the human element."[32] They argued that apartments were unsuitable for the Chinese family system, in which members of several different generations lived together. The expropriation and clearance program under Project No. 2 would force Chinese families to move out of Chinatown and would destroy four Chinese schools and sixty-eight Chinese fraternal associations. In response to the protests, Mayor Bill Rathie asked the Chinese community to present Council with a Chinatown redevelopment proposal which would suit their family life.[33] Acting on the mayor's request, the Vancouver Chinatown Redevelopment Association (VCRA), formed by a group of Chinatown homeowners and headed by Dean and Fay Leung, immediately commissioned Arthur Mudy to draw up an experimental redevelopment plan for one block.[34] The plan was typically Oriental in design, consisting of small houses, mostly duplexes, with moongate windows, pagoda roofs, and small interlocking courtyards. The city block would be subdivided into thirty-six small lots on which seventy-two units would be built. The VCRA was pleased with the design, which would preserve Chinese culture and make the redeveloped block a tourist attraction. But when the plan was presented to Council on 29 March 1963, it was criticized by Gilbert Eng as being anti-social and commercial.[35] He claimed that he had been asked by another Chinese interest group, headed by Harry James Agencies Ltd, to design another redevelopment plan based strictly on rehabilitation.[36] Council considered this as another example of dissension within the Chinese community. Nevertheless, it approved the VCRA plan in principle in July 1963 and permitted the association to repurchase the cleared block for a private housing development.[37] However, the Technical Planning Board did not recommend subdivision of the block into small lots, since this kind of tight subdividing would have resulted in too many buildings on small parcels of land, the reason for the blight in the first place. Its alternative scheme was to have fifty multiple units on eight lots, but the VCRA objected because large units were expensive and did not meet the social needs of Chinese who wished to own individual lots.[38] In March 1964, the VCRA plan was scrapped after being rejected by the federal government.[39]

Property acquisition began immediately in March 1965, soon after Redevelopment Project No. 2 was approved.[40] Within a year and a half, twenty-four properties, including the comparatively well-maintained Christ Church of China, had been appropriated for demolition.[41] Redevelopment

Project Numbers 1 and 2 had displaced about 1,000 people, of whom more than half were Chinese. In the implementation of these two projects, attention was paid only to physical improvement of Strathcona District; no consideration was given to the social problems of residents. A survey conducted by United Community Services revealed that the reaction of the people, particularly the Italians, whose homes were to be destroyed, bordered on panic.[42] Some people who had been moved into Maclean Park Housing Project led solitary, isolated lives and had little contact with the area's recreational agencies. The survey also revealed that most people in the district, particularly Chinese residents, would have liked to own their own homes, continue to raise home-grown vegetables and flowers, and maintain a sense of security by living close to Chinatown.[43]

In the summer of 1965, the city's Planning Department started to work on the details of the third and final stage of clearance of remaining housing in Strathcona District and relocation of its 3,000 residents. This clearance program was part of Urban Renewal Scheme No. 3, which covered five subareas: Strathcona, False Creek Flats, Clark Drive, Kingsway, Main, and Mount Pleasant.[44]

Meanwhile, another project drew the attention of the entire Chinese business community. On 17 October 1967, City Council approved construction of an elevated freeway above Carrall Street to link the new Georgia Viaduct with the proposed waterfront expressway along Burrard Inlet.[45] The route, running on the eastern side of Carrall Street, would require the acquisition and demolition of a strip of commercial buildings in Chinatown. The displaced business would be offered space beneath the elevated freeway. When Council's approval was known, the Chinese community protested angrily, complaining that the freeway would create a "Great Wall" on the western side of Chinatown, severing it from the city's commercial centre. On 18 October, about fifty architecture and community planning students from the University of British Columbia marched along Pender Street in Chinatown to protest Council's decision.[46] Black banners were hung up in Chinatown as a symbol of its death.[47] At the public meeting, the one hundred people in attendance expressed their desire to preserve Carrall Street as an historic site and elected a committee to spearhead the protest campaign. In January 1968, mainly because of these organized protests, City Council finally rescinded its previous decision about construction of the Carrall Street freeway.[48]

The successful fight against City Council had given much encouragement and confidence to residents of Strathcona District; if they were organized, persistent, and firm, they might be able to make Council abandon clearance of the district under Scheme No. 3. Civic organizations and over sixty residents submitted briefs to Council, complaining that "urban renewal is actually urban destruction" and that they were not given enough information about renewal plans.[49] In December 1968, the entire Strathcona community was mobilized. About 600 persons, Chinese and non-Chinese, owners and tenants, old and young, men and women, came to a meeting at which the Strathcona Property Owners and Tenants Association (SPOTA) was formed.[50] They sent briefs and petitions to City Council, demanding the right to continue to live in the Strathcona area and suggesting that Council should lend residents money to improve their homes instead of buying them and tearing them down. Residents were not against physical improvements in the Strathcona area, but they disapproved of how the city had handled the whole process. Urban renewal had created more social problems than it had solved; it had resulted in unaccountable psychological and socioeconomic costs, such as the anxiety and uncertainty of the residents, disruption of a familiar environment, financial loss resulting from inadequate compensation, and destruction of many structures of high heritage value. In his study of displaced Chinese families, Richard Nann found that "reactions of anger, bitterness, frustration, resignation and relief were constantly encountered by the researchers"; in some instances, relocation meant major disruption and frustration of life achievements and aspirations.[51] Some Chinese people perceived the redevelopment programs as an attempt by the government to remove Chinese families from valuable real estate.[52]

Coincidentally, the federal Task Force on Housing and Urban Development, led by Paul Hellyer, Minister of Housing, arrived in Vancouver while SPOTA was presenting its petitions to City Council. Strathcona residents immediately appealed to Hellyer to save their neighbourhood from demolition. After its tour of forty communities across Canada, the task force concluded that urban renewal schemes produced adverse and disrupting conditions in Canadian communities, and that public housing, in addition to being extremely costly, had negative social and psychological effects on its residents.[53] Accordingly, in early 1969, Hellyer froze federal funding for all urban renewal projects other than those currently implemented.[54] This was the beginning of a change in attitude by the federal government towards the urban renewal programs of inner city neighbourhoods.

Meanwhile, SPOTA had become one of the strongest community action groups in Vancouver. It continued to pressure Council to abandon its Urban Renewal Scheme. Its representatives also sought the support of Robert Andras, who had succeeded Paul Hellyer as Minister of Housing. On 14 August 1969, Andras announced that the renewal scheme planned for the Strathcona area would not proceed as originally planned.[55] He also told SPOTA that the federal government would not participate in any renewal scheme in an area which aimed at clearance rather than rehabilitation, which did not involve its residents, and which was opposed by them. In October 1969, for the first time, SPOTA was invited to join with the three levels of governments to form the Strathcona Working Committee, which was to chart a course for the future of Strathcona.[56] The most significant policies adopted by the committee were that no large-scale acquisition and demolition of property would be undertaken in the Strathcona area and that the general goal for the area was rehabilitation: to retain, repair, and improve individual properties, public works, and community services.

Vancouver's Chinatown entered the period of revival in July 1971 when the Strathcona Rehabilitation Program (SRP) was finalized. The SRP covered twenty city blocks, in which about 390 out of 557 structures might be suitable for rehabilitation.[57] It proposed a system of "grant-loans" to property owners for upgrading their property and extending the useful life of buildings. Between 1971 and 1976, a total of $925,800 ($513,700 contributed by the government and $412,100 by private property owners) was spent on rehabilitation of 229 properties in Strathcona District.[58] In addition to the housing rehabilitation, various neighbourhood improvement projects had been carried out, such as the development of a Linear Park and MacLean Park Housing Project Adventure Playground, the installation of bilingual English and Chinese street signs, and Gore Avenue landscaping.[59] After the SRP was completed in 1976, SPOTA continued to be a protector of the Strathcona community. In February 1971, for example, the city decided to locate a fire hall in one and a half city blocks of vacant land on the south side of Pender Street between Gore and Jackson avenues. After a strong protest by SPOTA, City Council rescinded its decision in early 1973. This incident demonstrated to the Chinese community that if they were united, their protests would be stronger and they would be able to change Council's decisions.

Two other important organizations playing an important role in revitalization of Vancouver's Old Chinatown are the United Chinese Community Enrichment Services Society (SUCCESS) and the Chinese Cultural Centre (CCC); both organizations were established in 1973. SUCCESS, a non-profit social service agency, initially aimed to help new immigrants in Vancouver. However, within a few years, it had become a high-profile social organization because of its dynamic presidents. For example, Dr. K.C. Li was instrumental in making SUCCESS a member of the United Way in 1979 when he was president, and later he assisted the organization to establish in 1986 a permanent charity fund, the Vancouver Foundation. Today, most of SUCCESS's activities encourage members of the Chinese community to overcome language and cultural barriers, to assume greater responsibility towards achieving self-reliance, and to contribute fully to Canadian society.

The CCC was established on 11 February 1973 by representatives of fifty-three Chinese organizations and registered with the provincial and federal governments as a non-profit organization in April 1974.[60] Dr. Wah Leung, former dean of the Faculty of Dentistry at the University of British Columbia, was elected chairman. The CCC's objectives were to promote understanding and friendship between the Chinese community and other ethnic groups; to preserve and interpret Chinese culture and traditions to Chinese Canadians and others; and to promote interchange between Chinese and other ethnic communities. On 27 July 1976, City Council agreed to lease the CCC about one hectare (2.5 acres) of land on the south side of Pender Street between Carrall and Columbia streets at one dollar a year for sixty years for construction of a $4.5 million cultural centre complex.[61]

The following year, the CCC approached the federal government for funding. The CBA of Vancouver regarded the CCC as a "leftist organization" challenging its leadership in the Chinese community. Accordingly, it created a rival organization known as the Chinese Canadian Activities Centre and also applied to the federal government for a grant to build a cultural centre, thereby putting both applications in limbo.[62] Supporters of the CCC were forced to respond to this challenge by forming the "Committee to Democratize the CBA," to counter charges that the CBA was controlled by Kuomintang members and that its board of directors was not elected according to its constitution.[63] The committee, established on 4 December 1977, was supported by over thirty Chinese organizations and 1,000 individuals who signed the petition calling for a general meeting and election in accordance with the CBA constitution. The committee took the CBA to court and won the case after a lengthy legal battle. On 12 May 1978, the BC Supreme Court decided that the CBA should abide by the terms of its constitution and hold an open election of the board of directors on 24 September 1978 (the date was postponed later to 29 October 1978), and that every Chinese Canadian or Chinese landed immigrant, paying a one-dollar membership fee, would be permitted to vote.[64] Victor Lee, an engineer, and chairman of the Committee to Democratize the CBA, won the election. The CBA, under the leadership of a democratically elected board of directors, gave their full support to the CCC for the cultural centre project.

Since the late 1960s and early 1970s, Vancouver's Old Chinatown had been booming, mainly because of new investments from Hong Kong entrepreneurs. The Chinese population was nearly 60,000 in 1981, doubling that in 1971.[65] As commercial spaces on East Pender and Keefer streets were not available, new immigrant investors began to set up businesses on Main, East Hastings, and other streets near Chinatown. In February 1971, the Gastown-Chinatown Area was designated a special protected area by the Archaeological and Historic Sites Protection Act.[66] The Chinatown Historic Area Planning Committee was established in 1975 to preserve and protect the heritage and character of the Chinatown area and to work with all city departments in the development and implementation of area policies and programs.[67] Throughout the second half of the 1970s, various beautification projects such as the installation of new streetlight fixtures, the creation of bilingual street signs, and tree-planting were carried out. Other new construction projects were undertaken and new sociocultural activities organized in Chinatown under the leadership of Dr. Wallace Chung, President of the CCC, Mr. Victor Lee, President of the CBA, Mr. Ronald Shon, President of Dr. Sun Garden Society, and many other community leaders of numerous associations.

The most significant construction projects were the Chinese Cultural Centre, which opened on 14 September 1980; the Dr. Sun Yat-Sen Park and Garden, which opened on 16 September 1986; and the Multipurpose Hall, which opened on 22 September 1986.[68] These projects depended not only on government subsidies but also on generous contributions of many corporate and individual donors. For example, one of the significant contributors was Dr. David See Chai Lam, who donated one million dollars to the garden project, and $250,000 to the Multipurpose Hall; the hall was then named the David See Chai Lam Auditorium by the CCC in appreciation.

One project to commemorate the centennial celebration of the City of Vancouver was a bronze plaque installed in February 1986 at the entrance of Shanghai Alley to identify its historic significance.[69] Another plaque was installed on the oldest surviving building in Chinatown, the Wing Sang Building, built in 1889, at 51-69 East Pender Street (Plate 12). The building was owned by Yip Sang, a contractor who had recruited Chinese labourers for the construction of the CPR.[70] In May 1987, the Chinese Cultural Centre acquired the Chinese arch from Expo 86, which was installed at 50 East Pender Street in front of the centre. The transfer and installation cost $65,000, of which the provincial government paid $20,000, the B.C. Enterprise Corporation $8,000, and the balance was raised by the China Gate Committee and Vancouver's Chinatown Lions Club.[71] Vancouver Chinatown's gate is the sixth Chinese arch in Canada, the first being in Victoria's Chinatown, the second and third in Montreal, the fourth in Winnipeg, and the fifth in Edmonton.

In 1986, the B.C. government considered redeveloping the North Park area on the north shore of False Creek. A preliminary North Park Development Plan was prepared, and five public meetings were held to provide opportunities for feedback from the public. On 12 March 1986, the director of planning and other city officials submitted to Council a North Park Development Concept report which stated that the North Park development was intended to respect and extend the character of the Chinatown and downtown areas to create new, distinct, yet compatible character areas, thus adding to the quality and diversity of the inner city.[72] The report summarized public opinion about the development and revision of the plan. For example, it reported that Chinatown community interest groups were concerned primarily with parking, historical preservation and reconstruction, and opportunities for senior citizen housing. To respond to these concerns, city officials recommended extending Shanghai Alley and creating Canton Court, allowing for parking on Block 17 and including senior housing in the plan.[73] Unlike its previous approach to the urban renewal of Strathcona District, City Council gave the public ample opportunities to review any development plan before a final decision was made. However the development plan probably has been shelved or rescinded because of the sale of the Expo 86 site to Concord Pacific Development Ltd.

CALGARY'S CHINATOWN

During the 1950s, Calgary's Chinatown encompassed about ten city blocks on the south bank of the Bow River, bisected by Centre Street (Plate 13). It was vulnerable to destruction because of the popularity of shoreline beautification programs. As early as 1945, the Reverend C.E. Reeve, secretary of the Calgary Ministerial Association, suggested to City Council that Chinatown should be removed and replaced by lawns and trees. He felt that Chinatown was "a disgrace to the city."[74] The real threat occurred in 1966 when the Downtown Master Plan included a proposal to construct a parkway road between 2nd and 3rd Avenues South, which would have resulted in elimination of half of Chinatown.[75] Accordingly, a group of Chinese businessmen, professional people, and concerned Chinese citizens formed the Sien Lok Society (SLS) in November 1968, mobilizing the Chinese community to oppose the plan. After the proposal was shelved, the SLS temporarily lost its main raison d'être. Some Chinese community leaders felt that the society did not represent the interests of the entire community; thus, the United Calgary Chinese Association (UCCA) came into existence in August 1969. It was organized along the lines of the CBA in other Chinatowns and represented twenty-four Chinese associations, such as the Chee Kung Tong, the Mah Association, and the SLS itself. The objectives of the UCCA were to achieve unity in Calgary's Chinese community, to present a common front against further threats to the existence of Chinatown, and to promote Chinese cultural and social activities.

A 1971 survey of Calgary's Chinatown revealed that it had 811 residents, of whom 492 (333 males and 159 females) were of Chinese ethnic origin.[76] Nearly 30 per cent of the Chinese residents in Chinatown were over sixty-five years of age, and most lived cheaply in the dilapidated "tong houses" owned by various Chinese benevolent societies.[77] In 1973, the Chinatown Development Task Force, headed by George Ho Lem, prepared a design brief for Chinatown. The following year, City Council delineated an area of about twenty hectares (forty-nine acres) as "Chinatown," designating it a Neighbourhood Improvement Program (NIP) area.[78] The NIP plan, which was subsidized jointly by the city, provincial, and federal governments, was intended for moderate-income, predominately owner-occupied areas that required rehabilitation. As most of Chinatown's old homes were owned by absentee landlords and were too derelict to be worth saving, plans to rehabilitate them were dropped. Instead, they were to be torn down to make way for new apartment and office buildings.[79] In 1974, two city blocks in the eastern part of Chinatown were levelled for construction of a federal government office complex.[80] The development resulted in the demolition of about thirty housing structures and the dislocation of about 200 Chinatown residents.

The Calgary Chinatown Design Brief, prepared by the task force, was approved by Council in 1976. This marked the beginning of the period of revival of Calgary's Chinatown. The brief recommended that Chinatown be retained and developed as a residential community and as a focal centre for both Chinese and non-Chinese.[81] The brief proposed different land uses on the western and eastern sides of Chinatown: family residential units, a multipurpose community centre, and other local services would be on the west side of Centre Street; on the east side, non-family housing projects, restaurants, ethnic food stores, and professional services would be encouraged. The brief also called for an immediate provision of low-cost housing.

Within four years after approval of the brief, many old houses in Chinatown had been demolished to make way for redevelopment projects. The Calgary Chinatown Development Foundation (CDF), formed by George Ho Lem, was a non-profit organization of about 170 professional and nonprofessional Chinese people. The SLS left the UCCA and joined the CDF as its "sister" organization. The CDF received funding from the Canada Mortgage and Housing Corporation (CMHC) to construct Oi Kwan Place, a senior citizens' home, in 1976 (Figure 25). Bowside Manor, another project financed by the CMHC, was built jointly by the CDF, the Lee Association, and the Chinese Public School. The main floor of the building is reserved for the school, the Lee Association's office, and commercial use, and the upper floors are residential.[82] Next to Bowside Manor is the new Calgary Indian Friendship Centre, completed in 1979.[83] Other projects included the demolition of old tong houses to provide space for new association buildings (Plate 14). Multi-storey commercial and residential apartment buildings, such as Bow Central Plaza (now known as Five Harvest Plaza) and Ng Tower Centre, had also been built.[84]

As these redevelopment projects were in progress, dissension between the CDF and the UCCA surfaced. The UCCA argued that the CDF was controlled by one or two persons and had misled City Council to think it represented the Chinese community. The CDF countered that the UCCA was not representative of the Chinese community in Calgary. The discord between the two organizations probably originated from construction of the Bowside Manor Project. The Chinese Public School had given up its land in return for use of

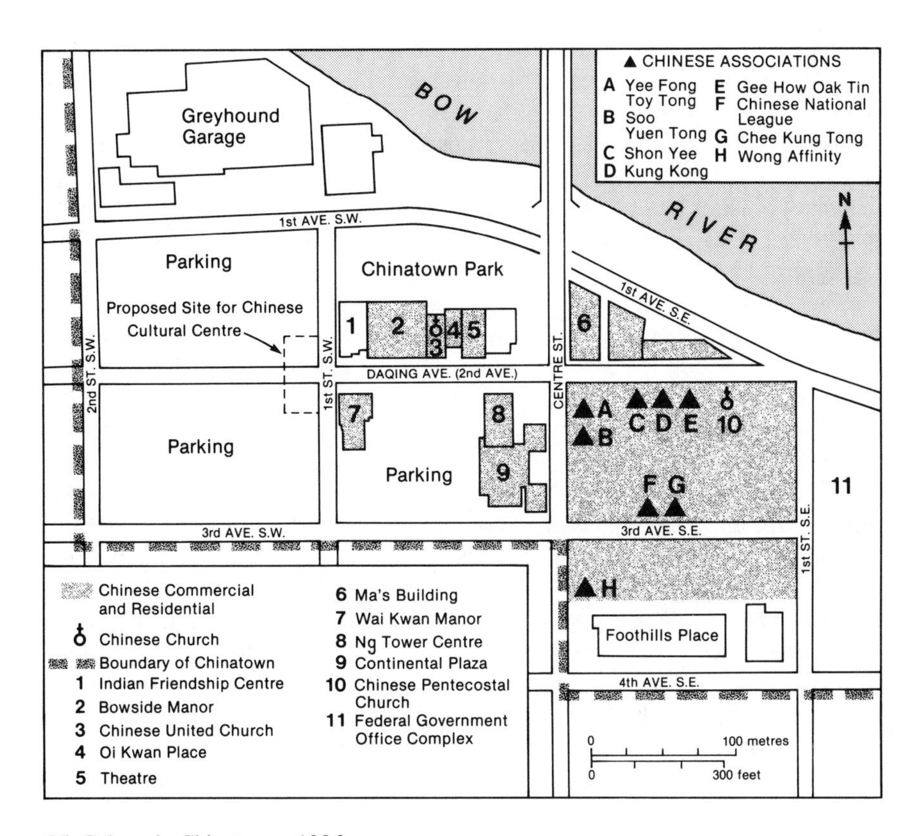

25 Calgary's Chinatown, 1986

the manor's main floor as a school. The school board wanted to buy more space for the school, but its request was turned down by the CDF because the project's space had to be reserved for shops and residential suites. The dispute exemplified the division of the Chinese community that characterizes many Chinatowns in Canada as well as in the United States.

In 1982, the CDF asked City Council to permit high-density development in Chinatown, but the UCCA informed Council that Chinese people wanted Chinatown to remain a low-density area, as originally recommended in the 1976 design brief.[85] The UCCA also charged that the CDF request represented the self-interest of a small group of people rather than the whole Chinese community. Faced with the CDF on one side and the UCCA on the other, City Council could not obtain a unified opinion about Chinatown redevelopment. The dispute was complicated further in 1982 when some Chinatown landowners and businessmen established another organization known as the Chinatown Ratepayers Association of Calgary and commissioned an architect to design a new redevelopment plan for Chinatown.[86] The plan called for high-density development for both residential and commercial use so that property owners would get full value from their land. The proposal was supported by the CDF, but was strongly opposed by the UCCA because it was contrary to the original design brief. The UCCA produced another plan for Chinatown which called for smaller buildings and more medium-sized apartment complexes. City Council thus had two plans which disagreed sharply about land use and intensity of redevelopment.[87]

The Calgary pattern of dissension in the Chinese community was repeated in Toronto, Winnipeg, and other cities when their Old Chinatowns were being revitalized. The internal community contention, complicated by politics, often delayed the revitalization program and forced the city government to look for a mediator. In Calgary, for example, City Council decided in October 1982 to hire an outside consultant to conduct a Chinatown design workshop, intended to provide a forum for identifying the major areas of controversy among various rival factions and to suggest realistic possibilities for achieving a vibrant Chinatown.[88] Based on the results of the workshops conducted by Gerald Forseth in March 1983, the Planning Department established a Chinatown Corporate Committee and a Chinatown Community Committee to resolve the outstanding issues of land use and density.[89] In consultation with the two committees, the Planning Department produced the Chinatown Area Redevelopment Plan in 1984 as a framework to guide future development. The guidelines of the plan allowed medium- to high-density projects in the Chinatown core and higher density commercial land use at its perimeter.[90] The plan also proposed construction of the Chinese Cultural Centre at a site right across 2nd Avenue Southwest (Daqing Avenue) on the west side of the intersection of 2nd Street Southwest and Daqing Avenue.[91]

In March 1986, Calgary's Chinatown had a residential population of about 800, mostly senior citizens and recent immigrants; it had over thirty restaurants, sixty grocery stores, and other businesses. Wai Kwan Manor, which was built by the Oi Kwan Foundation for senior citizens, was officially opened in 1985, and Wah Ying Mansion, another senior citizens' home with 104 suites was completed in 1988 on a site next to the Chinese Freemasons' Building. Chinatown Park, also known as Sien Lok Park, was completed, and its entrance marked by a semicircular gate. The Centre Street Bridge underpass, joining 2nd Avenue West with 2nd Avenue East, links the park with the eastern section of Chinatown. In January 1988, the city began the street beautification project in Chinatown and is now considering a Chinese gateway project.

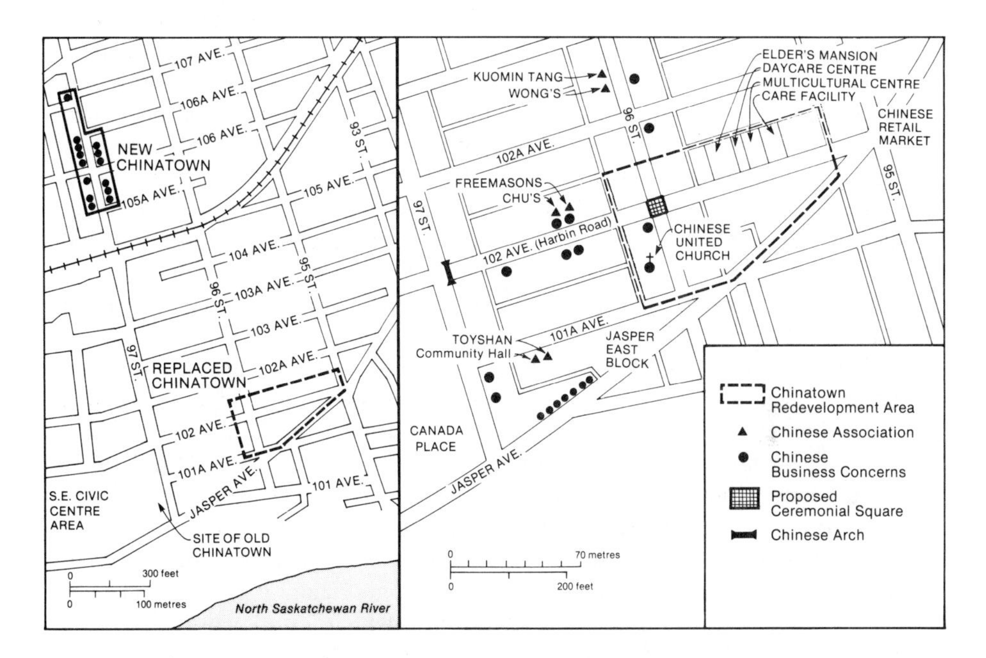

26 Edmonton's Replaced Chinatown, 1987

EDMONTON'S CHINATOWN

Like Calgary's Chinatown, the future of Edmonton's Chinatown was initially uncertain during the 1960s. Its commercial section was concentrated on the west side of 97th Street between Jasper and 102 avenues, within the Southeast Civic Centre area (Figure 26). In 1968, the federal government was planning to consolidate its offices in the area, which would have resulted in elimination of a few Chinese stores, restaurants, and the Lee Association building.[92] The CBA immediately appealed to both the federal and municipal governments to keep Chinatown alive within any urban renewal project.[93] For financial reasons, the federal government postponed its consolidation project in the Civic Centre area, and the Chinese community did not pursue the matter, since the immediate threat to Chinatown had been removed.

The survival of Edmonton's Chinatown was threatened again in July 1973, when the city considered widening the west side of 97th Street by 2.5 metres. Like the federal government's proposal, this project would have wiped out the remaining commercial strip of Chinatown. The CBA immediately called a public meeting, at which it was decided that if efforts to save the existing Chinatown failed, it might be relocated on two city blocks on the east side of 97th Street between 101A Avenue and 102 Avenue.[94] The CBA would apply to the provincial and federal governments for funding to build a senior citizens' home and a cultural centre in Chinatown as a focal point for the widely scattered Chinese community.[95] Accordingly, the Edmonton Chinese Community Development Committee was formed in April 1975 to look at possible sites for a new Chinese business strip and a Chinese cultural centre.[96] In the meantime, Council decided that widening 97th Street between Jasper and 125th avenues would not be undertaken until the impact of such widening had been assessed. Thus, the threat to Chinatown was again removed temporarily.

In 1976, the Alberta government started to construct an eleven-storey senior citizen home known as the Chinese Elders Mansion (Plate 15). It was located on 102nd Avenue, about a city block east of Old Chinatown. Adjoining the mansion was the Edmonton Bilingual Daycare Centre. In April 1977, local Chinese businessmen and representatives of fifteen Chinese associations formed the Edmonton Chinatown Planning Committee, the main objective of which was to co-ordinate preparation of a Master Development Plan for relocation of Old Chinatown.[97] In 1978, the Planning Department published a report on the future of Chinatown in which it outlined five possible relocation solutions. These included the dispersal of Old Chinatown's business or the creation of a Replaced Chinatown in one of four areas: the Southeast Civic Centre Complex, the City Market, on the Jasper Avenue Infill, or at 102nd Avenue between 96th and 95th streets (where the Chinese Elders Mansion is situated).[98] The first alternative was unacceptable to the Chinese community, since dispersal of Chinese businesses would lead to the loss of Chinatown. The second alternative depended on the interest and financial resources available for investment to dislocated businesses of Chinatown. The third alternative was more feasible because much land in the City Market area was owned by the city so land consolidation there was easier. However, parking and future expansion in the area would be restricted. The fourth alternative was also more feasible because the historic buildings in the Jasper East Block consisted of vacant premises to which dislocated Chinese businesses could be moved, but any major future expansion would be restricted as well. The last alternative site was the Elders' Mansion area, which was the most suitable because it had open space for expansion and was closer to many Chinese families living on 92nd and 93rd streets. The Chinatown Master Plan called for the demolition of Old Chinatown, and redevelopment of a four-block area around the intersection of 102nd Avenue and 96th Street as a Replaced Chinatown. The plan was approved in principle by City Council in September 1979.

While the Replaced Chinatown was being planned, Chinese business owners in Old Chinatown received notices to vacate their premises before April 1981 (Plate 16). A few operators closed their businesses, some relocated to the nearby Jasper East Block, and other proprietors planned to move to other parts of the city. By 1986, the Old Chinatown in Edmonton was extinct; its site is now occupied by Canada Place.

LETHBRIDGE'S CHINATOWN

Like other Chinatowns, a large part of Lethbridge's Chinatown was demolished by land speculators. In 1971, the city began the Downtown Development Scheme Phase 1, south of Chinatown.[99] Phase 1 development saw construction of the Lethbridge Centre and Provincial Building complexes; it was essentially complete by 1979. Phase 2 development, which includes Chinatown, is still being planned (Figure 27). During the early 1970s, some developers expected that there would be a demand for parking after the Phase 1 development, so they began to buy up nearby buildings, demolishing them to provide temporary parking and waiting for a rise in land value as the Downtown Development Scheme Phase 2 was implemented.

A survey of Chinatown in March 1986 revealed that Lethbridge's Chinatown was basically a ghost town, with only three Chinese stores on 2nd Avenue between 3rd and 4th streets still in operation (Plate 17). Most of its earlier buildings have been demolished. Only six pre-war Chinese buildings remain: the Chinese National League, the Chinese Freemasons, San Man Sang, Wing Lee, Bow On Tong, and Kwong On Lung buildings.[100] Although the Phase 2 development has not been finalized, one of its basic objectives is to preserve some of the historic buildings in Chinatown and give small businesses and landowners the opportunity to upgrade their properties in much of the Phase 2 area, bounded by 1st and 4th avenues South and 1st and 5th streets South.

WINNIPEG'S CHINATOWN

Like other Chinatowns on the prairies, Winnipeg's Chinatown had been physically deteriorating since the 1940s. In 1967, City Council engaged a consulting firm to prepare an Urban Renewal Plan to revitalize its stagnant downtown area. The following year, the plan—known as the Urban Renewal Plan for Area No. 2—was published. Since Chinatown was included in Area No. 2, it qualified for government funding for renewal. The plan proposed that Chinatown be demolished and developed as a moderate-priced shopping precinct with a strong ethnic flavour.[101] Not only would the present Chinese community remain in the area with its restaurants, specialty food stores, and other shops, but an effort also would be made to attract similar activities from

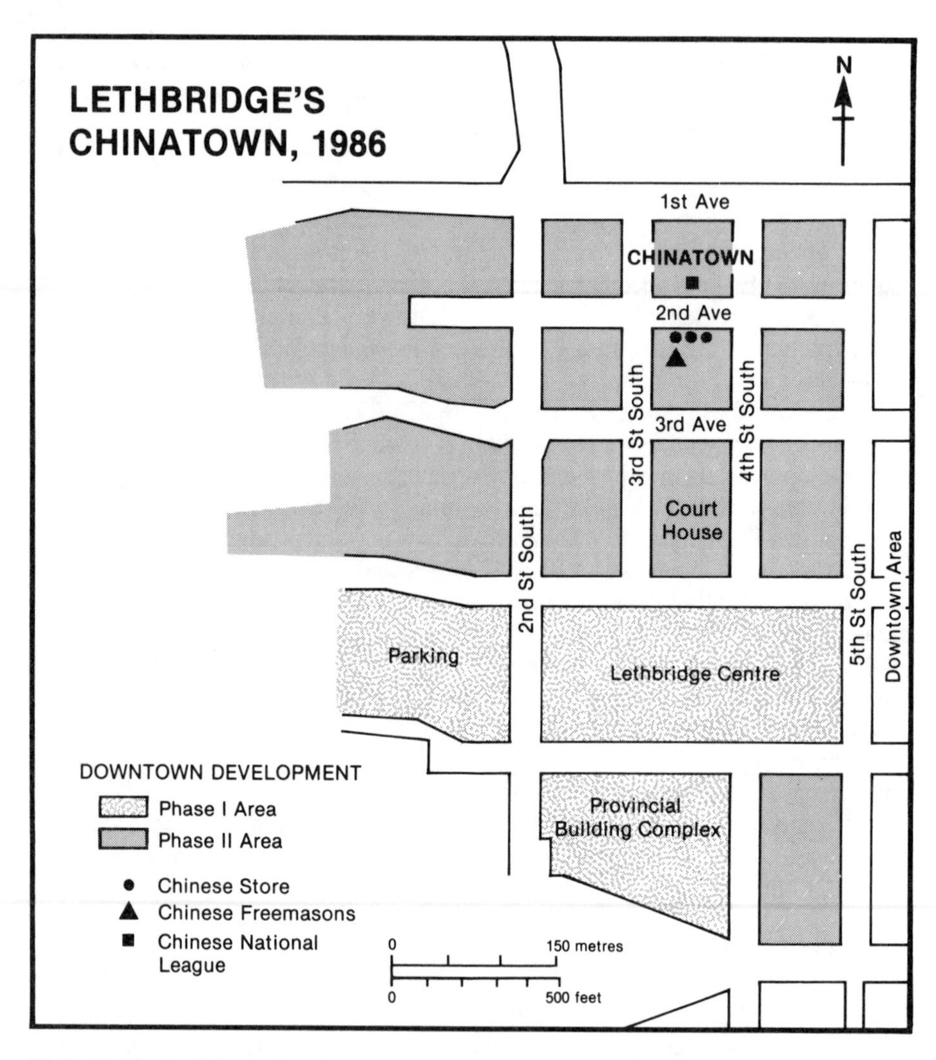

27 Lethbridge's Chinatown, 1986

a variety of ethnic groups. The Chinese Benevolent Association of Manitoba, headed by Hung Yuen Lee, welcomed this urban renewal plan and appointed a nine-member committee for redevelopment of Winnipeg's Chinatown.[102]

In 1971, a few Chinatown property owners and businessmen formed the Winnipeg Chinese Development Corporation, chaired by Dr. Martin Yeh.

With government funding of $95,00, the corporation engaged Gustavo da Roza, a Winnipeg architect, to draw up a Chinatown redevelopment plan.[103] The plan, completed in 1974, called for a closed mall with shopping, living, and recreational areas connected by a sidewalk.[104] They would be built in stages and eventually would cover the entire Chinatown area. The total cost of redevelopment was estimated at $10 million, of which 20 per cent had to be raised by the Chinese community. Some merchants felt that the proposal was too ambitious and impractical; they could not afford to close their businesses, tear down their buildings, and wait for new ones to be finished.[105] The plan also placed too little emphasis on the commercial aspect of Chinatown. Some people complained that the proposal had little or no input from the Chinese community, particularly from the elderly people whose lives would be directly affected by any developments in Chinatown.[106] As a result, the plan was never presented to Council by the corporation. The architect blamed a few parties who owned the land in Chinatown and allowed their personal interests to overcome their initial support of redevelopment of Chinatown. However, Dr. Yeh felt that some businessmen found it more attractive to purchase land for themselves.[107] Thus the plan was abandoned in 1975. Meanwhile, two more realistic projects were under way in Chinatown. The Sek On Toi Senior Citizens' Home was built by the Chinese United Church and completed in 1978.[108] Another project was Chinatown Plaza at the northeast corner of King Street and Alexander Avenue.

The Chinese population in Winnipeg had been increasing rapidly since the early 1970s, particularly between 1978 and 1981, when a large number of Indochinese refugees arrived.[109] The Chinese population in the city jumped from 2,535 in 1971 to 6,000 in 1981. This sudden expansion induced many local Chinese businessmen to buy more Chinatown properties. They expected that Chinatown would be redeveloped sooner or later, even though da Roza's redevelopment plan was dead. Their expectation came true in 1980 when City Council announced the $96 million Core Area Initiative (CAI), to be shared equally by the federal, provincial, and city governments, each contributing $32 million over five years from 1981 to 1986. The ten programs to be financed by the CAI were designed to revitalize the socioeconomic and physical development of Winnipeg's inner city. Programme 10, known as the Neighbourhood Main Streets Programme, encouraged new investment and the strengthening of focal points of commercial activity in the following neighbourhoods: Chinatown (King Street), Provencher Boulevard, Selkirk Avenue, Osborne Street, and Main Street.[110] The program marks the beginning of the period of revival of Winnipeg's Old Chinatown.

In 1981, a new corporation known as the Winnipeg Chinatown Development Corporation (CDC) was formed, headed by Dr. Joseph Du. It immediately presented to City Council a Chinatown improvement plan and a budget which was soon approved. With an initial funding of over $1 million from the CAI Main Streets Programme, the CDC carried out an Attitude and Need Study, a Housing Study, and a Cultural and Community Facilities Study.[111] At the nominal price of one dollar, the CDC purchased a city block in Chinatown from the provincial and city governments. The west half of the block was used for the construction of Harmony Mansion, a 111-unit, low-cost family housing complex. The east half was used for construction of a Chinese garden and the Dynasty Building, which would house the Chinese Cultural and Community Service Centre (Figure 28).

The various Chinatown projects were carried out by three corporations: the CDC, the Winnipeg Chinese Cultural and Community Service Centre Corporation (CCC), the the Winnipeg Chinatown Non-Profit Housing Corporation (CHC). The CDC was responsible for the overall Chinatown redevelopment, bearing the cost of King Street beautification and construction of a Chinese gate, a Chinese garden, and half the underground parkade.[112] The Chinese gate would be topped by a pedestrian overpass across King Street, linking the Dynasty Building with the Mandarin Building. By September 1985, the CAI Main Streets Programme had allocated $2.53 million to redevelop Chinatown: $820,000 was for the Chinese garden and streetscaping of King Street; $710,000 for half of the underground parkade; $500,000 for the Chinese Cultural Centre; and $500,000 for the Chinese gate.[113] The CCC, also headed by Dr. Du, was responsible for the construction of the Dynasty Building, a six-storey commercial complex with commercial and office space, a multi-purpose auditorium, a library, daycare, and other services. The CHC, headed by Ken Wong, received grants from the Canada Mortgage and Housing Corporation, the CAI Main Streets Programme, and the Manitoba Housing and Renewal Corporation to build the $7 million Harmony Mansion, and it shared construction with the CDC of the underground parkade.[114]

Like some other Chinatowns, there is discord in the Chinese community in Winnipeg, with the CHC on one side and the CDC and CCC on the other. The CCC was criticized for involving Metropolitan Equities International Inc., a private land developer, in building the Dynasty Building and for selling preferred income funds as tax shelters.[115] On the other hand, the CHC was criticized for breaking away from the CDC, which was the co-ordinator for all Chinatown redevelopment projects.[116] However, some people also speculated that Canadian politics might play a part in the discord.[117]

The rift between the two groups was deepened with the issue of roof overhangs. The curving roof of the Dynasty Mansion extended three metres over the Harmony Mansion. The CHC complained that the Dynasty Mansion had been redesigned without consulting its board of directors.[118] The increase in height had already caused some suites of the Harmony Mansion to be dark, and the overhanging roof also would cut off sunlight on some suites on lower

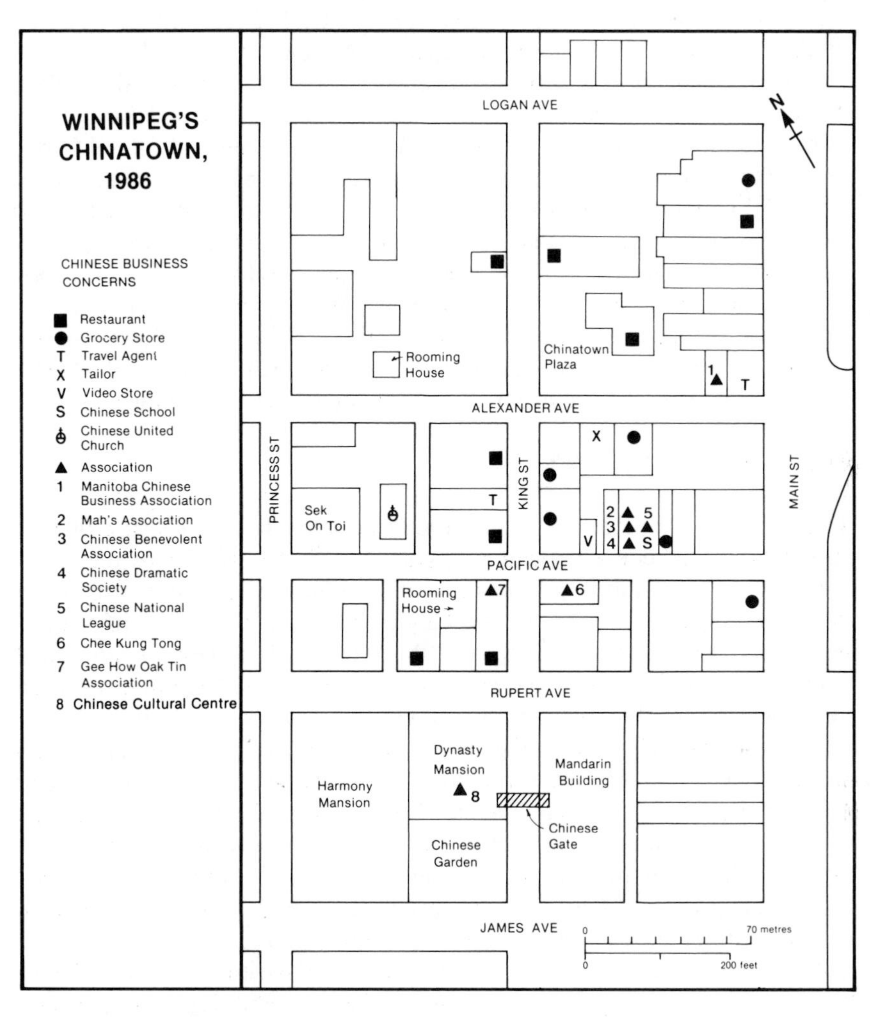

28 Winnipeg's Chinatown, 1986

floors and would cause other problems, such as dripping water and snow on balconies. Without an easement from the CHC, the CCC had to remove the curving roof on the west side of Dynasty Mansion, destroying the symmetric beauty of the roofline.

In spite of all the discord, Chinatown projects were progressing mainly due to the untiring efforts of Dr. Du. In August 1986, City Council approved another CDC project: the Mandarin Building and reconstruction of the historic police court building. This project, estimated at $9.5 million, would include a retail store, a dinner theatre, and office space.[119] On 13 September, Harmony Mansion was officially opened.[120] On 15 October, the official dedication of the Chinese Heritage Garden, China Gate, and King Street Beautification took place (Plate 18).[121] As well, the Chinese Cultural Centre was completed and officially opened on 1 August 1987.[122] The second CAI (1986–91) might provide additional funds for other Chinatown projects such as a sports centre and a second Chinese gate across Alexander Street.[123] This revitalized Chinatown in Winnipeg has attracted Hong Kong investors, notably the million dollar investment of the Wah Loong (Winnipeg) Ltd, a wholesale seafood distribution company.[124] With an estimated 20,000 Chinese (including over 5,000 Indochinese) in Winnipeg, Chinatown undoubtedly will again be the focal point for their social and cultural activities.

TORONTO'S CHINATOWN

Toronto's Old Chinatown was much reduced after the war because of land speculation and redevelopment projects. During the late 1950s and early 1960s, nearly two-thirds of Chinatown was obliterated to provide space for Nathan Phillips Square and the new City Hall (Figure 29). All Chinese businesses and residences on Elizabeth Street between Hagerman Street and Queen Street moved out; some businesses and residents were re-established on Dundas Street West between Bay Street and University Avenue and other businesses were relocated further west along Dundas Street beyond University Avenue. By 1965, the shrinking Old Chinatown, covering an area of about four hectares (ten acres) along Dundas Street West between Bay and Chestnut streets, was all that remained.[125] About 58 per cent of Chinatown's land was already in the hands of speculators, and only 42 per cent was owned by Chinese people.[126] City planners conducted many informal meetings with leading Chinese businessmen and invited them to present their own plans for Chinatown redevelopment. However, the businessmen had been disillusioned by the city's expropriation policy in the late 1950s and were reluctant to spend time and money on what they felt was a futile venture.[127] Furthermore, some Chinese owners, specially those approaching retirement, did not hesitate to sell their land whenever attractive prices were offered. To some Chinese residents and city officials, Chinatown's extinction was inevitable.

In the spring of 1967, Walter Manthorpe, the city's development commissioner, recommended to City Council that Chinatown be relocated to provide space for a northern extension of the civic square, since Chinatown residents could not agree about an overall development scheme and the city needed the area in Chinatown.[128] This proposal suggested that Chinese restau-

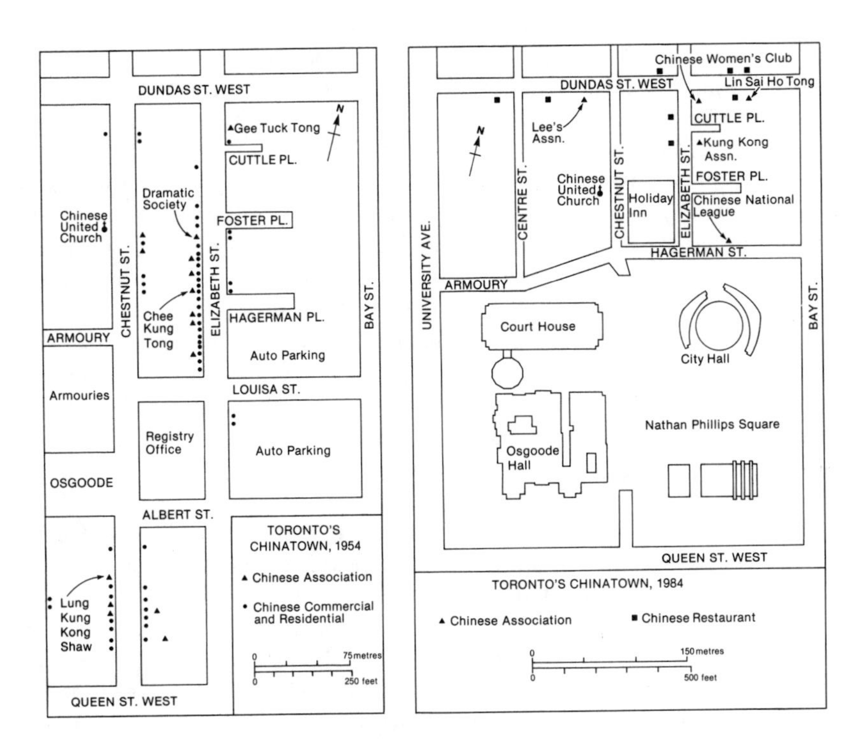

29 Toronto's Old Chinatown, 1954, 1984

rants and stores should be razed and replaced with office towers to complement the new City Hall. While the proposal was being studied, land speculators immediately assembled as much land as possible in Old Chinatown, thus inflating land prices from about $12 to more than $30 per square foot.[129]

So far, the Chinese community had acquiesced in all these proposals, feeling that it could do nothing about them. In February 1969, Alderman Horace Brown, chairman of the city's Buildings and Development Committee, objected to a suggestion that the Elizabeth-Dundas streets area be designated Toronto's Chinatown.[130] He felt that a Chinatown encouraged development of a ghetto. Another threat to Chinatown was Metro Council's proposal in 1970 to widen Dundas Street, forcing all Chinese businesses on Dundas Street and those remaining on Elizabeth and Chestnut streets to move. These various threats finally prompted some concerned community leaders such as Mrs. Jean Lumb to form the Save Chinatown Committee in March 1969 in a desperate effort to save Old Chinatown from obliteration.[131] At first, the committee rallied little support from the community. For example, only four Chinese turned up at the meeting on 14 April when the Buildings and Development Committee met to discuss the city hall expansion project.[132] The committee decided to defer its extension program until June when the Save Chinatown Committee would be able to present a brief.

In the following two months, the committee worked very hard to get support from various Chinese associations and concerned community leaders. On 16 June, about 350 Chinese attended the council meeting and succeeded in having Council endorse the principle of keeping Chinatown in its present location.[133] After this assurance from the city, three new restaurants and other businesses appeared within a few months.[134] In May 1970, the seventy-five business concerns in Chinatown included seventeen restaurants, fourteen grocery stores, four bakeries, and other stores and services.[135] Although Old Chinatown had been saved, its size had been further reduced by large redevelopment projects such as the Holiday Inn and the Meridian Group's commercial complex.

The city had always regarded the deteriorating Elizabeth Street in Chinatown as an eyesore and, in June 1971, it proposed to convert the street into a pedestrian mall at an estimated cost of $151,000.[136] The Chinese community was pleased with the project, which would create a car-free Oriental marketplace in the heart of Old Chinatown. To test the feasibility of the project, Elizabeth Street from City Hall to Dundas Street was closed between 29 August and 6 September, and a temporary mall known as the Dragon Mall was created.[137] The mall was a success, but the businessmen in the street were not pleased with it because of the loss of vehicular access and parking facilities.[138] As a result, the idea was dropped for several years.

The survival of Old Chinatown was threatened again in May 1975 when Metro Council's planning committee recommended that Dundas Street between Jarvis Street and University Avenue be widened to six lanes.[139] Both City Council and the local Chinese and non-Chinese communities opposed the plan. The threat was averted after Metro Council abandoned the proposal in June.[140]

In 1978, the city's Civic Design Group proposed another plan to improve Elizabeth Street which would narrow the Elizabeth Street roadway, widen its sidewalks, eliminate all its on-street parking places, develop an eighteen-space parking area on a city-owned lot, and build a pedestrian walkway system to connect Elizabeth Street and Bay Street.[141] Elizabeth Street merchants again opposed this revised plan at a public meeting on 27 March 1979 because the plan still restricted vehicular access to the street.[142] While the design was being revised, a number of land assemblies were active on Elizabeth Street. Several landholders had consolidated their properties into a larger redevelopment scheme. The city anticipated that these developers would also carry out related improvements on Elizabeth Street at their own expense, rather than at the city's. As a result, the improvement plan was not implemented. By 1984, Toronto's Old Chinatown had fewer than ten restaurants and four association buildings; the Chinese United Church on Chestnut Street and the Chinese National League building on Hagerman Street were the only remaining significant landmarks of Toronto's Old Chinatown.

MONTREAL'S CHINATOWN

Montreal's Chinatown is sandwiched between Old Montreal and the city's downtown business district. During the late 1950s and early 1960s, properties around Chinatown were increasing in value and speculators purchased old buildings in Chinatown, demolished them, used vacant sites as parking lots, and then sold them when attractive prices were offered. As a result, large numbers of old buildings were demolished without being replaced; low-rental housing in Chinatown became difficult to find. During the 1960s, a plan for Chinatown redevelopment was proposed, but it was soon shelved, not because there was lack of money and official support but because the Chinese Benevolent Association of Montreal "could not arouse the necessary interest within Chinatown and could not unite the various factions."[143] Like some other Chinese communities, the Chinese community in Montreal was so politically and religiously divided that it was difficult to carry out any community project. For example, the deteriorating Chinese Hospital on Lagauchetière Street was condemned in 1963 by the fire department.[144] The fund-raising campaign for a new hospital was bogged down for nearly a year, until Dr. David Lin became campaign chairman and mobilized the entire community to make contributions.[145] The new Montreal Chinese Hospital, located at the corner of St. Denis and Faillon streets, was completed and opened in June 1965; the hospital was built there because there was no suitable land available in Chinatown at that time. Another significant community project was construction of a small multicoloured pagoda in a tiny park at the northeast corner of Lagauchetière and St. Urbain streets. Pagoda Park was built to celebrate Canada's hundredth anniversary in 1967 and dedicated to "the cause of peace and harmony among all Canadians."[146]

Throughout the 1970s and early 1980s, Montreal's Chinatown was reduced by expropriation and redevelopment. The Chinese United Church and other structures were demolished and many residents, both Chinese and non-Chinese, had to move as Dorchester Boulevard West was expanded and Place du Complexe Desjardins and Hydro Quebec on the north side of the boulevard were constructed.[147] A large city block in Chinatown was expropriated, after the federal government announced in 1975 construction of

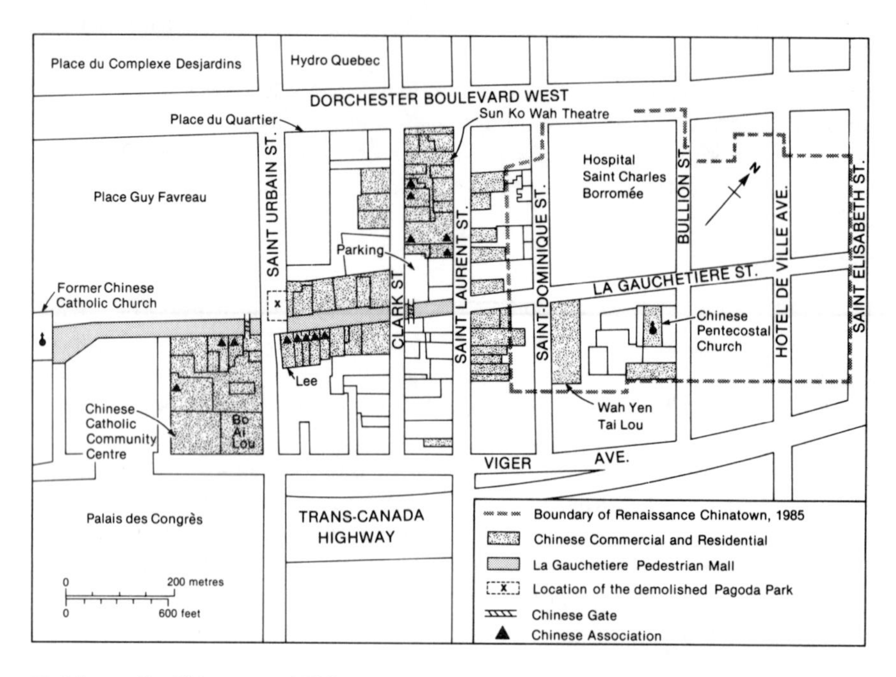

30 Montreal's Chinatown, 1986

Place Guy Favreau, a massive complex consisting of high-rise offices and apartment buildings (Figure 30). The expropriation resulted in the demolition of the Chinese Presbyterian Church, Chinese Pentecostal Church, a Chinese food factory, and a few Chinese grocery stores, as well as other non-Chinese structures on the block. The Chinese Catholic Church was the only structure that escaped demolition because it was designated an historic monument in 1977. Construction of the Place du Quartier, another high-rise commercial and apartment building project in Chinatown, resulted in displacement of many residents and businesses (Plate 19). At the southwest corner of Chinatown, properties were expropriated for construction of the Palais des Congrès (the Montreal Convention Centre). City officials regarded the massive projects of Complexe Desjardins, Complexe Guy Favreau, and the Palais des Congrès as the catalyst and incentive for the revitalization of Chinatown, although they were only marginally related to the Chinese area.

While these overbearing projects threatened the survival of Chinatown, the Chinese community in Montreal was still torn between pro-China and pro-Taiwan factions and could not form a united front to fight against these projects. Abraham Cohen, a city councillor, once warned the Chinese community that "as long as the community remained broken in rebellious and irreconcilable elements, not only would Chinatown continue to stagnate, but it would be extirpated by a gradual incursion of speculators."[148] After 1975, the divided Chinese community was further alienated by the English-French language dispute. Many people in the Chinese community did not care about the survival of Chinatown and were moving out of Quebec as the Parti Québécois made French the only official language in the province. About 1,000 Chinese families left Montreal between 1975 and 1977 because of the language bill.[149]

In spite of the general apathy towards the future of Chinatown, several concerned community leaders such as Kenneth Cheung and Dr. Kwok B. Chan succeeded in forming the Montreal Chinese Community United Centre (CCC) in September 1980, which was to be an umbrella organization for most dissident groups in the community. Soon after its formation, the CCC faced the problem of the imminent expropriation of another Chinese property for redevelopment. In September 1981, the city wanted to expropriate the Lee's Association Building and demolish it to widen St. Urbain Street. The CCC mobilized the community to fight to preserve the building, and more than 2,000 people signed a petition to that effect. Although the building was saved, tiny Pagoda Park, the only one in Chinatown, was eliminated to make way for road expansion (Plate 20). Meanwhile, the city engaged Henry Ng, a Chinese architect, to study the redevelopment and revitalization of Montreal's Chinatown. One recommendation in the report was conversion of Lagauchetière Street into a pedestrian mall, a plan later adopted by the city government.[150]

The revival of Montreal's Chinatown began in 1982 with installation of street signs throughout Chinatown and construction of the Catholic Community Centre and Bo Ai Lou, a senior citizens' home. Throughout 1983, the Montreal Chinatown Development Association (CDA), headed by Father Thomas Tou of the Chinese Catholic Church, worked closely with the city in implementation of the $3.5 million Chinatown facelift plan.[151] The plan called for the conversion of Lagauchetière Street between St. Lawrence Boulevard and Jeanne Mance Street into a tree-lined, brick-paved pedestrian mall and the construction of two Chinese arches spanning the mall and a Pagoda Park at the northwest corner of Lagauchetière and Clark streets (Plate 21). The CCC complained that the city planners consulted the CDA in the beautification project without involving the CCC which, formed by various Chinese organizations, was the elected representative. In their defence, city planners and officials said that they found it difficult to deal with so many organizations in Chinatown. Each organization seemed to have its own project. In August 1983, for example, the CCC started construction of its Chinese United Building (Wah Yen Tai Lau), a nine-storey low-cost housing complex with eighty-two units for senior citizens and low-income families, an indoor gymnasium *cum* auditorium, and underground parking.[152] The project did not involve the CDA. The same year, another project, organized by another group of people in the Chinese community, was started outside Chinatown—the addition of a new wing, named Sung Pai Pavilion, to the Chinese Hospital.

Two other significant projects were also completed in 1983: the Complexe Guy Favreau and the Place du Quartier. Both projects have attracted the families of many middle-class professionals and businessmen to Chinatown. These residents will play an important part in the revitalization of Chinatown because they can afford to dine out and shop more often than traditional Chinatown residents. On the other hand, the projects have engulfed extensive areas of Chinatown's land, hindering its future expansion. Furthermore, these projects do not benefit the displaced Chinese residents who, for example, could not afford the $460–$565 monthly rent for a one-bedroom apartment, $640–$840 for a two-bedroom apartment, and $760–$880 for a three-bedroom apartment in the Complexe Guy Favreau. Nor could they purchase Place du Quartier's condominium apartments, priced between $74,000 and $118,000.[153]

By 1984, Chinatown, wedged between the high-speed cross-town Dorchester Boulevard and the Trans-Canada Highway, was enclosed by Place Guy Favreau and the Palais des Congrès at the southwest end. The only possible area for Chinatown's future expansion was at the northeast end, where new Chinese businesses are being established along St. Laurent Boulevard and Lagauchetière Street east of the boulevard. However, on 22 October 1984, the city adopted Bylaw 6513, by which most of the areas on Lagauchetière Street on the east side of St. Laurent Boulevard were zoned residential.[154] The Chinese Professional and Businessmen's Association (CPBA), representing about half Chinatown's merchants, immediately protested against the bylaw, which would choke off the only avenue left for commercial growth of their much-reduced Chinatown.[155] In March 1985, Montreal's Chinese community, led by the CPBA, mounted an intensive campaign against the bylaw, staging a protest march, holding a day-long public hearing, and delivering a petition with 630 names to Mayor Jean Drapeau and other city officials.[156] At the meeting a group called the Centre for Research Action and Race Relations suggested the organization of an economic boycott and that Hong Kong investors should be told not to invest in Montreal. A report of the public hearings was later sent to City Council, with a request for the immediate lifting of the bylaw that banned commercial development on Lagauchetière Street East.[157]

Initially the Chinese protests and requests fell on deaf ears, as Yvon Lamarre, chairman of the city's Executive Committee, regarded the protests as the work of people with vested interests in the Chinese community.[158] However, because of persistent protests from the Chinese community, the city in August 1985 amended the bylaw, and the west side of St. Dominique Street between Dorchester Boulevard and Viger Avenue was zoned instead for commercial use.[159] Although the amendment gave Chinatown about half a city block for commercial expansion, the commercial section still could not expand along Lagauchetière beyond St. Dominique Street.

Since the spring of 1985, Kenneth Cheung, Chairman of CPBA, and Calude Touikan, a Chinese architect, had been consulting with Chinese community associations and Chinatown merchants to design a plan for redevelopment of Chinatown east of St. Laurent Boulevard as far as St. Elizabeth Street.[160] The plan, known as Renaissance Chinatown, proposed to establish a mixed commercial and residential district bounded by St. Laurent and Dorchester boulevards, Viger Avenue, and St. Elizabeth Street.[161] The redeveloped district would consist of 179 new housing units, arcades of shops, an indoor Chinese bazaar, a children's playground, a hospital, and other amenities and services.[162] In October 1985, the plan, estimated at $250 million, was presented to Mayor Jean Drapeau for consideration, but it was rejected by the Council because the plan did not conform to zoning Bylaw 6513. In June 1986, the city organized an economic conference of federal, provincial, business, labour, and city officials to discuss the development of downtown Montreal. At the conference, the Montreal Board of Trade strongly criticized the Drapeau administration for its treatment of Chinatown.[163]

After the civic election in November 1986 Jean Doré replaced Jean Drapeau as Mayor of Montreal, and the new administration was more responsive to the needs of the Chinese community. For example, according to Bill 59 (1984, Chap. 17) of the provincial legislature, Sunday shopping was prohibited in Chinatown because it was not a tourist district. In response to the request of the Chinese community, council declared in January 1988 that Chinatown was a tourist area, so its shops were permitted to open on Sundays.[164]

Throughout 1987 and 1988, the CCC had been fighting against Bylaw 6513. In 1987, it conducted two questionnaire surveys of the Chinese community's opinions about the future of Chinatown. The first survey in July revealed that of about 3,000 respondents 98 per cent opposed the bylaw and wanted the area east of St. Laurent Street to be developed for commercial uses.[165] The second survey in November revealed that 84 per cent of about 500 respondents wanted to abolish the bylaw and that 85 per cent wanted the area east of St. Laurent Street to be zoned for commercial and residential uses.[166] The surveys probably influenced the city's drawing up of a Master Plan for the Development of the Downtown Core; it was published in March 1988, one of its recommendations being that Lagauchetière Street east of St. Laurent Street be classified as "Le reseau d'ambiance," in which commercial uses were permitted.[167] If this plan is approved by the Council, Bylaw 6513 will be repealed.

In 1988, three organizations launched several development projects to help revitalize Montreal's Chinatown. The Chinese Catholic Community Centre built Ren Ai Lou, a twenty-two-unit senior citizens' home, at the corner of Lagauchetière and St. Elizabeth streets; the building was officially opened on 18 April 1988. The centre also began to renovate the old Chinese Catholic Church, which was sold back to the centre by the federal government at a nominal price of a dollar.[168]

The Montreal Chinese Hospital Foundation is now planning to build a 160-bed hospital in Chinatown to provide long-term care as well as daycare facilities.[169] Although the three levels of government would subsidize most of the funding for this $12 million project, the foundation fund-raising committee, chaired by Lewis Chow, had to raise $2.5 million for its construction; by April 1988, the committee had raised $2 million. The Hospital Foundation is now considering various sites near Chinatown for the new hospital. After it is completed, patients from the Chinese Hospital outside Chinatown will be transferred there.

Throughout 1987 and 1988, the CCC, led by its dynamic chairman L. Tom, launched two important development projects in Chinatown: the $2 million Community Centre Project and the $2.3 million Cultural Centre Project. The CCC received $2.9 million grants from the city and federal governments, but it had to raise $1.4 million for the two projects.[170] The Community Centre Project will involve renovation of the heritage building next to the Chinese Catholic Church on Lagauchetière Street. When completed, the first floor of the building will house a daycare centre; the second and third floors will comprise twenty-nine small apartments for senior citizens, and the top floor will consist of meeting rooms and office space for community organizations. The Cultural Centre Project will involve an extensive alteration of the warehouse at 1086 Clark Street. When completed, the building will consist of commercial units on the ground floor; the second floor will have an exhibition hall, a library, and conference rooms; and the third floor will provide space for an art school and other cultural activities. A new floor will be added onto the present structure and be used as an auditorium *cum* gymnasium. By April 1988, the CCC had reached half its fund-raising target.[171]

EDMONTON'S REPLACED CHINATOWN

A Replaced Chinatown is a new Chinatown which is planned to replace an Old Chinatown. It is a residential, commercial, and institutional neighbourhood, but unlike an Old Chinatown, its structures are modern designs and its businesses are more diversified. So far, the sole example in Canada is in Edmonton. Soon after City Council decided in 1979 to develop a Replaced Chinatown around the intersection of 102nd Avenue and 96th Street, Ip Investments Ltd., a private company, designed a plan for a new Chinatown which would cover the triangular block bounded by 96th Street and Jasper and 102nd avenues.[172] The plan recommended a ceremonial square at the intersection of 96th Street and 102nd Avenue, the conversion of 102nd Avenue to a pedestrian mall, the erection of a Chinese gate spanning 96th Street, and the construction of multistoreyed apartments, condominiums, hotels, and office buildings (see Figure 26). The investment company agreed to spend $138,000 for improvements such as landscaping and street furniture in the proposed Chinatown area.[173] The plan was immediately opposed by Pat O'Hara, owner of an auction shop inside the proposed Chinatown area. His opposition was based on his fear that Edmonton's proposed Chinatown development could become a racial ghetto full of crime and vice, not in the best interests of Edmonton. His stereotyped image of Chinatown was rooted more in fancy than in fact and was not shared by many other people in the city. In March 1980, Ip Investments applied to rezone the proposed Chinatown area to construct its fourteen-storey office and retail commercial complex on 102nd Street.[174] The application was rejected because the Chinatown Master Plan called for buildings of six to eight storeys and the preservation of a village-like character in Chinatown.[175]

In August 1980, the Chinatown Master Plan was revised, entitled the Chinatown Direct Development Control District; it was part of the proposed Boyle Street / McCauley Area Redevelopment Plan.[176] The revised plan designated the Chinatown redevelopment area for a mixture of high-density residential and commercial uses, as well as ancillary institutional uses, and proposed construction of a ceremonial square at the intersection of 102nd Avenue and 96th Street, and two Chinese gates marking the south and east entrances to Chinatown.[177] This Revised Chinatown Plan was included in the Final Report of the Mayor's Task Force on the Heart of the City, approved in principle by City Council in September 1984. One significant landmark in the Replaced Chinatown was the Edmonton Chinatown Multicultural Centre, built next to the Daycare Centre and officially opened in 1984. In 1988 a new Chinese senior citizens home will be built on the east side of the multicultural centre and a nursing home north of the centre. A Chinese garden, next to the Chinese Elders Mansion, is planned.[178] This garden not only will provide a unique recreational amenity for downtown Edmonton but also will serve as a catalyst in Chinatown redevelopment.

The boundary of the Replaced Chinatown has not yet been defined, as many new structures have been built outside the Chinatown Redevelopment

Area. For example, several new Chinese restaurants and other businesses, the Toyshan Association Building, and the Chinese Benevolent Association Community Hall were established on Jasper East Block. The extensive Chinese Retail Market was built at Jasper Avenue and 94th Street. Even the Chinese gate was built outside the Chinatown Redevelopment Area.

The Chinese gate in Edmonton's Replaced Chinatown is the fifth Chinese arch in Canada. In February 1985, the mayor of Edmonton wrote to Zhang Yongping, Head of the Architectural Arts Exhibition Delegation of China, requesting supplies and technical assistance for design of a Chinese gate for Chinatown. Two months later, Zhang informed the mayor that China would design, construct, and assemble the China-Canada Friendship Gate for $200,000, providing that all the decorative materials for the gate were shipped at the same time as the materials for the Chinese Gate for Expo 86 in Vancouver were delivered. In July 1985, the Chinese Benevolent Association informed the mayor of their commitment of $25,000 towards the cost of the gate. But, since no decision was made about the offer from China, and the materials for the Chinese Gate for Expo 86 had already been shipped, the Gateway Project was not started.

In March 1986, I lectured on Chinese gates and Chinatown revitalization to the Chinese community and staff of the Planning and Building Department in Edmonton.[179] I pointed out the ramifications of a Chinatown gate project, offering ideas on designs, materials, planning, engineering, and maintenance. Based on my advice, the Planning Department decided to construct one arch instead of two arches.[180] In December 1986, City Council approved a capital budget of $350,000 for construction of a Chinese arch spanning 102nd Avenue (Harbin Road) at 97th Street.[181] The gate was designed by Shenzhen Gardens Design and Decorative Engineering Co., and Harbin, Edmonton's sister city, donated the cladding materials, which were valued at $98,000.[182] The gate was officially dedicated on 24 October 1987 by Mayor Laurence Decore and Harbin's Vice-Mayor Hong Qipeng.[183] The Alberta Cultural Heritage Foundation contributed $50,000, the Alberta Department of Tourism $25,000, and the Chinese Benevolent Association $25,000.[184]

In addition to the Replaced Chinatown, Edmonton has a New Chinatown. As early as 1980, the Tai Fat Grocery and Hing Lung Store were set up on 97th Street between 106th and 107A avenues, about six city blocks north of Edmonton's Old Chinatown. Owners of the two stores listed the following reasons for their choice of location: available space at the right time, low rent, and proximity to Old Chinatown. While the future of Old Chinatown was uncertain, and the Replaced Chinatown was still being planned, an increasing number of Chinese businesses began to move out of Old Chinatown to 97th Street between 105th and 107A avenues, where a New Chinatown began to emerge. A survey of New Chinatown in March 1986 revealed that it consisted of twenty-two Chinese business concerns, including six restaurants, five grocery stores, and eleven firms such as importing companies, bakeries, and bookstores. In addition, there were several Vietnamese stores. The Chinese businessmen of New Chinatown would like to have the Chinatown gate built across 97th Street near 105A Avenue, although this commercial strip of Chinese firms has never been considered by City Council or the planning department as Edmonton's New Chinatown.

CHARACTERISTICS OF NEW CHINATOWNS

Unlike Old Chinatowns, New Chinatowns are not Chinese residential areas and rarely have Chinese institutions. They are neither historic districts nor tourist attractions. They may be sections of a street, or one or more shopping plazas where Chinese-operated restaurants, stores, and offices are concentrated, catering particularly to Chinese residents in nearby neighbourhoods. Basically, New Chinatowns are commercial areas, differing in origins, locations, and structures. For example, the birth of a New Chinatown in Saskatoon, Ottawa, or Windsor is mainly a result of the increasing demand for Chinese merchandise and services from a rapidly growing Chinese community. The Chinese population in Saskatoon increased rapidly from 225 in 1951 to 499 in 1961.[185] After the mid-1960s, a few Chinese restaurants and stores were set up on 20th Street West between Avenue B and Avenue D South. By November 1986, Saskatoon's New Chinatown consisted of about twenty Chinese business concerns, widely spaced along 20th Street West between Avenue A and Avenue H, and on Avenue C between 19th and 20th streets west.[186] In Edmonton, however, the New Chinatown resulted from a need to replace the Old Chinatown. There were other reasons for the emergence of New Chinatowns in large metropolitan cities such as Toronto and Vancouver. Real estate in Toronto's Old Chinatown, for example, was expensive and rarely changed hands; as new entrepreneur immigrants continued to set up businesses, the shortage of land for further development became acute. The problems of parking, traffic congestion, high rents, and throat-cutting competition in Old Chinatown forced some new businessmen or investors to look for properties in other parts of the city where there were lower rents or land prices, more parking spaces, and proximity to new Chinese residential areas. An examination of some new Chinatowns will illustrate their characteristics.

OTTAWA'S CHINATOWN

After the 1950s, Ottawa's Old Chinatown on Albert Street was on the way to extinction. By 1961, only four Chinese restaurants and one grocery store re-

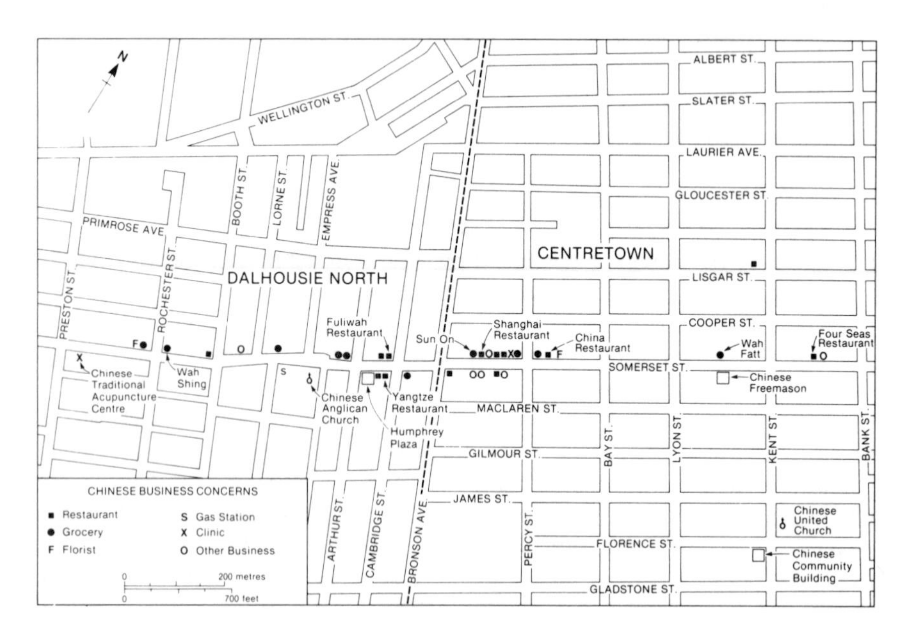

31 Ottawa's Chinatown, 1986

mained in operation.[187] As the Old Chinatown was near Ottawa's town centre, old buildings on Albert Street were gradually replaced by high-rise offices. By the 1970s, Ottawa's Old Chinatown had completed the cycle of its growth and no longer existed. However, a New Chinatown began to emerge at about seven city blocks southwest.[188]

During the late 1960s, a few Chinese merchants began to operate businesses on Somerset Street West, between Bronson Avenue and Kent streets in the Centretown area (Figure 31). Dalhousie North, a run-down lower-income residential neighbourhood, was inhabited by people with many ethnic origins. In 1971, for example, 54 per cent of its residents spoke English as their mother tongue, 22 per cent spoke French, 17 per cent spoke Italian, and the remaining 7 per cent spoke Portuguese, Chinese, East Indian dialects, and other languages.[189] Most were recent immigrants who lived in rented accommodation. Numerous small family-operated corner-store businesses were concentrated along Bronson Avenue, and Somerset, Booth and Rochester streets.[190] By 1974, for example, four Chinese business concerns—Sun On Grocery, China Restaurant, Shanghai Restaurant, and Wah Fatt Barbecue Shop and Grocery—operated on the north side of Somerset Street West.

Throughout the late 1970s, a large number of new Chinese immigrants came to Ottawa; its Chinese population increased from 3,060 in 1971 to 8,205 in 1981. Many lived in Dalhousie North mainly because of cheap rents and proximity to low-paid, menial jobs. In addition to Chinese immigrants, other immigrants such as Vietnamese, East Indians, and Arabs had moved into Dalhousie North and set up businesses there. Meanwhile, an undetermined number of Chinese merchants and investors from Toronto and Montreal moved to Ottawa. They purchased old houses in commercially zoned streets such as Somerset Street West, converting the houses into commercial premises for their own businesses or to rent (Plate 22). By 1980, about ten Chinese businesses had spread along Somerset Street West in the Dalhousie and Centretown districts and Bay Street. However, Somerset Street was not yet perceived by the public, including the Chinese themselves, as a New Chinatown, because there was no grouping of Chinese business concerns.

On 16 April 1980, City Council designated Dalhousie North as a redevelopment area; two million dollars were to be spent to improve the area, funded by the Community Services Contribution Programme.[191] One project was to improve Somerset Street between Bronson Avenue and Preston Street and a portion of Booth Street north and south of Somerset Street.[192] This resulted in unprecedented commercial growth of Chinese businesses on Somerset Street during the early 1980s. In 1982, for example, a group of Chinese entrepreneur immigrants purchased a few old buildings, demolished them, and constructed Humphrey Plaza, a two-level commercial building. The same year, the eight-storey Ottawa Chinese Community Building was officially opened, consisting of forty-six apartment units, a community hall, a Chinese library, and an office for the Chinese Community Association of Ottawa.[193] In the next few years, scores of Chinese restaurants, grocery stores, gift shops, florist shops, beauty parlours, and offices were established in new commercial buildings or renovated houses on Somerset Street West between Arthur and Bay streets. Thus, this section of Somerset Street West soon became known as Chinatown.

On 2 February 1986, representatives of eight Chinese organizations and several Chinese community leaders set up a Chinatown Development Committee to develop a Chinatown project proposed by Rose Kung, an urban planner.[194] The committee requested City Council to undertake a study of Somerset Street West, with the objective of developing a more distinctive Chinatown in the area; the request was approved readily since such an idea had been part of Mayor Jim Durrell's campaign platform in the fall of 1985.

On 1 April 1986 Somerset Street merchants and Dalhousie Ward residents met with city planners to discuss the proposed study. At the meeting, the Somerset Street Citizens' Committee (SSCC) was formed by representatives of the Ottawa Chinese Business Association, Dalhousie Community Association, Cambodian Association, Vietnamese Community Association, and other organizations.[195] Mayor Durrell supported the idea of creating a New Chinatown as a tourist attraction, but some Dalhousie merchants and residents preferred development of Somerset Street into a multicultural area instead of a New Chinatown because of the multi-ethnic nature of the district.[196] In 1981, for example, Italians accounted for about 11 per cent, Chinese people 4 per cent, and Arabic-speaking people nearly 3 per cent of the residential population of Dalhousie Ward.[197] Some local residents and merchants wanted Somerset Street West to retain its variety of small shops, restaurants, and services; they wanted new businesses to be compatible and not push out those already established.[198] Furthermore, local residents insisted that affordable housing be available within Dalhousie Ward for families displaced by development projects.

On 27 May the SSCC, chaired by Frank Ling, reviewed the draft of the Terms of Reference for the proposed Somerset Street Planning Study.[199] The SSCC felt that the proposed study, prepared by the Ottawa Planning Committee, put too much stress on development of a more distinct Chinatown on Somerset Street between Rochester and Percy streets. It wanted the study to cover a larger range of community needs to reflect a more multicultural approach, including other ethnic groups such as Vietnamese, Japanese, and Laotian.[200] On the other hand, Y.C. Lee, a Chinese-Canadian architect, considered a Chinatown on Somerset Street as a regional tourist mecca. Joe Pavone, co-owner of Pavone's Authentic Italian Cuisine, worried that his restaurant would suffer because people would assume that there could not be an Italian restaurant in a Chinatown, but Birinder Ahuja, owner of the India Foods Centre, felt that his store would stand out in a better-defined Chinatown and that his business would boom because people coming to Chinatown would want to try something different.[201] Within City Council there was also a division in opinion about the proposed study. Alderman Darrel Kent felt that if all ethnic groups were included in the redevelopment plan, Somerset Street would become just another commercial business strip and lose its potential to become a regional tourist attraction like the Byward Market and the Rideau Centre. However, Alderman Diane Holmes opposed the Chinatown idea because many businesses along Somerset Street were not Chinese, and Alderman Mac Harb felt that a Chinatown should not be created without considering other ethnic groups.[202] Eventually, on 6 August, City Council approved a $43,000 feasibility study of Somerset between Preston and Bay streets, with the section from Rochester to Percy streets as the core area; the study was to consider all possible themes for Somerset Street West.[203]

In July 1987 Rose Kung was hired as a Somerset Street Planning project co-ordinator to carry out the tasks of getting public participation, responding to public inquiries, and gathering information and ideas from citizens on issues such as parking, garbage removal, and marketing the region as a tourist attraction.[204] Haigis-MacNabb-De Leuw Ltd. was engaged as a consulting firm to carry out and integrate the technical tasks of data collection, analysis, impact studies, and recommendations. In October, it released an interim report which included a series of marketing options, such as maintaining the street as a low-scale shopping area; the addition of building facades to reflect the ethnic character of the stores behind them; and the amendment of city bylaws to allow larger and more colourful signs on the street.[205] The report did not recommend an increase in parking space and stated that an above-ground parking garage would not be economical unless it could accommodate at least 400 cars; such a garage would cost more than three million dollars.[206] Residents wanted the study to focus on problems such as traffic, parking, and cleanliness, whereas merchants preferred high-rise development and the transformation of Somerset Street into a multicultural commercial avenue of Dalhousie.[207] Thus, the interim report pleased neither residents nor merchants.

In January 1988 I gave lectures on the development of North American Chinatowns to the Chinese community, politicians, planners and other professionals in Ottawa.[208] Having met and talked with the Chinese community leaders and people from the Dalhousie community, I identified five major immediate problems to be addressed:[209]

(1) Some people in Dalhousie Ward do not like the name Chinatown, and others do; some wish to have the multi-ethnic theme, some do not. During my presentation at a public meeting on 15 January, a man shouted that he did not want the name Chinatown, Italian Town, or any other names at all, nor did he want any theme for development of Somerset Street West.

(2) The relationship between the Chinese business community and the Dalhousie community is not bad, but it is not very amiable. At present, possibly only a few white residents blame the restaurants and grocery stores, particularly those owned by Chinese people, for the problems of parking, cleanliness, and so on. The prejudice may be deepened if the Chinese community does not establish rapport with the Dalhousie community.

(3) The boundary of the study area of Somerset Street West has not been clearly defined and made known to the public. There is no committee claiming to represent the merchants, owners, and residents in the study area, which presumably covers a section of Somerset Street between Rochester and Percy streets.

(4) At present, the parking problem is not acute, but a shortage of parking will become a serious problem and will cause conflict between merchants

and residents; worse, it may lead to a racial problem like that in Scarborough, a New Chinatown in the suburb of Metro Toronto.

(5) The last major problem is residents' apprehension about the area's cleanliness, especially after development.

I suggested the following possible solutions to these problems:

(1) The name for the economic development of the study area is not critical; in fact, it is unimportant. In recent years, Somerset Street West has been growing, being transformed gradually from a stagnant neighbourhood shopping street to a vibrant commercial core with no definite plan, name, or theme. A Chinatown was budding, but not officially called a Chinatown; an Italian town was budding, but not officially called an Italiantown. Other clusters of ethnic business establishments had also emerged but had no specific ethnic names.

For the sake of harmony in the community, I suggested that Chinese, Italian, and other ethnic groups' businesses in Somerset Street West be promoted. The street will be known to some people as a Chinatown, to some people as an Italian town, and to some other people as an Arabic town, an Indian town, a Vietnamese town, or some other ethnic town. The idea is to develop a Harmonious Multi-Ethnic (HOME) town. Like the International District in Seattle, a HOME town is suitable in a neighbourhood where various ethnic businesses and residents have been firmly established.

(2) The Chinese Community Association of Ottawa, the Ottawa Chinese Business Association, and the Eastern Ontario Chinese Restaurant Association should form a Somerset Street Liaison Committee (SSLC) to promote mutual understanding between the Chinese community and other ethnic groups in Dalhousie. The SSLC should support the social functions and other activities of other ethnic groups.

(3) An information sheet should be printed. It should show the boundary and land uses of the study area and list all businesses and owners in the area. A Study Area Committee (SAC), consisting of representatives of merchants, owners, and residents in the area should be formed, representing all interests in the study area which will be affected by any project there.

(4) The most important project for the study area is construction of a multilevel parkade. There are only a few possible sites in the study area, such as the open space south of the Dalhousie Community Centre and some spaces being used for gas stations.

(5) A scheme should be developed by which garbage collection will be more effective and the study area kept clean and tidy.

My suggestions have been acted upon or are being considered. For example, the three Chinese associations sent a joint letter to Mayor Jim Durrell stating that "in the interest of multicultural unity and community harmony and the long-term development of the area, we would respectfully like to suggest that a specific name at this time is not critical," and that "the issue [of a name or theme] be dropped to let the urgent problems, such as parking, be dealt with immediately."[210] The planning department had decided that the controversial issue of creating a New Chinatown on Somerset Street be dropped and the street be left to develop on its own.[211] The consulting firm is now looking at several potential sites where parking lots could be built; one proposal was to construct a forty-eight-car underground garage under the Dalhousie Community Centre.[212]

WINDSOR'S CHINATOWN

The birth of Windsor's New Chinatown was different from Ottawa's. The Chinese community in Windsor had grown from about 500 people in 1961 to over 5,000 in 1981, including about 1,000 Chinese Vietnamese boat people and many Chinese students from Hong Kong and Southeast Asian countries studying in private schools or at the University of Windsor.[213] To cater to these students, a few Chinese restaurants and grocery stores were set up at the west end of University Avenue and Wyandotte Street West, where the university campus is situated. Gradually more stores were set up on Wyandotte Street West, widely spaced eastward toward Windsor's central business district. The comparatively compact cluster of Chinese businesses is on one city clock on Wyandotte Street West between Rankin and Partington avenues; in April 1985, there were three restaurants, two groceries, one bakery, and one bookstore. East of this block, a few Chinese restaurants and stores were widely spaced along the street for more than twelve city blocks. Two or three Vietnamese stores were set up between Church and Pelissier streets, specializing in Chinese and Vietnamese foodstuffs for the nearby Chinese and Vietnamese patrons. A temporary Chinese cultural centre, located at 1671 Church Street, was opened on 24 January 1982.[214] The Essex County Chinese Canadian Association and the Chinese Benevolent Association of Windsor plan to establish a permanent community centre as the basis of Windsor's Chinatown.

RICHMOND'S NEW CHINATOWNS

Unlike the New Chinatowns in Ottawa and Windsor, the New Chinatown in Metropolitan Vancouver was established in the suburb of Richmond. During the late 1970s, a grocery store was first set up in Park Village Plaza in Park Road between No. 3 Road and Buswell Street, and two Chinese restaurants emerged in Richmond Plaza on No. 3 Road between Cook Street and the

Westminster Highway (Figure 32). The two plazas are readily accessible to several thousand Chinese residents in Richmond (Plate 23). In the first half of the 1980s, several Chinese grocery stores, restaurants, a seafood market, and a bakery were set up in Park Village, Times Square, and Richmond Plaza. Together they form the embryo of a Chinatown in Richmond, attracting upper middle-class Chinese customers from South Vancouver. On Steveston Highway, about four kilometres south of Richmond's Chinatown, a Kuan Yin Temple built by the International Buddhist Society was officially opened on 23 October 1983.[215] In the early 1980s a large Chinese restaurant and a sizeable grocery store were established in Cambie Plaza on Cambie Street near No. 5 Road. Since the plaza is too far from the town centre of Richmond, it has not attracted many other Chinese businesses.

In 1987, a shopping plaza was built at a cost of about ten million dollars on Westminster Highway between No. 3 and Cooney roads.[216] This new plaza is called Johnson Centre (Xiang Shun Center) but is more popularly known as Liezhiwen Xin Tangren Jie (Richmond's New Chinatown). Several Chinese business concerns, notably the Top Gun Restaurant, have been established there. Today, Richmond has two New Chinatowns, the earlier one consisting of a group of nearby plazas, and the recent one having a single plaza.

TORONTO'S CHINATOWN WEST

After the 1960s, four New Chinatowns were established in Metropolitan Toronto: Chinatown West, Chinatown East, Scarborough's Chinatown, and Willowdale's Chinatown (Figure 33). Chinatown West, located in the South-East Spadina District, was the first New Chinatown in Toronto. Before the Second World War, the district, bounded by Spadina, College, McCaul, and Queen streets, was a low-density residential area, containing two- and three-storey wooden houses built in the late nineteenth century. In 1951, the Jewish community predominated in the district, and Chinese households accounted for only 20 per cent of the population.[217] However, in the late 1950s the Jewish community began to relocate to newer residential areas on the northwestern fringes of the city, and a large number of houses in the South-East Spadina District were available for sale or rent.[218] In 1951, the Chinese population in the City of Toronto was less than 2,900, but it soared to nearly 18,000 by 1971 because of the influx of new immigrants, including many from Hong Kong and Taiwan. Many moved into the South-East Spadina District, and by 1971, nearly half of the 8,000 residents in the district were Chinese in origin.[219] Chinese restaurants, grocery stores, beauty parlours, real estate and insurance agents, medical offices, and other service and retail establishments occupied most of the commercial properties on Dundas Street.[220] Thus a new Chinatown emerged on Dundas Street between Spadina Avenue

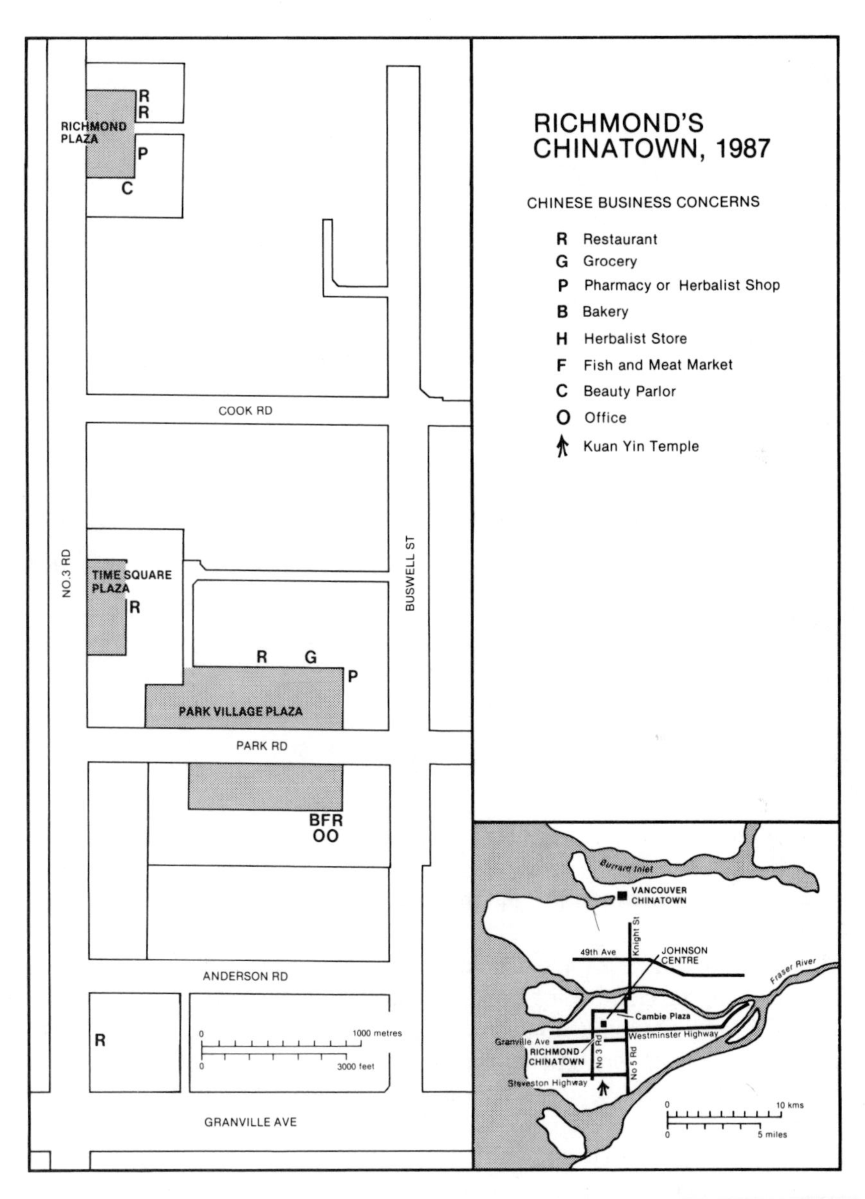

32 Richmond's Chinatown, 1987

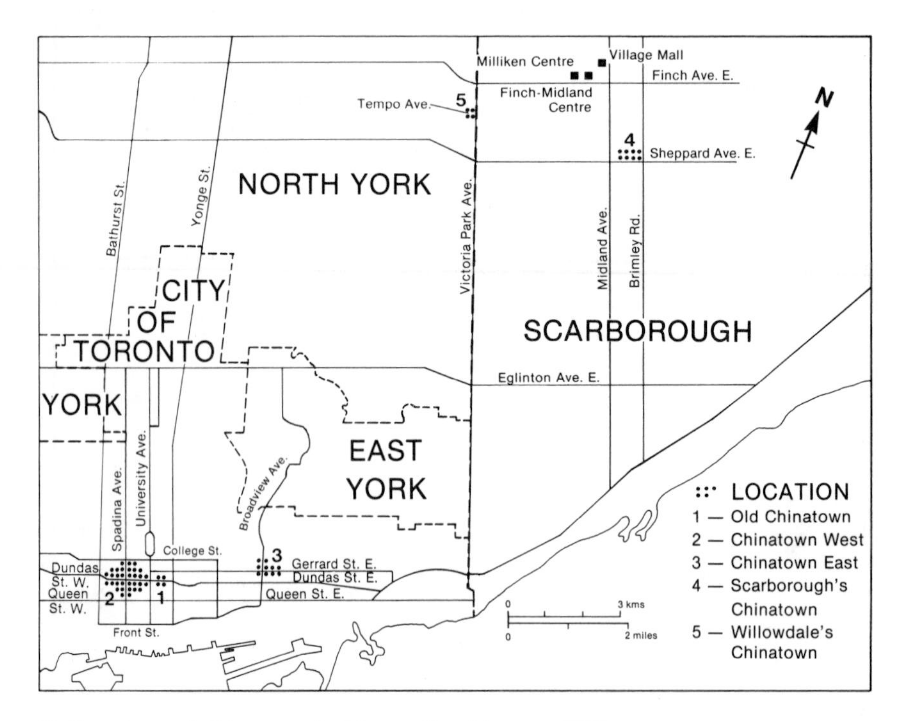

33 Locations of Chinatowns in Metropolitan Toronto, 1987

and Beverley Street and became known as Chinatown West (Plate 24). Outside Chinatown West, there were also two small clusters of Chinese businesses: one on Baldwin Street and the other on Dundas Street east of McCaul Street.

Initially the future of the Chinese community and their budding New Chinatown in the South-East Spadina District was uncertain. In 1969, the City of Toronto Planning Board recommended to City Council that the district be zoned as an institutional area in order to accommodate the expected expansion of the University of Toronto.[221] In June 1970, the University of Toronto abandoned its plans to expand south of College Street because its student enrollment had declined.[222] If the plan had been carried out, a large number of Chinese residents would have been displaced. Another threat to the Chinese community came from Ontario Hydro, which bought up an entire city block facing Chinese businesses on Baldwin Street, and all the block's residents, half of them Chinese, had to move out. In 1971, Ontario

Hydro boarded up the block, intending to build a high-rise terminal station on the site, a structure completely out of place in the low-rise residential area.[223] A committee formed by residents of the area, supported by City Council, appealed to the provincial government to build cheap housing on the Hydro block instead of constructing a station there.[224] Eventually the provincial government decided to transfer the block to the Ontario Housing Corporation for construction of a low-cost housing project.[225] While the fight against the Hydro project was going on, the Chinese community was trying to raise funds for construction of the five-storey Mon Sheong Home on D'Arcy Street. This senior citizens home, completed in 1974, was financed by the CMHC and provided accommodation for sixty-five elderly Chinese.[226]

In 1970, the Planning Board recommended to Council that the low-density residential and commercial uses in the South-East Spadina District be replaced by high-density residential, commercial, and institutional development.[227] The Chinese Canadian Association, led by Doug Chin, Doug Hum, Duncan Lang, and Wes Lore, and the United Action of Chinese Canadians led by Fred Kan, strongly opposed the recommendations because the high-density development would alter the physical characteristics of the district and threaten the existence of the Chinese and non-Chinese communities in the area.[228] Another organization, Grange Park Residents' Association, also opposed the high-density development. Accordingly, the Planning Department created a structure for citizen involvement in the planning process by forming the South-East Spadina Ad Hoc Steering Committee. It had neither formally designated membership nor decision-making power. Since its first meeting on 20 December 1972, it has met every two weeks. Attendance at meetings ranged between twenty and thirty people, who "represent, in a broad sense, their own interest groups [but] they generally speak as independent individuals rather than as delegates of specific organizations."[229] Gradually a core group of about twenty people, mostly social service leaders, emerged, and they attended practically all meetings. Local businessmen and larger property owners showed little public interest in the committee's deliberations, and their involvement in its work was sporadic.[230]

The stability of the South-East Spadina District remained uncertain until the Planning Board recognized its low residential characteristic.[231] In 1973, it recommended that most of the district be stabilized as low-density residential and commercial areas and that only its south and east sections be changed to a mixed-use, high-density residential area. This recommendation was a positive response to the steering committee's wishes, providing some stability for Chinatown West. Accordingly, when the Manbro Investment Company, a Chinese corporation, applied to the Planning Board in 1975 to develop a large commercial project on Spadina Avenue near Dundas Street, the project was rejected because it did not conform to the planning principle of the steering committee.[232] The corporation immediately allied itself with traditional merchant associations such as the Chinese Community Centre of Ontario and formed the Toronto Chinatown Community Planning Association (TCCPA), a coalition of twenty-five organizations.[232] The TCCPA argued that the steering committee, dominated by the social service leaders, was unrepresentative, but the social service groups were obliged to demonstrate how they represented their constituency. Because of dissensions in the Chinese community, the Planning Board reconstituted membership of the steering committee and directed the neighbourhood planning team to develop plans agreeable to both the TCCPA and social service groups.[234] Finally, in 1979, Chinatown West was designated an area of Special Identity, where development had to be compatible in form and character with the emerging Chinese motif and architectural details.[235] Intensive retail and service uses such as a split entrance to the commercial spaces below and above the street level were permitted in the area (Plate 25).

Throughout the late 1970s and early 1980s, the Chinese population continued to expand westward along Dundas Street and north and south along Spadina Avenue. According to the 1981 census, over 30 per cent of the population in South-East Spadina and nearby areas spoke Chinese as their mother tongue. By the summer of 1986, retailing activities had spread along Dundas Street as far west as Augusta Avenue and north along Spadina Avenue to College Street and south beyond Grange Avenue. The huge Dragon Centre, a multistorey complex of condominiums and shopping centre, was completed on Spadina near Dundas. A similar project, the Chinatown Centre, covering an area of 1.75 hectares (4.3 acres), is planned.[236] Hong Kong developers are particularly interested in Toronto real estate because its value is increasing. It was estimated that in 1986 Chinese investors put about one billion dollars into property in Toronto. In August 1987, I interviewed two Chinese land developers in Hong Kong who, declining to be identified, were considering investing more than two million dollars in real estate in Toronto or Vancouver.

TORONTO'S CHINATOWN EAST

During the 1970s, many lower-income Chinese immigrants found it difficult to obtain cheap accommodation in Old Chinatown or Chinatown West. They looked for places where rents were lower. Riverdale, about three kilometres (two miles) east of Old Chinatown, was chosen because properties there were several thousand dollars cheaper than in Old Chinatown and Chinatown West, and rents were much lower. Furthermore, the area, dominated by an Anglo-Saxon working class, has a large park, public libraries, other amenities and services, and is on the street car and bus routes from Old Chinatown and Chinatown West. As a result, Chinese students, young couples, and restaurant workers began to move in.[237] In July 1972, Charles Cheung opened Hung Kee Store at 383A Broadview Avenue, probably the first Chinese store in the neighbourhood. Later that year, the Hsu Store on 597 Gerrard, and the Hai Fung Fish Market on 339 and the Ko Sing Restaurant on 341 Broadview Avenue opened. An embryonic Chinatown East thus emerged at the intersection of Broadview Avenue and Gerrard Street East (Plate 26). In the next three years, more stores were established along Broadview Avenue. In the early 1980s, Chinatown East expanded east along Gerrard Street East, many Chinese business concerns being established as far east as Jones Avenue.

Chinatown East has several advantages over Old Chinatown and Chinatown West. First, commercial premises at reasonable cost are more available in Chinatown East. Second, Chinatown East has more on-street parking spaces, whereas the two other Chinatowns have severe parking and overcrowding problems. Third, Riverdale has a large Chinese market created by many lower-income Vietnamese Chinese and Chinese from China. Last, Chinatown East is very well served by Toronto's public transit system, thus attracting customers from the two downtown Chinatowns.[238] Because of these advantages, Chinatown East began to boom in the late 1970s and its property value began to soar: a small store at Gerrard and Broadview which sold for only $30,000 in 1975 was four times higher in market value in 1980.[239] Today, Chinatown East is centred at the intersection of Broadview Avenue and Gerrard Street, expanding south along Broadview Avenue towards Dundas Street East and east along Gerrard Street East. Outside Chinatown East, a few Chinese businesses, including a Chinese theatre, have been established on Queen Street East. About 320 metres (350 yards) north of Chinatown East is a bronze statue of Dr. Sun Yat-sen, founder of the Republic of China, a memorial which led to a rift in Toronto's Chinese community.[240] Those who opposed erecting the statue preferred to install something that honoured the Chinese workers who had helped build the CPR. Later the choice of a location created a dispute, and the wording on the statue caused another argument.[241] It remained controversial for nearly five years before eventually being installed in 1985 at the south end of Riverdale Park.

According to the 1981 census, nearly 8,000 of the 60,000 residents in Riverdale spoke Chinese as their mother tongue.[242] Although there are no Chinese associations in Chinatown East, Woodgreen Community Centre on Queen Street East organizes a number of services and programs specifically to meet the needs of the Chinese community in Riverdale, such as the Chinese Outreach Programmes and Chinese Adult Programmes.[243]

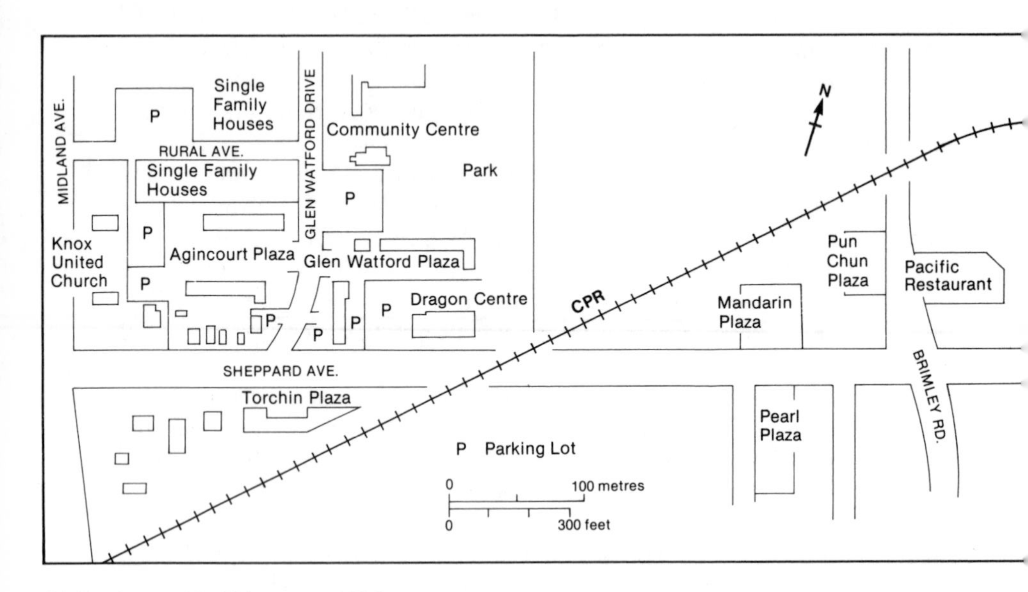

34 Scarborough's Chinatown, 1986

SCARBOROUGH'S CHINATOWN

Scarborough's Chinatown, the third New Chinatown in Metro Toronto, consists of a series of shopping plazas on Sheppard Avenue East between Midland Avenue and Brimley Road (Figure 34). During the 1960s, the Glen Watford commercial area, consisting of Glen Watford and Agincourt plazas, was established to serve the small community of Agincourt, but did not attract much business. Glen Watford Drive, a residential street running off Sheppard Avenue, was the main access road to Agincourt's commercial and residential areas. In 1977, a Chinese developer purchased a piece of land on Sheppard Avenue East and developed Torchin Plaza.[244] Chinese businesses such as the New City Oriental Cuisine and the Original Mandarin Restaurant were established. Two years later, Ching Kee Market and East Court Restaurant were the first Chinese businesses in Glen Watford Plaza. The market's owner listed low rents and proximity to Chinese customers as the two major reasons for his choice of location; the restaurant owner located his restaurant in the plaza because the two Chinatowns on Dundas offered too much competition and because most of the 45,000 Chinese residents in Scarborough, about one-third of Metro Toronto's Chinese population, were professional people from Hong Kong, Taiwan, and southeast Asian countries and could afford to dine out more often than other Chinese immigrants. Five years after the Ching Kee Market and the East Court Restaurant had been established, other Chinese restaurants, grocery stores, supermarkets, hair-stylists, real estate agents, travel agencies, bakers, doctors, dentists, and even an acupuncturist had set up businesses in the Glen Watford Plaza or Agincourt Plaza. Thus, Glen Watford, Agincourt, and Torchin plazas were collectively referred to as Agincourt Chinatown or Scarborough's Chinatown. Business depended mainly on the Chinese living in Scarborough and Markham. According to the owner of the Maple Garden Café, about 70 per cent of his customers were Chinese, 25 per cent Filipinos, and the remaining 5 per cent of other ethnic origins.[245]

New Chinese investment resulted in prosperity in the Glen Watford commercial area, but also led to traffic congestion and a shortage of parking spaces in the area. The two-lane Glen Watford Drive was too narrow to accommodate increased traffic in the area. The influx of Chinese businesses and customers had also left an uneasy feeling among some long-time residents who complained that "the masses, the bustle and the activity" had disturbed the once quiet residential community in Agincourt.[246] Some white residents were annoyed further by a Chinese developer who talked to a reporter about his vision of creating Chinatown Two, in Agincourt, building decorative Oriental gates on Sheppard Avenue and adding Chinese characters on street signs.[247] One local white merchant blamed the developer for building an ever-expanding Chinatown in Scarborough and threatening the "well-being of its residents." Partly in response to the dispute between the Chinese and non-Chinese merchants over the parking problem, Mayor Gus Harris set up a Task Force on Multiculturalism and Race Relations in May 1983.[248] The Task Force, composed of the mayor and six ethnic leaders, was not very active, mainly because Mayor Harris did not think that there was a racial problem in Scarborough.

In 1984, a Chinese entrepreneur purchased a failed roller rink south of Glen Watford Plaza and converted it into a shopping mall known as the Dragon Centre (Plate 27). It consisted of a 350-seat restaurant and more than twenty stores selling mostly Chinese goods. Soon after it opened in April, it attracted many Chinese customers not only from Scarborough and Markham but also from North York and the City of Toronto. Although the centre had 130 parking spaces, that was not enough, particularly on weekends.[249] White merchants in Agincourt and Glen Watford Plazas were annoyed and angry when the centre's shoppers, mostly Chinese, occupied their parking spaces, driving away their own customers.[250] White residents of Agincourt also were annoyed that the increased traffic, particularly on weekends, caused back-ups on Glen Watford Drive, clogging side streets and endangering pedestrians.

Initially, some white merchants and residents blamed the city for poor planning and the Dragon Centre for aggravating traffic and parking problems, but soon they blamed a Chinese realtor for building an ever-expanding Chinatown in Agincourt; finally, the target of anger was the Chinese people. For example, a white merchant told a local radio station that she did not think the Chinese fitted into the community of Agincourt because "they are a type of people that shop in their own stores, and they cling with their own. Canada is like a salad bowl, and all these people coming into the country do add a bit of spice, but if you're going to have a concentration of them in one area, it's going to create hatred."[251]

On 28 May 1984 more than 500 people, nearly all white, held a public meeting in the Agincourt Collegiate Institute auditorium to discuss the problems of traffic congestion and inadequate parking.[252] Chinese merchants in the area were not invited, although they were not barred from attending. Throughout the meeting, all sorts of fears other than the traffic problem were expressed. For example, one person asked "How is it morally or ethnically possible for politicians to change street names to Chinese?"[253] Alderman Bob Aaroe immediately appeased the audience by saying that bilingual signs were not yet considered. "Never!" roared the crowd. The meeting, chaired by Ross Rennie and Dr. Doug Hood, soon turned into a shouting match. "Let 'em [Chinese] learn English," a man yelled. Someone cried out "Teach 'em [Chinese] how to drive." One person proclaimed subtly that "we should not actively encourage any group to cling together as an enclave."[254] Some people suggested that restrictive zoning and a development freeze would be the best way to "defend our interests."[255] Organizers of the meeting probably had not expected it to turn into an anti-Chinese rally. Probably some racists exploited the traffic problem to fuel racism. Accordingly, Ronald Kelusky, president of the Central Agincourt Ratepayers Association, appealed to its members to keep the organization from becoming a haven for racial bias because racist attitudes would ruin its credibility.[256]

In early August 1984, a pamphlet written by Margaret Hunter (probably a fictitious name) was distributed in Agincourt. The author charged that of Hong Kong's three million inhabitants, one in six was a member of the criminal Triad Societies and that the change in the immigration law in January 1984 had left "the door wide open for wealthy drug traffickers from the Orient to establish Chinese businesses as fronts of criminal activities . . . . A large part of the money invested in this [Agincourt's] real estate could be from organized crime."[254] The writer urged her readers to write to their MP and local newspapers to ask the federal government to stop its present "open door" immigration policy. The distribution of Hunter's pamphlet prompted establishment of the Federation of Chinese Canadians in Scarborough, organized under the chairmanship of Richard Wong, to deal with problems concerning the Chinese community there. Meanwhile Scarborough Council passed a resolution condemning the hate literature aimed at the Chinese community.[258]

In the summer of 1985, a report prepared by the Task Force on Multiculturalism and Race Relations revealed that a racial problem did exist in Scarborough and was most serious in three "hot spots": the Glendower Housing Project at Birchmount Road and Finch Avenue, where there had been minor flare-ups between black and white youth gangs; the high-rise and townhouse community near Danforth and Warden avenues, where two-thirds of the tenants were black, unemployment was high, and recreational facilities were inadequate; and Agincourt Chinatown, where hate literature was distributed.[259] The report recommended a permanent mayor's committee on community, race, and ethnic relations to advise council and alleviate race tensions. It also recommended hiring an executive director and a secretary to facilitate the work of other subcommittees.[260] However, Alderman Scott Cavalier, who was asked by the mayor to review the task force's report, merely recommended that council set up a special committee to revise the mandate and membership of the city's multicultural and race relations committee so that it could better meet the needs identified by the task force.

In January 1985, the planning department completed the Glen Watford commercial area study and recommended several courses of action for improving traffic conditions, such as extending parking restrictions on the east side of Glen Watford Drive from the Dragon Centre driveway to Sheppard Avenue and installing "Watch for Pedestrian" signs north and south of the commercial area.[261] The study also revealed that at least 110 out of the 403 parking spaces in Dragon Centre, Glen Watford, and Agincourt plazas were not available to customers because they were used by employers and employees. Accordingly, the report recommended that the city assist employers and employees to develop a system to use available parking spaces in nearby parking lots such as those of Agincourt Community Centre and Knox United Church. However, some residents were not pleased with these recommendations and suggested that if Chinese merchants would take their trade elsewhere, all the problems would be solved.[262]

In January 1986, the Monarch Development Company proposed to build a shopping mall beside Chartland Plaza at Huntingwood Drive and Brimley Road, about 1,000 metres north of Scarborough's Chinatown on Sheppard Avenue.[263] The mall was to include a 440-seat Chinese theatre, a 130-seat restaurant, and a number of fast food outlets. Local residents strongly opposed the plan, fearing that expansion of Chartland Plaza would create parking problems and traffic chaos in Brimley Road and Huntingwood Drive, as Dragon Centre had done to Glen Watford Drive. Some opponents did not want another Chinatown in their neighbourhood and others objected to the

Chinese theatre because it would result in traffic congestion. At the same time, a spokesman for the Federation of Chinese Canadians in Scarborough felt that the area might not be the right location for a Chinese theatre and that the Chinese might not want it. Council rejected the proposed Chinese theatre plan and the issue died.

Throughout the 1980s, Chinese businesses had sprung up in another group of plazas near the intersection of Sheppard Avenue and Brimley Road. Nearly all the businesses in Pun Chun Plaza, for example, were Chinese-run stores; opposite the plaza was the glamorous Pacific Restaurant. By the summer of 1986, Scarborough's Chinatown consisted of two groups of shopping plazas on Sheppard Avenue East separated by the CPR track. Meanwhile, another group of Chinese shopping plazas began to emerge near the intersection of Finch and Midland avenues, north of Scarborough's Chinatown. In June 1987 this new group included the Village Mall, Finch-Midland Centre, and Milliken Shopping Centre, where a large number of Chinese businesses were concentrated (see Figure 33). The grand Hsin Kuang Restaurant dominated the "plazascape" of Milliken Shopping Centre and might become the seed for another budding Chinatown in Scarborough.

WILLOWDALE'S CHINATOWN

Willowdale's, or North York's, Chinatown is a small shopping plaza on the eastern border of North York. Local people commonly refer to it as the "New World" because the New World Oriental Cuisine, established in the late 1970s, was the first restaurant in the plaza (Plate 28). This budding Chinatown, the fifth New Chinatown in Metro Toronto, is located at the corner of Tempo and Victoria Park avenues, about six kilometres (or four miles) northwest of Scarborough's Chinatown. In July 1986, the fourteen Chinese business concerns in Willowdale's Chinatown included two restaurants, two trading companies, one coffee shop, and other stores and services such as real estate and travel agencies.

PROPOSED CHINATOWNS IN REGINA AND HAMILTON

In the past few years there has been occasional talk of creating a New Chinatown in Hamilton and Regina, and both cities are trying to attract investments from Hong Kong business magnates. In Regina, for example, after an economic mission to Hong Kong in the spring of 1985, the City's Economic Development Department proposed a Regina Chinatown in one city block, bounded by Saskatchewan Drive on the north, 11th Avenue on the south, Broad Street on the west, and Osler Street on the east.[264] The Broadway Theatre would be converted into a Chinese Cultural Centre for social and cul-

1 Barkerville's Chinatown, 1860s

2 Kamloops' Chinatown, 1890s

3 Bay windows of Sam Kee Building, Vancouver's Chinatown

4 Cheater Floor and Recess Balcony of Association Building, Vancouver's Chinatown

5 Cumberland's Chinatown, 1910s

6 Nanaimo's Chinatown, 1910s

7 Dupont Street, Vancouver's Chinatown, c1887

8 Carrall Street, Vancouver's Chinatown, c1887

9 Elizabeth Street, Toronto's Chinatown, 1937

10 Huang Geng's cabin, Cumberland's Chinatown, 1971

11 Duncan's Chinatown, 1970

12 Wing Sang Building, Vancouver's Chinatown

13 Calgary's Chinatown, bisected by Centre Street, 198

14 Tong House replaced by new Association Building, Calgary's Chinatown, 1986

15 Elders' Mansion and Multicultural Centre, Edmonton's Replaced Chinatown, 1986

16 Edmonton's Old Chinatown, 1981

17 Lethbridge's Chinatown, 1986

18 Chinese Cultural Centre, Chinatown Garden, and Chinatown Gate, Winnipeg's Chinatown, 1986

19 St. Urbain Street between Place Guy Favreau and Place du Quartier, Montreal's Chinatown, 1986

20 Lee's Association, Montreal's Chinatown

21 Chinese Gates on Lagauchetière Street, Montreal's Chinatown

22 Conversion of residential building into restaurant, Ottawa's Chinatown

23 Park Village, Richmond's Chinatown

24 Dundas Street, Toronto's Chinatown West

25 Split entrance to commercial building, Chinatown West

26 Broadview Street, Toronto's Chinatown East in Riverdale

27 Dragon Centre, Scarborough's Chinatown

28 New World Oriental Cuisine, Willowdale's Chinatown

29 Chinese Girls' Rescue Home on Cormorant Street, Victoria, 1906

30 Tam Kung Temple, Victoria's Chinatown, 1900s

31 Chinese Methodist Church on Fisgard Street, Victoria, 1960s

32 Chinese Consolidated Benevolent Association Building on Fisgard Street, Victoria's Chinatown, 1880s

33 T and L Co. on Government Street, Victoria, 1890s

34 Cormorant Street, Victoria's Chinatown, 1890s

35 Victoria's Chinatown on northern bank of Johnson Street Ravine, 1890s

36 Chinese Public School, 1982

37 Chinese Girls' Drill Team, established in 1950

38 Fisgard Street, before and after rehabilitation

39 Gate of Harmonious Interest on Fisgard Street

40 Fan Tan Alley, before and after rehabilitation

tural activities. The proposed New Chinatown would include a residential facility, traditional businesses such as stores selling women's and men's clothing, computer software, and Chinaware, and new services such as a Chinese health clinic, a bank, and a hotel/motel. Should this New Chinatown be built, it would be like the Replaced Chinatown in Edmonton.

On 19 February 1986, the Economic Development Department called a meeting with representatives from ten Chinese social, cultural, and business organizations to discuss the initial ideas for a New Chinatown in Regina.[265] The response from the Chinese community was positive and a Regina Chinatown Steering Committee, consisting of five Chinese members and a representative from the city administration, was formed. After a few months, the steering committee, headed by Dennis Fong, completed a conceptual plan for a Chinatown, and presented it at a public meeting on 19 November. The proposed Chinatown, covering an area of one city block, would house a senior citizens apartment complex, a cultural centre, a garden with skylight, a emporium, a quality restaurant, a tea and coffee room, a bakery, a grocery, and a meat store.[266] As much as 80 per cent of the $25 million investment required for the project was expected to come from Hong Kong investors.

At the public meeting, several important questions were raised about the project which was, in fact, a private-sector development. For example, would this proposed Chinatown affect existing business concerns in other parts of the city? Since the overseas investors/developers might control the project, what guarantee would there be that they would include non-profit facilities such as a cultural centre? What were the opportunities for the local community to become participants in the proposed development process? Did the steering committee have any control over what the local community would like to see in the Chinatown?[267] Although most of Regina's Chinese community of three to five thousand members would like to have a kind of Chinatown or cultural centre, their support of the steering committee's proposal was not overwhelming. Choon Yong, president of the Regional Chinese-Canadian Association, expressed reservation about the project and wanted more information.[268] Bill Lim, a Chinese resident, wrote to a local newspaper that "the whole Chinatown project was simply an investment proposal by the [Economic] Department in the pursuit of off-shore and possibly other investor/developers, in which local input can only serve as suggestions to those investors/developers for their consideration."[269] He considered the steering committee to be "an investment-scouting and brokerage outfit in the service of the economic development department's investment-enticing scheme, and masquerading as a publicly-supported community agent," and suggested setting up a brand-new community-based Regina Chinatown Development Council to seek public input on the design, planning, and development of the proposed Chinatown. Since there was no assurance that Regina's Chinese community was strongly behind the project, City Council decided in January 1987 to take another look at the proposal before offering further support to its backers.[270]

In Hamilton, there is also a desire to attract Hong Kong money. Since the disappearance of their Old Chinatown, Chinese businesses in Hamilton have been scattered through the city. The Hamilton Chinese Alliance Church on Breadalbane Street is now the focal point of cultural and religious activities of the three to four thousand Chinese people in Hamilton. In 1985, Alderman Paul Cowell expressed a desire for a new Chinatown which could be created as "a potent sales pitch to Hong Kong investors" and expanded to become an international village.[271] The proposed Chinatown should be a planned commercial centre and different from "the sprawling, commercial jungle that exists in Toronto's Dundas-Spadina Avenue area." Clemant Chan, a director of Hamilton's Chinese Cultural Association, supported the idea but he wanted to include a community centre in the project. On the other hand, Michael Ng, president of the Chinese Community Centre, wanted to have a Chinese senior citizens home.[272] There is no suggestion what the proposed Chinatown should be or where it should be located.[273] Thus there is a concept for a New Chinatown but no action has been taken.

RECONSTRUCTION OF BARKERVILLE'S CHINATOWN

In addition to the proliferation of New Chinatowns across Canada after the Second World War, a reconstructed historic Chinatown also emerged. In 1958, the government of British Columbia embarked on development of a provincial historic park at Barkerville as one of its centennial projects.[274] The objectives were to portray the townscape of Barkerville between 1869 and 1885 and to develop it into an open museum to attract tourists.[275] Barkerville still had over 120 buildings, of which forty structures had been built before the turn of the century.[275] Many of these buildings were carefully restored, and some destroyed buildings were reconstructed (Figure 35). Each building was refurbished with exhibits from the Cariboo Gold Rush. Today the Reconstructed Historic Chinatown inside Barkerville Provincial Historic Park is a tourist attraction. In 1986, it consisted of eleven restored or reconstructed buildings: Sin Hap Laundry, Kwong Lee Wing Kee House, Wa Lee Store, Yan War Store, Min Yee Tong (gambling house), Kwong Sang Wing Store, Lung Duck Tong Restaurant, Chinese Miner's Cabin, Tai Ping Fong (a hospital), Chee Kung Tong (Chinese Masonic Hall), and Sing Kee Herbalist Store.[277] In May 1987, for example, Foundation Group Ltd. was engaged to develop a plan for display of Chinese artifacts in the Chee Kung Tong building.[278] I suggested to the consulting firm a plan for display of signboards, flags, altar, and ceremonial artifacts in the society building (Figure 36).

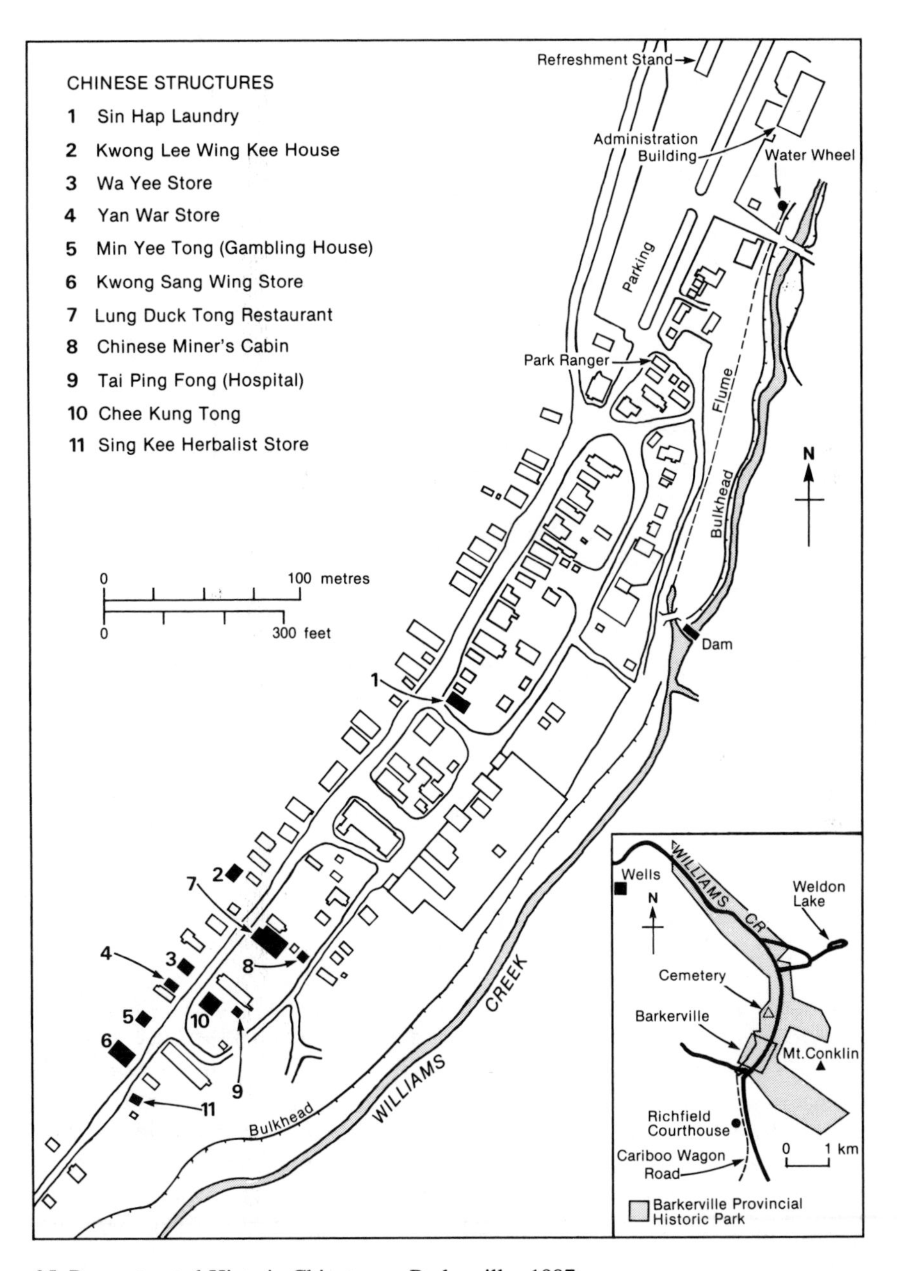

35 Reconstructed Historic Chinatown, Barkerville, 1987

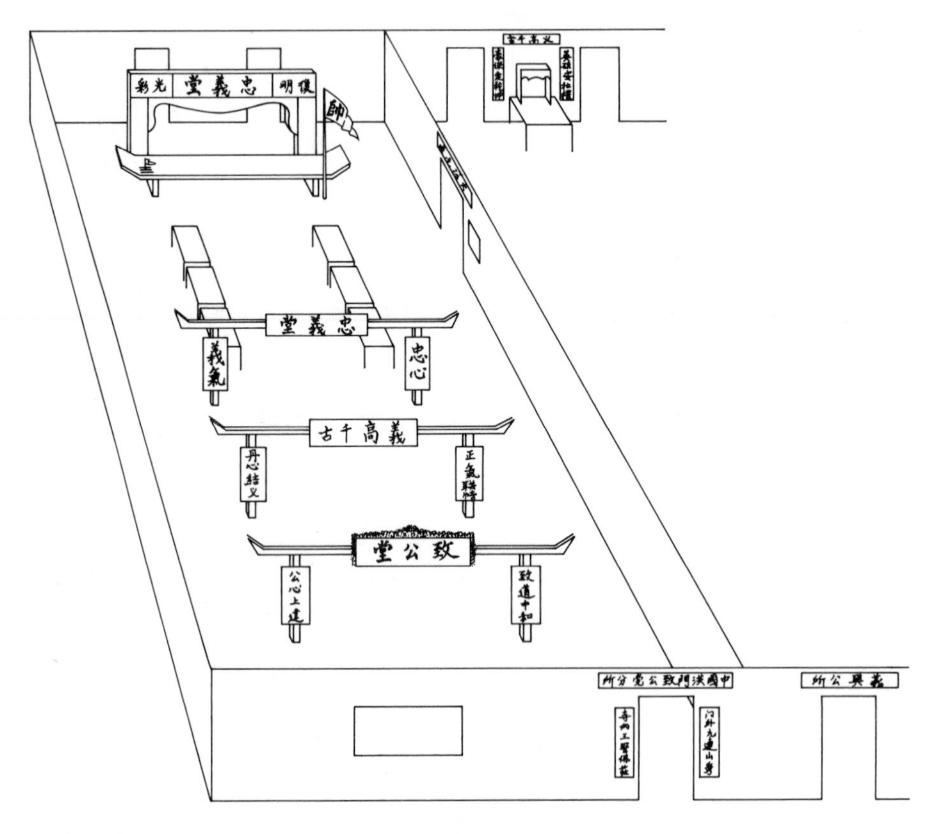

36 A Proposed Plan for the Chee Kung Tong Hall, Barkerville

CUMBERLAND'S CHINATOWN RESTORATION PROJECT

The success of the Barkerville Project inspired the people of Cumberland, B.C., to restore their Chinatown as a tourist attraction.[279] In 1962, a committee for the Chinatown restoration was formed. After its tour of Cumberland's Chinatown, the committee estimated that restoration would cost more than a million dollars and the idea had to be dropped when it was impossible to raise the funds.[280] In 1984, the restoration idea came up again. The Cumberland Museum constructed a scale model of Chinatown in its heyday during the early 1900s, and Jumbo's cabin, the sole surviving cabin of Cumberland's Chinatown, was relocated and restored as an historic building in 1984.[281] The following year, Brian Woodason, a mortgage appraiser, proposed to Cumberland Council that a true-to-scale reproduction of Cumberland's Chinatown, at a cost of about $125,000, be built as a tourist attraction on the former site of Chinatown.[282] He felt that this miniature Chinatown would be much cheaper than reconstruction of a Chinatown like Barkerville's. However, his proposal was not accepted, partly because such a model might be expensive to maintain. Instead, Council applied in June 1986 to the Expo 86 Legacy Fund concerning a Cumberland Chinatown Restoration Project, which, at the cost of $447,272, would involve reconstruction of three Chinatown buildings to scale—the Cumberland Chinese National League Building, the Masonic Hall of the Chee Kung Tong, and the Sun On Wo Company Building.[283] In February 1987, the application was approved, and a subsidy of $202,600 was granted.[284]

Thus the Cumberland Heritage Conservation Society, headed by William Moncrief, Jr., Mayor of Cumberland, was formed to carry out the project. The project is to construct a Comox Valley cultural centre consisting of a new interpretive centre and the reproduction of three historic Chinatown buildings, namely, the Cumberland Chinese National League Building, Sun On Wo Company Building, and the Chee Kung Tong Building. The interpretive centre will include logging and mining displays, models of railways, and other exhibits, whereas the three Chinatown buildings will house a museum, a tea house, a concert/theatre hall, and an information centre. The total cost was estimated at nearly $675,000, of which about 30 per cent was to come from the Expo 86 Legacy Fund, 36 per cent from the Village of Cumberland, and 34 per cent from fund-raising campaigns.[285] The Conservation Society is now trying to raise funds for the project.

PART TWO

Victoria's Chinatown

8

The Budding Period, 1858–1870s

The evolution of Victoria's Chinatown can be the model for the growth patterns of Old Chinatowns in Canada and the United States because physically and socioeconomically most of them tend to have had a similar ecological process, that is, having had budding, blooming, withering, and dying or reviving stages. A detailed study of Victoria's Chinatown will reflect similar characteristics of other Canadian Chinatowns.

Fort Victoria, on the southernmost end of Vancouver Island, was established by the Hudson's Bay Company in 1843 as a Pacific coast trading post.[1] In addition to fur-trading and trapping, the company engaged in farming, salmon-curing, and lumbering on the island. In 1849, recognizing the economic importance of the area, the British government proclaimed Vancouver Island a Crown colony; three years later, a townsite for Victoria was laid out, with an orderly grid pattern of streets bounded by Johnson Street on the north, Government Street on the east, Fort Street on the south, and the harbour on the west.[2] By 1854, Victoria was still a small town, with a population of about 200 people.[3] It was the Fraser River gold rush that transformed Victoria from a quiet frontier town into a busy commercial centre; between March and June 1858, about 20,000 people arrived at Victoria on ocean steamers from California.[4] Hundreds of makeshift tents were set up to accommodate the influx of gold-seekers. Having obtained their mining licences and purchased their equipment, many of these men took the first available boats to the Fraser River but many other immigrants remained in Victoria, established businesses, and engaged in land development and speculation. Indeed, in a short time land prices soared: a standard 12.3 m x 36.6 m (60′ x 120′) lot, for example, bought at £10 to £15, might sell for £300 to £600 within a month.[5]

In January 1860, Governor James Douglas proclaimed Victoria a free port, which meant it had to rely on business and property taxes for its revenue. Each business in the city had to pay a tax in accordance with the assessment of its business volume, and the property tax was fixed at one quarter of 1 per cent of the current real property value.[6] As the real estate business prospered, Victoria boomed. However, it was still a shack town when it was incorporated in 1862, and had a residential population of about 5,000, exclusive of native Indians.[7]

THE SITE

In the summer of 1858, certain sections of Victoria had already been taken over by particular groups. Professional Englishmen such as doctors and lawyers resided on the hillsides fringing the city, former employees of the Hudson's Bay Company occupied Humboldt Street, and most of the Jewish merchants had opened shops on Johnson Street.[8] When the Chinese arrived, they had little option but to set up tents or shacks on mudflats on the north bank of the Johnson Street ravine, which ran along the northern fringe of the city's business centre, filling in the potholes before they could pitch their camps on Cormorant Street, which ran parallel to the ravine (Figure 37). A budding Chinatown on this street began to emerge, accessible from the south bank of the ravine via three narrow footbridges which spanned the ravine at Store, Government, and Douglas streets. The plank sidewalks of Government and Douglas streets ended at Johnson Street and were not built beyond the ravine to Chinatown.

As the Chinese community began to grow, Chinatown gradually expanded northward. The first reason for this was that the Johnson Street ravine, which was not reclaimed until after 1890, was a physical barrier to southern expansion. Second, a Jewish community was already well established on Johnson Street, constituting another barrier. Third, there was a vast stretch of undeveloped land north of Cormorant Street, which was available for settlement. Thus, by the end of the 1870s, Chinatown had been firmly established on Cormorant Street and was beginning to expand northward to Fisgard Street.

POPULATION GROWTH AND AGGREGATION

On 23 June 1858, a Victoria newspaper reported that the first "batch of Celestials have landed," camped in the vicinity of the sign "Chang Tsoo."[9] During the second half of the year, hundreds of Chinese miners came to Victoria's Chinatown, using it as the base camp in preparation for their trip to the Fraser Valley. By 1862, the city's Chinese population was estimated at only 300, or about 6 per cent of the city's total population; almost all the Chinese living in Chinatown were barbers, tailors, cobblers, or other tradesmen whose services were required by the Chinese miners.[10] During the winter and spring, Chinatown would have many temporary residents, mostly miners or

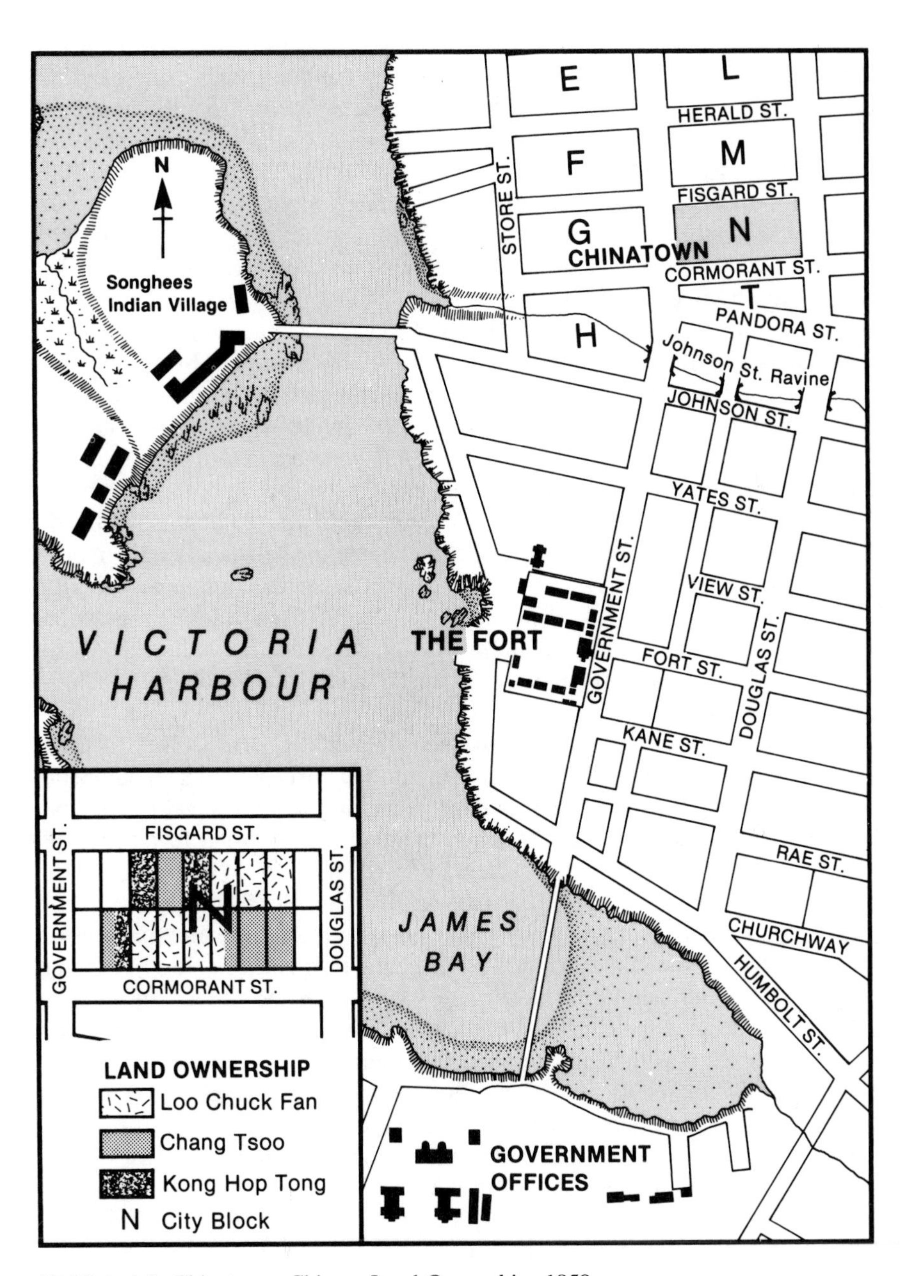

37 Victoria's Chinatown: Chinese Land Ownership, 1859

labourers from the mainland. They drifted back to Victoria's Chinatown when construction of trails or wagon roads stopped during the winter or when the high water level of the Fraser River during the late spring made panning on river bars impossible.[11] If the influx of these labourers from the mainland coincided with the arrival of new immigrants from China and San Francisco, the residential population in Chinatown would double or triple. On 18 May 1965, for example, the ship *Fray Benton* from Hong Kong brought 408 Chinese to Victoria[12] and, the following day, another ship arrived with 600 Chinese from China and Hong Kong;[13] within two days, Chinatown's population soared by over a thousand.[14]

During the 1860s and early 1870s, the number of Chinese in Victoria fluctuated between two and four hundred. In 1871, for example, it was 211: 181 males and 30 females.[15] As Chinese immigrants continued to arrive, the population of Chinatown increased accordingly. In 1874, for example, the Chinese population rose to 493, of which 453 were males and 40 females.[16] Near the end of the 1870s, Chinatown's population had more than doubled. Except for a very few wealthy merchants who could afford to come with their families, nearly all Chinese immigrants had come to Canada as single males, leaving their parents, wives, or children in China. According to the city directory's enumeration of 1874, the sex ratio of Chinatown's population was about ten males to one female. In reality, the ratio was well over 200 to one most of the time. This imbalance caused many social problems, including prostitution.

The Chinese had always been accused of confining themselves to their Chinatown and refusing to be assimilated into the host society. The linguistic barrier was probably the most important factor for minimal social interaction between the Chinese and the host community. Except for a few educated merchants, most early Chinese immigrants were ignorant of the English language, not to mention English customs and culture. Since they had to work long hours to make a living, they could not afford the time and expense to study English. Even if they could, no mechanism was available to acquaint them with Western culture unless they were willing to attend the church school and accepted the Christian faith. After all, prejudice against them was so strong that it was difficult to assimilate them into the host society.

Once a city councillor in Victoria remarked that the Chinese "did not eat our grub, wear our boots, nor did they even leave their bones in the country."[17] They did not go to church nor did they patronize local industries, arts, or science.[18] They did not conform to the English pattern of life and customs in Victoria; they were opium addicts, inveterate gamblers, morally and mentally inferior to white people, and their taste in clothing and food was unbearable. On the other hand, the Chinese themselves knew that they were unwelcome and insecure in the white society. They were hissed at in the streets, and called "Chinks," "yellowbellies," "yellow pagans," or other offensive names. Occasionally they were physically abused outside Chinatown. Thus, they preferred to work and live in Chinatown unless their businesses or occupations, such as laundry and market-gardening, required them to live in the host society.

LAND OWNERSHIP PATTERNS

Chinese ownership of land was an important factor for the location of the budding Chinatown in Victoria. Some people asserted that during gold rush days all dry land in Victoria was at a premium, and since the Chinese could not afford a premium, they had to settle down in the mudflats, where real estate values were low.[19] This assertion is unsubstantiated; some Chinese merchants from San Francisco participated in real estate speculation as soon as they arrived in Victoria. On 12 July 1858, for example, seven Chinese merchants went to a land auction and bought seven 9.1 m x 30.5 m (30′ x 100′) lots on Esquimalt Harbour at prices ranging from $600 to $1,450.[20]

The same month, Chang Tsoo and Loo Chuck Fan purchased thirteen lots in Victoria at $100 per lot.[21] Loo Chuck Fan and his brother, Loo Chew Fan, set up the headquarters of their Kwong Lee & Company and built many wooden shacks to house their imported labourers on their properties, where the embryo Chinatown was thus conceived (Figure 38).

During the 1870s, the Loo brothers were the largest landholders in Chinatown, owning half the land on City Block N and two lots in another part of Chinatown. Other Chinese property owners in Chinatown were Tsay Ching and Dong Sang, who together purchased a small lot at Government Street, on which Tam Kung Temple was built.[22] Chun Tan bought a lot in Block G, where he set up an import and export company.[23] Some Caucasians also purchased or leased properties on Cormorant Street, on which they set up small shanties to rent to Chinese immigrants.[24]

ECONOMIC GROWTH

During the heyday of the gold rushes, the Chinatown economy was monopolized by an established Chinese mercantile class from San Francisco. A search of the Victoria district assessment rolls, city directories, Chinese archival materials, and other records reveals the economic activities of the pioneer Chinese merchants and their import and export companies. For example, the first assessment roll in 1861 indicates that there were three Chinese import and export companies (Kwong Lee, Tai Soong, and Yang Wo Sang) and one Chinese restaurant (Ah Chong Restaurant) in Victoria. The assessment on their business amounted to £8,250, from which they to-

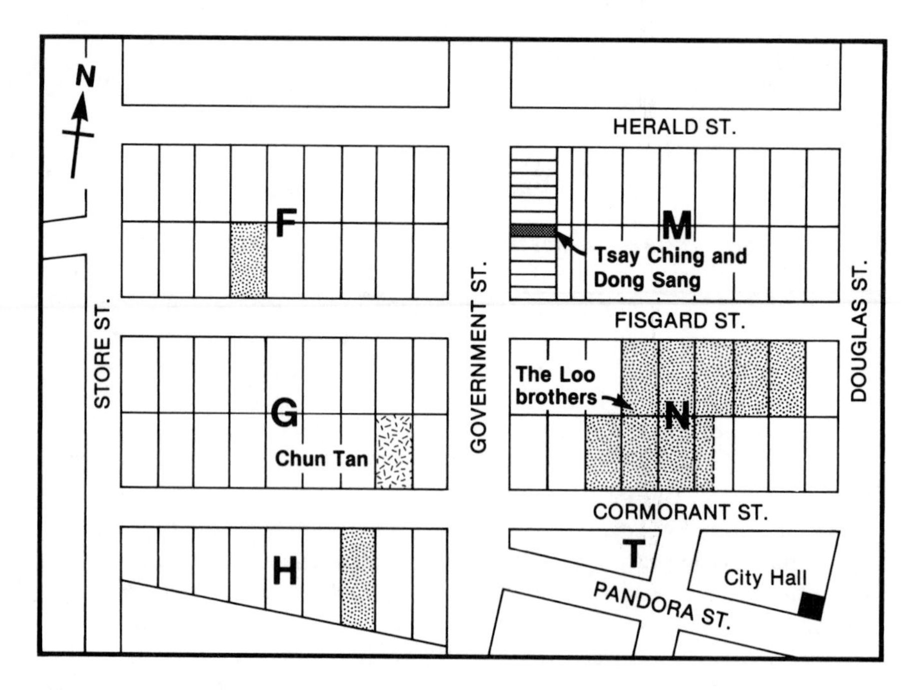

38 Victoria's Chinatown: Chinese Land Ownership, 1870s

gether paid £22.50 in business taxes.[25] The head offices of the three companies were located either in San Francisco or in Hong Kong, where they had commercial connections with China or Chinese communities in Southeast Asia.

Yang Wo Sang & Company, for example, was a branch of a San Francisco company owned by Chang Tsoo. He was among the first group of Chinese to set up a company at Cormorant Street.[26] Tai Soong & Company was a large Hong Kong-based company with branches in both San Francisco and Victoria.[27] The Victoria branch at Johnson Street was managed by Wong Shiu Chiu and later by Tong Kee.[28] The company had its own vessels to ship Chinese merchandise directly from Hong Kong to Victoria and San Francisco and to take lumber and other products from North America to China on return trips. Kwong Lee & Company was by far the largest one in Victoria. Its headquarters were in San Francisco, and branches were set up in Victoria, Yale, Lillooet, Barkerville, and other towns throughout British Columbia.[29] The Victoria branch was on Cormorant Street, managed by Lee Chong and Tong Fat.

During the gold rushes, these big companies not only provided Chinese gold-seekers with goods, equipment, and other daily necessities, but also arranged for Chinese labourers to come to Canada. In April 1860, for example, Kwong Lee & Company arranged for the first group of 265 Chinese workers and nearly $12,000 worth of goods to come from China.[30] Once these workers arrived in Victoria, the company provided them with room and board in Chinatown. The three big companies in Chinatown also functioned as banks for the workers, keeping their savings when they were out of town and remitting money on their behalf to their families in China. In October 1860, when a shipping agent charged the miners five dollars more than the regular fifteen-dollar fare from Victoria to San Francisco, the companies chartered a sailing vessel for their miners rather than pay the excessive rate.[31] Without a fixed place of work, the miners or labourers had to use the address of a company for correspondence with their families in China, so Chinese miners had to rely totally on the few Chinese companies in Victoria not only for employment, room and board, but also for maintaining family contact.

Owners of these companies were also land developers and speculators. They invested heavily in real estate which, then as now, was one of Victoria's chief trading commodities. In 1858, for example, Chang Tsoo purchased two and a half standard lots for $300; within half a year, he had sold the land to Kong Hop Tong at $800, making a profit of 200 per cent.[32] About two years later, Kong Hop Tong sold the lots to a European for $1,800, more than double his purchase price.[33] The land boom, which lasted for several years, was over in 1865 when Victoria faced an economic recession; properties which once sold for thousands were worth little because there were no buyers.[34] Many speculators lost their land through default or tax sales. In June 1867, for example, Chang Tsoo and Li Hing lost all their lots to George Hills, a developer and speculator, after they failed to pay him $10,000 in loans and interest.[35] The Loo brothers held onto their land and managed to survive the slump in real estate, mainly because of their strong financial backing from San Francisco. By the end of the 1870s, they still owned ten and a half lots, and remained the largest landlords in Chinatown (Figure 39).

The three big Chinese companies were also opium importers and manufacturers. Opium was grown in India by the British, manufactured in the British colony of Hong Kong, and then shipped to Victoria and other places. In March 1865, Dr. John S. Helmcken introduced in the legislature a motion to levy a tax or licence fee of $100 on opium sellers, although he "had not the remotest idea of what amount of revenue would be raised from the licence."[36] As all four druggists in Victoria were Caucasians who also dealt with the opium trade, the motion was passed with an amendment that druggists who sold opium for prescriptions were exempted from the tax. In other words, only Chinese opium dealers would be taxed. In August 1865, after a Chinese

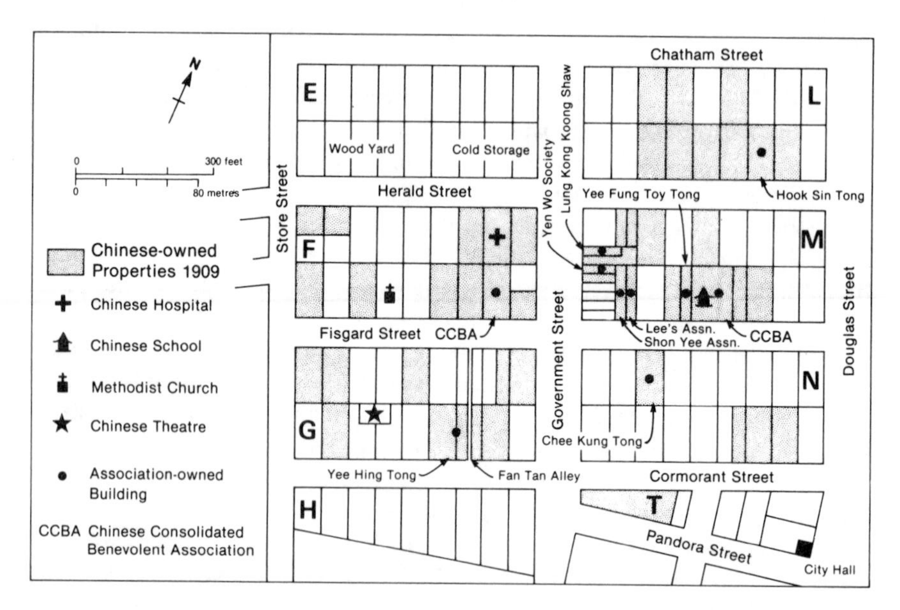

39 Victoria's Chinatown: Chinese Land Ownership, 1909

was found selling opium without a licence, his seventy-four cans of opium were confiscated and sold for $290.[37] Thus, revenues were generated not only from the fines but also from customs duties on opium and re-exports. Records of the Customs House in Victoria, for example, show that nine cases of opium valued at $4,050 and weighing 1,440 lbs. were being held in bond in 1871, meaning a tax had to be paid before the goods could be released.[38] During the 1870s, opium was the third largest export item to the United States after coal and furs.[39]

The economy of the budding Chinatown depended much on the enterprise of the few pioneer Chinese merchants, particularly Lee Chong, who was frequently mentioned in local newspapers. He was popularly known as "Kwong Lee" and was described as "a gentleman of most polite manners and very intelligent. Speaks English fluently in ordinary conversation. Free from Yankee twang and slang."[40] He received his first publicity in August 1858 when a newspaper correspondent reported that Lee Chong, "a well-known and respectable commission merchant," had interpreted the charge to a Chinese offender who was accused of selling liquor to a native Indian.[41] On 29 February 1860, his wife and two children arrived in Victoria; she became the first Chinese female in Victoria.[42] On 7 March, Lee Chong and two other Chinese merchants went to see Governor Douglas after hearing a suggestion to impose a poll tax on Chinese immigrants into Canada.[43] When Governor Arthur Kennedy arrived in Victoria in April 1864, Lee Chong, Tong Kee, and Chang Tsoo called on him to express concern about the fair treatment of the Chinese in Canada and the government's plan to modify the colony's free trade policy.[44] Kwong Lee was mentioned again in the local newspaper after J.S. Drummond was elected mayor in 1875. The *Colonist* estimated that out of ninety-two Chinese votes, seventy-seven went to "Lummond," who promised them "plentee workee" and shoved them into the ballot stalls.[45] When City Council met on 21 January, Councillor Hayward said that one day the Chinese voters might be able to elect Kwong Lee as the mayor and fill the council with gentlemen wearing pigtails.[46] These newspaper reports about Lee Chong indicate that he must have been the most influential leader of the Chinese community in the early 1860s.

The 1870s saw the advent of a class of merchants of moderate means. An increasing number of small Chinese stores were established in Chinatown, although its economy was still dominated by the Kwong Lee and Tai Soong companies on Cormorant Street. Ninety-six Chinese were listed in the city directory in 1871. A study of their occupations and places of residence reveals that over two-thirds lived in Chinatown. Five small stores—Ah Taie, Ah Click and Wo Chin, Quong, Hop Sing, and Tai Chong—were located on Cormorant Street. A cook, butcher, doctor, barber, and several labourers lived there as well. On Hing was the only company on Fisgard Street. Outside Chinatown, the Chinese worked as domestic servants, cooks, market-gardeners, washermen, and labourers. The directory listed a few Chinese market-gardeners scattered throughout the Victoria suburbs: Ah Chew lived on Cedar Hill Road; Ah Hau on Mount Tolmie Cross Road; How Chong in Fairfield Estate, and so on. Similarly, most laundrymen lived outside Chinatown because their clientele was the white community. Mee Hing lived on Fort Street; Mee Wah on Yates Street, Own Sin on Humboldt Street, and many other Chinese washermen lived in different parts of the city's business centre. A hand laundry was the most popular business among the Chinese because it could be started with little capital, little skill, and without their needing much English to communicate with the host society.

The city directory did not list many other Chinese labourers since their whereabouts could not be located easily. For example, it did not list many store assistants, domestic servants, and labourers who did odd jobs. Many Chinese were employed by white families to cook and do other housework. Cooking, washing, and ironing were considered women's jobs, left to women or coloured people. There was a great shortage of maids, and many white families employed Chinese cooks and servants, paying them about thirty dollars a month. Employers felt that they overpaid their servants and were enraged when Chinese cooks struck for high wages in 1872. "Like Oliver Twist, these yellow pagans cry for more. The demand ought not to be acceded to."[47]

The economy of Chinatown and the livelihood of Chinese labourers would have been very difficult if the Chinese Tax Act of 1878 had not been declared unconstitutional. Soon after the act was passed by the provincial legislature on 2 September 1878, the City of Victoria hired Noah Shakespeare as the tax collector to force the Chinese people to pay ten dollars every three months for a licence to reside in British Columbia. Accompanied by a policeman, Shakespeare went to Chinatown to hunt for the Chinese people and seize their belongings if they did not have a licence.[48] Chinese merchants immediately organized a city-wide strike to protest the act. On 17 September, stores throughout Chinatown were closed; Chinese merchants refused to sell goods to any white person, and all Chinese cooks, wood-cutters, washermen, factory workers, and domestic servants declined to go to work.[49] Because of the strike, the Belmont Boot and Shoe Factory, which usually employed over fifty Chinese workers, had to close; several hotels were short of clean napkins and had difficulty in their kitchens; and some housewives were inconvenienced for the want of servants. The Chinese ended their strike five days later when the government agreed to return their belongings.[50] Meanwhile, Tai Chong & Company engaged Messrs. Drake and Jackson and Mr. A.R. Robertson, QC, to serve an injunction on John Maguire, the bailiff, restraining him from disposing of the merchandise he had seized from the company.[51] Disregarding the injunction, Maguire proceeded with sale of the goods and was sued for contempt of court. On 26 October, Mr. Justice Gray judged Maguire guilty of disobeying the court's order and fined him forty dollars. The judge also requested Maguire's lawyer to deposit $950, the value of the goods seized under the Chinese Tax Act, subject to further orders of the court.[52] The court case discouraged Shakespeare from enforcing the Chinese Tax Act in Chinatown.

INSTITUTIONAL ORGANIZATIONS

During the budding period, the few formal institutions in Victoria's Chinatown included a secret society, a few *fangkou,* a Chinese temple, and a missionary school. (A *fangkou,* meaning "rooms," was primarily an informal group of men banded together to share a room or cabin.) Chee Kung Tong, founded in 1876, was the first formal organization;[53] in 1877, it had only thirty members, but two years later its membership had increased to 300.[54] Since it was a secret society, its activities were little known to outsiders. However, it was not a secret that members of the society operated or controlled the Chinese brothels, opium dens, and gambling clubs in Chinatown.

There were a few *fangkou* in Chinatown. By living together, they could reduce expenses and help each other. Usually they came from the same village, spoke the same dialect, and often had the same surname. They coined a name for their *fang* and wrote it on the door of their lodging. When they corresponded with their families in China, they used the name of their *fang* as an address. These *fangkou* were predecessors of the more organized county or clan associations developed after the late 1870s.

Tam Kung Temple was the first Chinese temple in Victoria's Chinatown. According to legend, Tam Kung (Tangong) was a Hakka (Kejia), native of Huiyang County in Guangdong Province in South China.[55] During his lifetime he had repeatedly performed miracles, such as healing incurable diseases and pacifying storms at sea. He was deified after his death and became a patron saint of seafarers in South China, second only to Tianhou, the Queen of Heaven. Another legend claims that Tam Kung was an elderly Hakka villager who faithfully followed the last Song Emperor to Yashan in South China, where he died in 1279 during a sea battle between the Song force and the Mongol pursuers. After his death, Tam Kung was made a patron saint of the Hakka people, fishermen and lightermen in South China.

The origin of Tam Kung Temple in Victoria is not clear. Legend says that in the early 1860s, a Hakka gold-seeker brought with him a statuette of Tam Kung. When he left Victoria for the Fraser River, he placed the statuette near the mouth of the Johnson Street ravine for his fellow countrymen to worship. Later, Hakka residents of Victoria felt that their god should be properly housed and so raised enough money to purchase a small lot at Government Street, where they built Tam Kung Temple.[56] It was dedicated on 21 January 1876.[57]

Another important religious institution in Chinatown was the Chinese Mission School. In 1868, the Reverend A.E. Russ of the Methodist Church started missionary work among the Chinese community in an empty bar at the corner of Government and Herald streets.[58] For six years, lessons were taught by his wife and other volunteers. In 1873, Mr. W.E. Sanford, a wealthy merchant and senator from Hamilton, Ontario, visited Victoria. He was impressed by the missionary work among the Chinese people but not with the school environment. Accordingly, he promised to contribute $400 a year for several years to establish a proper Chinese mission school.[59] With the financial support of Sanford and other donors, the Methodist Church organized the Sanford Mission on Government Street. The school, the first Chinese mission school, was opened on 18 March 1874, attended by children and adults.[60] It was run by Miss Annie Pollard and later by a Mrs. Burgess; the religious service was conducted by the Reverend W. Pollard and the Reverend Derrick; the Sabbath School was supervised by Mr. W. McKay.[61]

Although the Chinese attending were more eager to learn English than to become Christians, the missionaries felt they would eventually convert the Chinese.

SOCIAL ACTITIVIES

Life in Chinatown was dull; there were few organized activities for the Chinese community. The most important social event of the year was celebration of the Chinese New Year, during which firecrackers were set off. All shops were closed, tidied, cleaned, and decorated with peach blossoms, narcissi, and pieces of red paper on which good luck verses were written. People dressed in their best and exchanged visits. Visitors were entertained with Chinese wine, nuts, cakes, fruits, and other delicacies, and the celebration ended with the Feast of Lanterns on the fifteenth day of the first lunar month. However, enjoyment of the Chinese New Year was frequently dampened by the city's regulation that no fireworks be exploded in public. On several occasions, Chinese were fined for setting off firecrackers in the street. Normally few Caucasians visited Chinatown, but during the Chinese New Year some availed themselves of Chinese hospitality and came to try the food and wine.

Other festivals celebrated in Chinatown were the Dragon Boat Festival on the fifth day of the fifth lunar month, the Mid-Autumn Festival on the fifteenth day of the eighth month; and the Chongyang Festival on the ninth day of the ninth month. These festivals were, however, less noticeable to the host community. A wedding ceremony was a rare event in Chinatown because it was expensive to get married. For example, when Mee Wah, a laundryman, got married in November 1871, a local newspaper reported that he had to pay $900 for his bride and the wedding feast.[62] That amount was equivalent to several years' income for a Chinese labourer.

Another excitement in Chinatown was a funeral procession. The white community was fascinated by the attire of the mourners, the variety of flags and banners, and the array of burning incense and candles, roast pigs, chickens, fruits, wine, and other offerings on the altar. Only rich merchants or prominent leaders of the Chee Kung Tong could afford luxurious funerals.

The shipment of bones of the deceased back to China also drew the attention of the host community. According to a traditional explanation, this practice of returning originated from the British recruitment of Chinese coolies to work on plantations in Malaya and Singapore in the early nineteenth century. Chinese employment agents, under whose auspices coolies were shipped abroad, were held responsible by the Manchu government for the safe return of Chinese coolies. If they died en route to the plantations or during employment, their remains had to be sent back to China to prove their death. Accordingly, deceased coolies were buried for seven years, the minimum time for bodies to decompose completely. The bones were then exhumed, cleaned and dried, and crated and stored until the number of crates was large enough to warrant chartering a vessel to ship them back to China.

This tradition was maintained by the Chinese in Canada. Every seven years, Chinese remains in Ross Bay Cemetery in the suburb of Victoria were exhumed. After washing, cleaning, and drying, the bones would be crated and stored in a hut in Chinatown. When several hundred crates were collected, they would be shipped back to Hong Kong on a chartered vessel and stored in Tung Wah Hospital's mortuary, where they would be collected by relatives or shipped back to the home villages. In Victoria's Chinatown, the shipment of bones was an important event. The farewell ceremony was held on the street, attended not only by the Chinese but also by many curious Caucasians or native Indians. On 26 January 1872, for example, the city's entire Chinese population assembled in Chinatown to give a parting feast to about 150 crates of bones before shipment to Hong Kong.[63]

SOCIAL DEVIATION

Generally speaking, most Chinese people were law-abiding citizens. Occasionally, Chinese storekeepers were fined for selling spirits to native Indians, and domestic servants were heavily punished for thefts. In the so-called "Daring Burglary" case, for example, Ah Sam and Joe were arrested for having stolen cash and gold pieces from their white master and were sentenced to three years' imprisonment with hard labour.[64]

Common offences in Chinatown were gambling and prostitution. When Chinese labourers were very busy, they did not have time to think of their families in China, but if they had some spare time, they felt lonely or forlorn. As there were no amenities in Chinatown, they resorted to prostitution, gambling, or opium-smoking. Many indulged in these vices, and so gave the host community the impression that the Chinese were immoral and degraded. Often gamblers were handled roughly by the police during raids: they were handcuffed in the gambling dens, taken to the police station, and jailed for a night before being charged the following morning. In one case, twenty-five Chinese gamblers were arrested; the two bankers were fined £20 each and the remaining gamblers 40s. each.[65] On several occasions, gamblers were reprimanded for using knives in disputes.[66]

The other source of trouble in Chinatown was prostitution. Brawls and fights as a result of disputes over ownership of women or prostitutes were common in Chinatown. During the early 1860s, Ah Gutt, a notorious procurer, owned a brothel on Cormorant Street and was in the news several times.[67] In May 1875, concerned Chinese residents sent a petition to City

Council asking it to suppress the nine Chinese brothels in Chinatown.[68] However, a bylaw to close the Chinese brothels would also have been applicable to many other brothels kept by native Indian women or Caucasian women.[69] Hence, Council did not respond to the request.

TOWNSCAPES AND IMAGES

During its budding period, Victoria's Chinatown consisted of several rows of rough wooden huts on Cormorant Street between Douglas and Store streets and a few isolated crude shacks on Fisgard Street. On the periphery of Chinatown were drinking saloons, which were vastly out of proportion to the needs of Victoria's population. The waterfront was dominated by warehouses and shipping offices. The most imposing landmark near Chinatown was city hall at the corner of Douglas and Pandora streets. Designed by John Teague, it was built in 1878 in an Anglo-Italian manner, with dark red brick and a tin mansard roof and recessed windows.[70]

In Chinatown the streets were unpaved and did not have plank sidewalks. There were no sanitary facilities and no sewers to drain off rain and waste. As Chinatown was situated on a muddy flat, heavy rainfall caused torrents of water to flow through the street, inundating the crude wooden sheds. Between the closely packed shanties and cabins were dark narrow garbage-clogged passageways which formed a forbidding labyrinth. Some alleys were so narrow that neighbours could touch each other through their bedroom windows. Most Chinese lived in the dirt-floored cabins and shacks and ate, cooked, washed, and slept together. Many wooden huts were divided into small cubicles, each measuring three by four metres (10′ x 14′). Several people would be crammed into a single cubicle, and two or three men slept on one bed. The white community had always complained about overcrowding in Chinatown, fearing it might breed pestilence. At one time, there was an attempt to introduce a Cubic Air Ordinance, similar to one in San Francisco, by which people were forbidden to rent rooms with less than fourteen cubic metres (500 cu. ft.) of air per person.[71]

Two structures stood out in the urban landscape of Chinatown. The first was Tam Kung Temple, a one-storey structure and one of the rare brick buildings in Chinatown. The other structure was a large frame building on Store Street. It was used as a mortuary for storing crates of bones sent to Victoria from other Chinese settlements in the province. The use of the building was unknown to outsiders until it caught fire on 3 October 1871.[72] Remains of fifty dead Chinese were found inside the building after the fire was put out, and there was fear that this "Chinese Dead House" was a health hazard. The mayor and a councillor inspected the building and ascertained that the bones stored in the building were clean.[73]

The streetscape of Chinatown displayed Oriental characteristics. Vertical signboards bearing Chinese characters were hung outside each shop. Virtually no Chinese women were seen outside their homes, and nearly all pedestrians were Chinese men. Peddlers carried their vegetables or other products in baskets on a bamboo pole, walking from one shack to another to look for customers. In front of Chinese stores, small groups of onlookers crouched, sat, or stood while chatting or smoking. Chinese men usually wore homemade cloth shoes and straight-cut short jackets which buttoned from throat to hem and hung out loosely over baggy trousers. In winter, their clothes were padded and quilted with cotton. Only a few merchants could afford to wear handsome silk jackets over brocade robes, and they had leather shoes or cloth shoes richly embroidered with flowery designs.

Occasionally, Chinatown storefronts were brightened by ornaments during the celebration of certain festivals such as the Chinese New Year or during the visit of a dignitary. For example, when Lord Dufferin, Governor-General of Canada, visited Victoria on 16 August 1876, Cormorant Street was colourfully decorated with three impressive evergreen arches, Chinese lanterns, and other paraphernalia. In those days, virtually no Caucasians strolled in Chinatown, which was perceived by them as an exotic, hellish, and mysterious enclave of "long-tailed, rice-eating aliens." Even policemen would not enter Chinatown unless there was a dire necessity to do so.

9

The Blooming Period, 1880s–1910s

In the late nineteenth century, Victoria was a prosperous and important seaport with a regular steamship service across the Pacific as well as along the Canadian and American coasts. It was the destination for cargoes and passengers from China and the shipping centre for goods to other parts of Canada and the United States. When the CPR was being completed in the early 1880s, Chinatown was booming, functioning as a distribution point for the infant Chinatown in Vancouver and other Chinatowns in British Columbia.

After the CPR was completed in 1885, many unemployed Chinese railway workers found themselves stranded in Victoria. Throughout the economic depression of the late 1880s and early 1890s, the smell of bitter poverty pervaded every shack and alley in Chinatown.[1] After this depression, Chinatown entered a short period of economic prosperity and territorial expansion; real estate boomed from the mid-1900s onwards, reaching its zenith in the mid-1910s. This prosperity was reflected in the emergence of a new landlord and merchant class and the construction of many new three- to four-storey brick buildings in Chinatown.

POPULATION GROWTH

The national census of 1881 listed 693 Chinese in Victoria, but three years later the Royal Commission Report on Chinese Immigration revealed that there were 1,767 Chinese in the city, nearly triple the census data.[2] This big jump in population had been caused not only by the influx of Chinese immigrants from China, but also by the arrival of many unemployed Chinese railway workers trying to smuggle themselves back to the United States. From Victoria, it was relatively easy to sneak back into the United States at night by sailing to Port Angeles, Washington, which took seven hours. A revenue officer at Port Townsend, Washington, estimated that at least 1,000 Chinese had crossed the strait into the United States during the first ten months of

1883.[3] This figure did not include those who left Victoria by steamers and were smuggled into Portland and San Francisco.[4]

Victoria City Council claimed that the national census tended to underestimate the Chinese population. After the 1891 census revealed that Victoria had a population of 16,841, of which 2,080 were Chinese, Council conducted its own census.[5] Its survey revealed that Victoria had a population of 22,981, about 6,000 more than shown on official returns, and that the Chinese population was 3,589, nearly 1,510 more than the census enumeration. Similarly, the 1901 census showed that Victoria had a Chinese population of 2,978. But in the same year, the Royal Commission on Chinese and Japanese immigration conducted a survey which revealed that the city's Chinese population was 3,282.[6]

The Chinese population was difficult to enumerate accurately partly because of seasonal labour conditions. After the winter season, labourers left Chinatown for farms, mines, lumber camps, or fish canneries outside Victoria. At the beginning of May, for example, fish-canning factories began to recruit Chinese workers to make tins and prepare for the fishing season. During the peak canning season in July and August, more Chinese workers were required; as a result, Chinatown was virtually depopulated in the summer, when the national census was usually conducted. But as winter approached, Chinese workers were laid off and returned to Victoria. Its Chinatown grew crowded in January when Chinese people from other parts of the province also thronged into town, waiting to return to China to spend the Chinese New Year with their families. As a result, Chinatown's population from late autumn to early spring was usually several times higher than in the summer.

SEX RATIO, MARITAL STATUS, AND NATURALIZATION

The 1911 census, the first to include the composition by sex of the Chinese in Victoria, revealed that the ratio was highly unbalanced: nearly thirteen males to one female (Table 31). As time passed, more Chinese workers saved enough money to return to China to marry. Then they would return to Victoria and work for years in order to save enough to have their wives and children join them. Gradually, the Chinese population grew less unbalanced; by the 1920s, the ratio was about six males to one female.

The national census did not record the marital status of the Chinese, but according to the two Royal Commission reports the number of married Chinese women in Victoria increased from forty-one in 1884 to ninety-two in 1901.[7] Intermarriage between Chinese and Caucasians was extremely rare; both communities objected to it. In 1907, for example, Leon Wing, son of a wealthy Chinese merchant in Victoria, fell in love with a white American girl, but his father prevented him from following her to the United States.[8] In

TABLE 31: Chinese population in Victoria, 1881–1981

Census year	Male	Female	Total	Ratio males to female
1881	–	–	693	
1891	–	–	2,080	
1901	–	–	2,978	
1911	3,205	253	3,458	12.7
1921	2,938	503	3,441	5.8
1931	3,192	510	3,702	6.3
1941	2,549	488	3,037	5.2
1951	1,399	505	1,904	2.8
1961	1,364	773	2,137	1.8
1971	1,105	950	2,055	1.2
1981	2,825	3,000	5,825	0.9

Source: Census of Canada, 1881–1981

1909, Amy Morris came from San Francisco with the intention of marrying Lee Barker, a Chinese merchant in Victoria, but she was soon deported by the police as "undesirable."[9] In one case, Amanda Clapton married Lee Land, owner of a grocery store on Fort Street, in September 1908. After they had spent their honeymoon in Vancouver and were seen on the boat returning to Victoria, the couple disappeared; thus the marriage ended mysteriously.[10] In nearly all cases of intermarriage, white girls married Chinese boys; in Victoria, a permit from the sheriff was required for intermarriage.[11]

Newspaper reports occasionally provided sketchy information about the naturalization of the Chinese. For example, it was reported in the *Colonist* that ten Chinese were naturalized in 1880 and ninety-one in 1899.[12] Although most Chinese did not want to give up their nationality and remained as "sojourners," there was an undetermined number of Chinese residents, particularly merchants, who considered Canada their home and strove for naturalization in the hope that they might be treated as Canadians and that their children would have the same privileges as other Canadians. Huang Tsun Hsian, Consul-General for China, testified to the Royal Commission on Chinese Immigration in 1884 that "I know a great many Chinese will be glad to remain here permanently with their families, if they are allowed to be naturalized and can enjoy privileges and rights."[13] The Chinese found it not only difficult to be naturalized but also to be treated as "Canadians." They were still "Chinaman," which was defined by the B.C. government as "any native of the Chinese Republic or its dependencies not born of British parents . . . [including] any person of the Chinese race, naturalized or not."[14] In other words, whether a Chinese was naturalized or not, he was still a second-class citizen. Thus there was no incentive for naturalization.

TABLE 32: Occupations of Chinese in Victoria, 1884

Occupation	No. of persons	% of total
CLASSIFIED		
Cooks and servants	180	10.2
Store employees	179	10.1
Boot-makers	130	7.4
Vegetable gardeners	114	6.5
Washermen	90	5.1
Fuel cutters	65	3.7
Brickmakers	60	3.4
Merchants	45	2.5
Farm labourers	40	2.3
Prostitutes	34	1.9
Others	274	15.5
UNCLASSIFIED		
Married women	41	2.3
Girls	31	1.6
Boys under 12 years	10	0.6
Boys between 12 and 17	92	5.2
New arrivals	382	21.6
Total	1,767	100.0

Source: compiled from Report of the Royal Commission on Chinese Immigration (Ottawa: Printed by Order of the Commission 1885), 363

OCCUPATIONS

The two Royal Commission reports described Chinese occupations in Victoria. The 1884 Royal Commission showed that about 20 per cent of the Chinese were store employees, cooks, or domestic servants; 50 per cent were labourers engaged in bootmaking, gardening, washing, and other labouring jobs; and 20 per cent were new arrivals who did not yet have fixed occupations (Table 32). This increase resulted partly from the economic growth of Chinatown, which is described later in this chapter. The remaining 10 per cent were merchants, women, and children. About one-third of the one hundred females in Chinatown were prostitutes, and less than half of all females were married. Seventeen years later, the second Royal Commission survey revealed that the occupation pattern of the Chinese in Victoria had not changed much since the 1880s. The labouring class was still predominant, but the number of merchants was six times larger and that of housewives had doubled (Table 33). This increase was also partly a result of the economic growth of Chinatown. A second generation was also emerging, as 145

TABLE 33: Occupations of Chinese in Victoria, 1901

Occupation	Number of persons	% of total
CLASSIFIED		
cannery men	886	27.0
labourers	638	19.4
cooks	530	16.1
merchants	288	8.8
gardeners	198	6.0
laundrymen	197	6.0
tailors	84	2.6
sawmill hands	48	1.5
interpreters	2	0.1
preachers	1	0.0
UNCLASSIFIED		
unemployed labourers	173	5.3
housewives	92	2.8
children (male), native-born	63	1.9
children (female, native-born	82	2.5
Total	3,282	100.0

Source: compiled from Report of the Royal Commission on Chinese and Japanese Immigration (Ottawa 1902), 12–13

Canadian-born Chinese children were reported in the 1901 Royal Commission.

Demographically, Chinese society in Canada was still a "bachelor" society, with few married families. Economically and socially, the society was a dual community: at one end was a small group of wealthy and comparatively educated merchants, and at the other end was a multitude of poor and illiterate workers. A few merchants could afford to have servants and lead a comfortable life in Canada. They were able to speak and read English and so were more westernized and more willing to be assimilated into the host society. They led the Chinese community fight against discrimination and abuses because the host society still ostracized them. Through their leadership in various Chinese organizations, they controlled Chinatown and dominated its people. On the other hand, the mass of labourers in Chinese society knew little or no English and found it difficult to communicate with the host society. They were regarded as undesirable and unassimilated "slaves," despised not only by the white community but also by some of their own wealthy countrymen.

TABLE 34: Home county and clan origins of Chinese in Victoria, 1884–5

Clan	Taishan	Kaiping	Xinhui	Enping	Panyu	Zhongshan	Other counties	Total
Li	70	5	9	3	6	9	7	109
Ma	45	0	0	0	0	0	1	46
Huang	13	0	3	3	15	2	7	43
Chen	5	0	4	5	5	0	15	34
Liang	4	1	2	8	6	0	6	27
Xie	0	1	0	2	21	0	0	24
Zhou	0	3	2	4	9	0	5	23
Xu	0	0	0	2	17	0	1	20
Other	46	34	33	45	74	24	90	346
Total	183	44	53	72	153	35	132	672

Source: compiled from CCBA donation receipt stubs, 1884–5

HOME COUNTY AND CLAN ORIGINS

Additional information on the Chinese population in Victoria can be obtained from non-census sources, from which data on occupations, age-sex profiles, clanship, and the like may be obtained or deduced. For example, the county and clan origins of the Chinese in Victoria can be deduced from the stubs of 1884 and 1885 CCBA donation receipts.[15] On each stub were recorded the full name and home county origin of the Chinese donor and the amount of his donation. Study of the 672 donation receipts issued to Chinese residents in Victoria reveals that over half originated in Siyi (Taishan, Xinhui, Kaiping, and Enping counties) in Guangdong Province (Table 34). These donors belonged to ninety-one clans, of which the Li, Ma, and Huang clans were the largest. Most had come from Taishan County.

Another source of county and clan origins is 104 Chinese Hospital donation receipt stub booklets, from 1892 to 1915.[16] Each stub shows the full name and home county origin of the donor and the two-dollar "donation." Seventy-three per cent of the 6,155 donors came from Siyi, 9 per cent from Panyu County, 7 per cent from Zhongshan County, and the remaining 11 per cent from other counties in Guangdong Province. The total donation from the 6,155 donors amounted to $12,310, of which one-quarter was from 446 people surnamed Ma, 420 surnamed Huang, and 303 surnamed Li from Taishan County, 220 persons surnamed Zhou of Kaiping County, and 185 surnamed Lin of Xinhui County (Table 35). Therefore the Li, Ma, and Huang clans were most numerous, and most came from Taishan County. Other clans such as the Zhou, Guan, Xie, and Situ came mainly from Kaiping County, the Lins from Xinhui County, and the Xus from Panyu County.

TABLE 35: Home county and clan origins of Chinese in Victoria, 1892–1915

Clan	Taishan	Kaiping	Xinhui	Enping	Panyu	Zhongshan	Other counties	Total
Li	303	24	142	1	20	57	98	645
Huang	420	30	75	3	25	24	62	639
Ma	446	1	4	0	0	6	4	461
Zhou	11	220	16	24	113	2	31	417
Chen	157	4	116	22	9	39	59	406
Lin	66	7	185	0	4	26	8	296
Liu	67	1	13	3	6	41	30	161
Zhang	6	76	46	2	5	3	13	151
Yu	78	34	22	0	5	5	2	146
Guan	2	128	0	0	0	1	7	138
Others	531	442	510	226	376	207	403	2,695
Total	2,087	967	1,129	281	563	411	717	6,155

Source: compiled from Chinese Hospital donation receipt stubs, 1892–1915

Although the above two Chinese non-census sources do not cover the entire Chinese population in Victoria, they provide substantial evidence that most early Chinese immigrants came from only a few counties on the Zhujiang delta (see Figure 4). The highly clustered origins of Chinese immigrants in Victoria had considerable influence on Chinatown's socioeconomic structure. For example, the Siyi dialect, markedly different from Cantonese, Hakka, and other dialects, was the most widely spoken language in Chinatown, in Victoria; also some type of foods and styles of cooking were characteristic of the Siyi cuisine. The county and clan associations formed by these people were powerful in Victoria's Chinatown because of their larger membership and greater wealth; leaders of these associations usually became the decision-makers in the community. This concentration of power was also apparent in other Chinatowns in Canada and the United States.

COUNTY, DIALECT, AND CLAN ASSOCIATIONS

During the blooming period of Victoria's Chinatown, a great variety of voluntary associations were formed. Many *fangkou,* each consisting of several people with the same county, dialect, or surname, merged to form a bigger association. In 1893, for example, Yuqingtang, a county association of people from Taishan and their descendants, was formed.[17] By 1902, ten county associations had been established in Chinatown: Yuqingtang (Taishan County); Changhoutang (Panyu County); Fuyintang (Nanhai County); Xingantang (Shunde County); Fushantang (Zhongshan County); Baoantang (Dongguan County); Renantang (Zengcheng County); Fuqingtang

(Xinhui County); Tongfutang (Enping County); and Guangfutang (Kaiping County).

Unlike a county association, a dialect association united people with similar dialects, although they came from different counties. In Victoria, Nanhuashun Sanyi Lianhui (the United Association of Nanhai, Hua, and Shunde Counties) was formed by people who spoke the Sanyi dialect, although they were from different counties. Another dialect association was Renhe Huiguan, formed by the people speaking Hakka dialect; they had come from Wuhua, Chixi, and a few other counties in Guangdong Province.

The clan association is founded on the assumption that all persons with similar surnames are of a common ancestry. In 1907, for example, the Chinese in Victoria surnamed Lee (Li) formed the Lung Sai Tong (Longxitang), and in following years, the Wong (Huang) people formed the Wong Kong Har Tong (Huang Jiangxiatang), and the Lum (Lin) established the Lum Sai Ho Tong (Linxihetang).[18] Other clan associations were also formed: the Fung Toy Tong (Fengcaitang) of the Yu clan, the Jinzitang of the Mahs or Mars (Mas), the Jingchuantang of the Chans (Chen), the Jiaoluntang of the Szetos (Situ), the Ailian Gongsuo of the Chows (Zhou), and the Lujiangtang of the Hos (He).

THE HONGMEN SOCIETY

Another association was the Hongmen Society, the earliest organization in Victoria's Chinatown. In 1886, ten years after its establishment, Chee Kung Tong had accumulated $6,000 to construct a wooden building at 22-24 Cormorant Street, next to the Chinese theatre.[19] The two-storey building was officially opened on 25 September; the first floor was used for lodgings and the upper floor contained the shrine of Guanyu, a patron saint of the Hongmen. During the early 1880s, Hip Sing Tong (Xieshengtang), another branch of the Hongmen Society, was established in Victoria in opposition to Chee Kung Tong. Little was known about the early activities of these two secret societies, but occasionally their activities were exposed by local newspapers. In 1886, for example, the abduction of a girl by Hip Sing Tong became known to the public after a reward of fifty dollars for her return was advertised in a local newspaper.[20] In 1898, Chee Kung Tong demanded fifteen dollars a month from the proprietors of the gambling dens in Chinatown on the pretext that imposition of the levies was an attempt to discourage members from gambling.[21] In another incident, a Hongmen member was accused of prostituting the society's secret after he converted to Christianity. Marked for death, he had to leave for eastern Canada under police protection.[22]

The rivalry between Hip Sing Tong and Chee Kung Tong was also little known to the public. According to a Hongmen elder, Chee Kung Tong eventually eliminated Hip Sing Tong in the late 1880s and became the only branch of the Hongmen Society in Victoria. In July 1899, Chee Kung Tong purchased a lot at 617-19 Fisgard Street and erected a two-storey brick building; the following year, the society moved its headquarters from Cormorant Street to the new building.[23] By then, Chee Kung Tong had become the most influential and powerful branch of the Hongmen Society in Canada, having lodges in all large Chinatowns in the country.

In 1904, Dr. Sun Yat Sen (Sun Yixian), a member of the Hongmen Society, assisted the Chee Kung Tong to change its constitution, transforming the secret society into an open revolutionary party against the Manchu government.[24] In 1911, Dr. Sun toured Canada and the United States to raise funds for the Huanghuagang Uprising in Canton.[25] In response to Dr. Sun's appeal, the Victoria chapter of the Chee Kung Tong mortgaged its building at Fisgard Street for $12,000 to support the uprising.[26] After the success of the revolution and establishment of the Republic of China in 1912, Chee Kung Tong was recognized by the Chinese government as a political party.

CHINESE GIRLS' RESCUE HOME

One of the most significant institutions in Chinatown was the Chinese Girls' Rescue Home, started by John Endicott Gardiner. Gardiner, born of missionary parents in China, could speak the Cantonese dialect fluently. In January 1885, he came to Victoria from San Francisco to act as an interpreter in a Chinese trial, and was shocked to discover the amount of the buying and selling of child prostitutes in Chinatown.[27] Touched by the sufferings of these small children, Gardiner decided to remain in Victoria to help them. After the trial was over, he accepted a post as interpreter offered by the Customs House, and, with the help of the Methodist Church, set up a temporary refuge on Frederick Street, employing Mrs. M.E. Hopkins as matron.[28] After the police raided the brothels in Chinatown and took away the young girls, Gardiner placed them in his home.[29] By June 1887, it housed six Chinese girls under the care of Mrs. Marion L. Fowler, Mrs. Hopkins's successor.[30] The home was maintained mostly at Gardiner's expense, and partially with subsidy from the Methodist Church.

Without much money to carry on the home, Gardiner appealed to the Missionary Board of the Methodist Church for financial help. Impressed by his rescue efforts, the board agreed to set aside $250 towards support of the home, provided that the Women's Missionary Society managed it.[31] At its annual meeting in October 1887, the Society agreed to do so. A special committee was formed to organize the Chinese Girls' Rescue Home, with plans to raise thirty-five dollars a month to support it.[32] In December 1887, the home was established in rented premises at 54 Herald Street, and Miss Annie Leake was employed as the first matron.[33]

The Chinese Girls' Rescue Home gradually had become known not only to child prostitutes but also to child servants and concubines in Chinatown. As well, many battered wives or maids tried to escape from their husbands or masters and found their way to the home. Local Chinese merchants had no objection to the home's rescue of child prostitutes, but they resented its involvement with their maids or concubines. In China, it was not an offence to purchase a girl as a maid or concubine or to mistreat her; such a practice was, however, unacceptable in a Christian country. In rescuing the battered children, Gardiner and the matrons often had to confront infuriated masters who often took the home to court.[34]

As the number of rescued girls increased, the home required more space, and in 1887, the Women's Missionary Society decided to build a large two-storey wooden house at 100 Cormorant Street (now 728 Cormorant Street) (Plate 29). In September 1888, the Chinese Girls' Rescue Home was moved to the spacious new building, where the girls were taught English, arithmetic, geography, and other subjects, including music and needlework.[35] All the girls were converted to the Christian faith and married into Christian families of their own choice.[36]

The Rescue Home was originally intended for the rescue of Chinese women and girls, but it began to take in Japanese prostitutes in the 1890s. In 1896, Miss Kate Morgan also opened a school for Oriental children bereft of their mothers.[37] Both Chinese and Japanese widowers soon brought their children to Miss Morgan for safekeeping and education, and school attendance rapidly increased and there was much interest in it. The following year, Miss Morgan employed a Miss Churchill, an experienced educator, to take charge of the school while she ran the Chinese night school. According to Miss Morgan's evidence given to the Royal Commission of 1901, eight Japanese and forty Chinese had been rescued from "houses of ill fame" and placed in the rescue home. Of the Chinese rescued, four girls were under the age of ten; nineteen were between ten and nineteen; fifteen were between twenty and twenty-nine; one was thirty-six, and one forty-five.[38] By 1901, twenty-two had married, mostly to Christians, and some had returned to China.

In time, the home and the school became overcrowded. In April 1908, for example, the home housed seventeen Chinese girls, and the school had a total enrolment of thirty-six Chinese and Japanese pupils.[39] Fourteen were residents of the home, the remaining twenty-two non-residents. To meet the need for more space for the home's diversified activities, the Women's Missionary Society built a new two-and-one-half storey brick building next to the wooden building being used. The brick building, designed by Hooper and Watkins, was begun in April 1908 and completed six months later.[40]

On 30 December 1908, the Chinese Girls' Rescue Home, renamed the Oriental Home and School, was officially opened at the new brick building at 732 Cormorant Street.[41] The basement was used for laundry, drying, and storage. The ground floor was occupied by a large reception hall, an office, and a drawing room. There were also two dining rooms, a kitchen, a children's playroom, and a classroom. Upstairs were two large dormitories for the girls, three bedrooms for teachers, and a separate bedroom reserved as a sickroom. There was also a large attic. On the day of its official opening, the Oriental Home and School had nineteen children (nine Chinese and the others Japanese). They were under the care of the matron, Mrs. Ida Snyder, and two teachers, Miss Annie T. Martin and a Miss Smith.

After 1908, the Oriental Home and School was not only a rescue home, but also a temporary shelter for sick women and a boarding school for motherless girls. Since there were many more Japanese women than Chinese women in Victoria, the home became more important to the former, particularly after the Chinese were excluded from entering Canada. Between 1924 and 1942, the home housed more Japanese than Chinese, but in May 1942, all Japanese girls in the home were evacuated to the interior of the province, leaving only six Chinese girls in the home.[42] The home was closed and changed to a service club for men in the forces; a small home was set up temporarily for remaining Chinese girls.

RELIGIOUS ORGANIZATIONS

There were no formal religious organizations in Chinatown. Tam Kung Temple was the only important temple and was frequented by Chinese people not only from Victoria but also from other towns on Vancouver Island (Plate 30). All Saints Temple, installed in the Chinese School, was mainly for pupils to pay respect to Chinese saints such as Confucius, Huatuo, and Guanyu. The Chee Kung Tong had a shrine to Guan Yu and other deities of the Hongmen Society. Other clan associations installed their own shrines. The host society called these temples and shrines "Joss Houses."

The two significant religious organizations in Chinatown were the Chinese Methodist Church and the Chinese Presbyterian Church. The Methodist Church had been running a Chinese Mission School on Government Street since 1876. Some white people strongly objected to this missionary work. At one public meeting, for example, a speaker warned the audience that the Chinese might one day become Christians, and cries of "No! No!" came from the crowd.[43] At another meeting, a speaker criticized the mission school for teaching English to the Chinese, enabling them to compete with the whites. "If this was the only way of Christianizing the Chinese, they would have to leave them to their own devices."[44] Immediately after this remark was made, cheers of "Hear! Hear!" resounded in the assembly hall.

Despite the resentment of some white people, John Endicott Gardiner continued to extend his missionary work in the Chinese community. He felt that more Chinese people would benefit if the mission school were held in Chinatown. In 1885, with the financial support from the Reverend Percival, pastor of the Methodist Church, Gardiner approached a number of Chinese merchants to obtain pledges for additional financial help. Many contributed, including Loo Choo Fan of Kwong Lee & Company, who offered fifty dollars and promised further assistance.[45] Gardiner then rented an upstairs room in a building on Cormorant Street and furnished it with desks and benches. On 3 February 1885, the Chinese Mission School was inaugurated. In 1887, Gardiner resigned his government post, entered the ministry of the Methodist Church as a clergyman, and devoted all his time to missionary work in Chinatown.

Gardiner was much impressed with the enthusiasm of the Chinese to learn English and become Christians; he also saw a need for a Chinese church in Chinatown. With a grant of $10,000 from the Methodist Mission Board, he purchased a lot at 526 Fisgard Street and built the Chinese Methodist Church.[46] The building was a two-storey structure, with the church auditorium on the ground floor and a school above (Plate 31). It was officially opened on 13 March 1891 by the Reverend C. Bryant, president of the British Columbia Conference.[47] Every day, girls from the Chinese Girls' Rescue Home attended a service, and a weekly Bible class, Sunday school, and morning and evening services were also conducted in the church.

The Chinese Presbyterian Church was the other Christian organization in Chinatown. In April 1892, the Reverend Alexander Brown Winchester started the Victoria Presbyterian Gospel Hall for Chinese at a rented flat at the corner of Government and Cormorant streets.[48] He was assisted by C.A. Coleman, who could speak both Cantonese and Shanghai dialects fluently. They started an evening school, sabbath service, and later, a boys' day school. The obstacles to Winchester's efforts to expand the missionary work were lack of a suitable place for the mission, his inability to speak the Cantonese dialect, and his inexperience in running a mission among the Chinese in Canada. For example, he rented the Chinese theatre for two months from about mid-December 1893 to early February 1894 to hold Sunday service, when the Chinese were busy in preparation for the coming New Year festival; services thus were frequently disturbed by actors and their helpers walking in and out of the theatre. Attendance at the service gradually dwindled from several hundred to about fifty.[49] Neither the day school nor the evening school attracted many boys. In September 1894, Winchester went to Canton to pursue his Chinese studies, leaving Coleman to look after missionary work in Victoria. The next year, Winchester returned with Ng Mon Hing, a graduate of the American Presbyterian Theological School in Canton. Ng was employed as a catechist to help develop the English night school and promote missionary work among the Chinese in Chinatown. On 22 January

1899, the status of the mission was upgraded, and it became known as the First Chinese Presbyterian Church.[50] After Winchester resigned in December 1900, Dr. J. Campbell, assisted by Ng Mon Hing and Ma Seung, was given charge of the church. In 1908, the church was moved to a building at the corner of Government and Pandora streets, and throughout the 1910s, Leung Moi Fong was in charge of the church.

POLITICAL ORGANIZATIONS

The Baohuanghui (Protecting the Emperor Association) was the first Chinese political party established in Chinatown. Known to Western society as the Chinese Empire Reform Association (CERA), it was formed by Kang Yu Wei (Kang Youwei), a brilliant Chinese statesman and scholar. In an attempt to save the declining Manchu Empire, Kang wanted to establish a constitutional monarchy like Britain's and to adopt Western science, technology, and education.[51] Following Kang's advice, Emperor Guangxu issued a series of reform edicts between 11 June and 21 September 1898 for the modernization of China. However, the Empress Dowager, the emperor's mother, opposed radical reform, imprisoned her adopted son, revoked the reform edicts, and executed the pro-British reformers.

Kang escaped from China, arriving in Victoria on 7 April 1899 en route to London.[52] He was regarded by the overseas Chinese as a patriotic hero, and while in Victoria he was well received by Lee Mong Kow (Li Mengjiu), a Chinese interpreter at the Customs Office, and prominent Chinese merchants such as Lee Folk Gay (Li Fuji) and Huang Xuanlin.[53] In July, Kang returned from London to Canada to mobilize the overseas Chinese to form a political party to save the Empire and the Emperor. Throughout the summer of 1899, he toured Victoria, Vancouver, and New Westminster, established the CERA in all three cities, and advocated an uprising to overthrow the rule of the Empress Dowager and to rescue Emperor Guangxu from imprisonment. (The emperor was not put in prison, but rather was confined to an island.) As a bounty of $60,000 had been placed on him, the Canadian government assigned a Northwest Mounted policeman as an escort during his tour. Some of Kang's supporters resorted to violence when they faced opposition in Chinatown. For example, after the Reverend Chan Sing Kai, pastor of the Methodist Chinese Mission, advised his followers not to be involved in CERA activities, a bomb was placed in the church, exploding on Christmas Eve and causing slight damage to the building.[54] Although CERA members denied their association with the explosion, many people believed that it was caused by them. Throughout the 1900s, the CERA in Victoria was very active under the leadership of Lee Folk Gay, Chu Lai, and Dong Tai, and had several hundred members. In 1905, the association purchased a building at 1715 Government

Street, establishing headquarters there.[55] After the downfall of the Manchu government, however, the CERA became virtually defunct.

Another political organization in Chinatown was Jijishe (Striking Oar Society), formed in 1907 by Wu Ziyuan, Situ Yingshi, and other supporters of Dr. Sun's revolution.[56] The society was dissolved two years later when Wu left Victoria; it was replaced by Tongmenghui, a revolutionary society led by Dr. Sun. After establishment of the Republic of China in 1912, Tongmenghui was called the Kuomintang (the Nationalist League). The Victoria branch of the Nationalist League published a Chinese newspaper called *Xinminguo Bao* (The New Republic), and recruited members from the Chinese community.

Rivalry among Chinese associations was not very obvious in Victoria. The battles, or "tong wars," notorious in San Francisco, rarely occurred. Most tong wars which did break out resulted from rivalry between one political group and another. In 1908, for example, when a service in memory of the death of Emperor Guangxu was conducted by the members of the CERA, it was interrupted by members of the Jijishe. The altercation might have ended in a brawl had the police not stepped in.[57]

A more serious and lasting confrontation was between the Nationalist League and the Chee Kung Tong. After establishment of the Republic of China, the Nationalist League became a powerful political party. It set up branches in Chinatowns across Canada, challenging the leadership of the Chee Kung Tong in every Chinese community. This resulted in a conflict between the two parties, although many people belonged to both. In 1915, the Chee Kung Tong in Victoria started a "party purification." A new organization known as Dart Coon Club was formed, accepting as members only Chee Kung Tong members who were dedicated Hongmen people and who had no relationship with the Nationalist League. In other words, Dart Coon Club was an exclusive organization of the Hongmen Society, whose members had been screened. On 8 October 1916, Loo Gee Guia (Liu Zikui), alias Charlie Bo, and his Chee Kong Tong members were attacked by Hoo Hee and other members of the Nationalist League.[58] This sparked off the tong wars between the two parties throughout the late 1910s not only in Victoria but also in other cities across Canada.

OTHER CLUBS AND SOCIETIES

Besides the above organizations, other societies also emerged in Victoria's Chinatown. In 1893, for example, a group of Chinese merchants formed the Zhaoyi Gongsuo, predecessor of the Chinese Chamber of Commerce. In 1907, the Gold Field Society was formed by a few community-minded people; its objectives were to improve and develop the mental, social, and physical conditions of young men and to promote the cause of temperance and moral reform.[59] In March 1908, an Anti-Opium League was established to petition to the Canadian government to stop the import of opium and outlaw its sale in Canada.[60] This organization was formed in response to the international opium suppression effort.

As locally born Chinese children increased in number, they began to form their own organizations. In 1909, a group of Chinese youth organized an athletic club which was so popular that, soon after it was formed, ninety members had enrolled.[61] Two years later, a group of Chinese students formed the Chinese Football Team, with fourteen members.[62] Then in the same year, a group of locally born Chinese established the Chinese Young Men's Progress Party, which was involved in activities such as organization of a Chinese library and the fight against gambling in Chinatown.[63]

CHINESE CONSOLIDATED BENEVOLENT ASSOCIATION

During the early 1880s, Victoria's Chinatown, the biggest in Canada, had the largest number of voluntary associations. Of them, the most important was the CCBA. Some prominent merchants in Victoria, such as Li Youqin, Huang Yanhao, and Lo Zhuofan, felt that since there was no Chinese consulate in Canada, the Chinese people did not have an official representative to act on their behalf in dealing with the Canadian government. Accordingly, in March 1884, they sent a letter to Huang Tsim Hsim (Huang Zunxian), Consul-General in San Francisco, requesting his approval of the establishment of the CCBA in Victoria and his endorsement of it as a representative organization for the Chinese in Canada.[64] Soon after the request was granted by the Consul-General, the organizing committee of the CCBA sent out copies of a notice to all the Chinatowns in British Columbia, urging the people to support the association by donating at least two dollars. The notice included a warning that anyone who did not contribute would have to pay ten dollars to the association before he was permitted to return to China. In those days, all vessels leaving Canada for Asia departed from Victoria, the largest port on the Canadian Pacific coast, and as most Chinese did not know English, they had to rely on Chinese shipping agents to arrange any journey. In addition, all Chinese agents were either founders of the CCBA or business partners or friends; the CCBA could make a special arrangement with agents not to sell tickets to any Chinese who failed to present a donation receipt issued by the association. As a result, nearly all Chinese had to make a contribution in order to get a donation receipt.

The CCBA registered with the provincial government on 9 August 1884 and raised nearly $30,000 within a year for construction of an association building and a Chinese hospital. The three-storey brick building, at 554-62

Fisgard Street, was completed in July 1885.[65] The ground floor held stores, the second floor was used for the association's office, and the third floor housed a temple (Plate 32).

The CCBA was a community-wide organization since its directorate consisted of representatives from county, dialect, and clan associations, political parties, church groups, and a great variety of other clubs and societies. It set a pattern of co-operation across the spatial, clannish, religious, and political lines, attempting to unite all factions of the Chinese community. Its objectives, outlined in its constitution, were to promote a close relationship among the Chinese in Canada, solve their disputes, eliminate internal troubles and vices, help the sick and poor, and fight against foreign oppression. The CCBA organized a great variety of activities for the benefit of Chinese not only in Victoria but also in other parts of Canada, such as fighting for the repeal of all kinds of discriminatory regulations and laws against the Chinese. The CCBA soon became a spokesman for Chinese interests in Canada and a de facto government in Chinese society: it made legislation, exercised jurisdiction, enforced regulations and orders, and protected the interests of the Chinese in the country.

THE CHINESE HOSPITAL

The CCBA operated two important institutions in Chinatown, namely the Chinese Hospital and the Chinese School. A small wooden hut was rented and used as the Taipingfang (Peaceful Room).[66] It functioned as a "hospital" for poor, sick Chinese men who had no one to look after them. Female patients were not sent to the hospital because they were cared for at home by their families or by their procurers if they were prostitutes. For fifteen years, Taipingfang was housed in a cold, dark wooden hut until the CCBA raised enough funds to purchase a lot at 555 Herald Street, behind the association building. The Zhonghua Yiyuan (Chinese Hospital), a two-storey brick building, was built; it was officially opened in the winter of 1899.[67] The hospital, with twenty beds, accepted male patients not only from Victoria but also from other cities in British Columbia.

The hospital was financed by donations, both compulsory and voluntary. The CCBA ruled that whenever a Chinese left Canada for China, he was obliged to give two dollars to the hospital. Donors would be given receipts by the association in Victoria or its representatives in Vancouver or Nanaimo. A special arrangement was made with all Chinese shipping agents whereby a boat ticket to China would be sold to a Chinese only upon his presentation of the hospital's donation receipt. On the day of departure, the hospital caretaker would go to the pier to collect the donation receipts. In other words, for every trip a Chinese made to China, he had to make a separate donation to the Chinese Hospital. However, poor, elderly Chinese returning to China for retirement were exempt from the compulsory donation.

During the first ten years of its operation, the Chinese Hospital received sufficient compulsory donations to pay for its expenses, but after the 1900s its revenues were drastically reduced for two reasons. First, it ceased to be the only Chinese hospital which issued donation receipts as "departure permits." In 1908, for example, the Chinese Hospital in Vancouver's Chinatown accepted donations and issued receipts which were equally valid as "departure permits." The following year, the Chinese Hospital in New Westminster followed suit. Thus donations to Victoria's Chinese Hospital were drastically reduced. Second, some organizations strongly opposed the hospital's compulsory donations. In 1915, for example, the United Chinese Christian Society in Vancouver wrote to the Customs Office in Victoria to complain that some members were prevented from boarding the ship for China because they refused to make a donation to the Chinese Hospital in Victoria.[68] Having received this complaint, Customs officers in Victoria did not permit the CCBA's representatives to come to the pier and collect donation receipts. Since the CCBA could not enforce compulsory donation, income for the hospital was thus greatly reduced.

THE CHINESE SCHOOL

The Chinese School was another important institution under the administration of the CCBA. In the early 1880s, there were few children in Chinatown, and most came from rich merchants' families and were taught at home by private tutors. But by the 1890s, there were about a hundred Chinese children in Chinatown, and the CCBA saw a need to provide them with Chinese education.[69] After several meetings, the CCBA decided on 8 January 1899 to operate a free school on the top floor of its building. With the enthusiastic support of the Chinese community, the CCBA raised more than $3,000[70] and the school, known as Le Qun Free School, was opened on 1 July 1899, attended by thirty-nine pupils.[71] They were grouped into two classes under two Chinese scholars recruited specially from Canton to teach the children Confucian ideals. The Four Books (*The Great Learning, Confucian Analects, Doctrine of the Mean,* and *Mencius*); and *Five Canons* (Canons of Changes, History, Rites, Odes, and the Spring and Autumn Annals) were taught to senior pupils in Class A, and the younger children in Class B studied the *Trimetrical Classic,* the *Book of Family Names,* and other elementary classics. Although all the pupils were Chinese, Lee Mong Kow, one of the school promoters, told a newspaper reporter that the school was also open to white pupils.[72]

In addition to Le Qun Free School, a few Chinese schools or classes were organized by other associations. For example, the Chinese Empire Reform

Association ran Aiguo Xuetang (Patriotic School); the Chinese Freemasons operated Qinge Xuexiao (Educating Youth School); and the Methodist and Presbyterian Church organized English classes for Chinese children as well as adults. A few wealthy merchants employed English tutors to give English lessons to themselves as well as to their children at home.

In the summer of 1907, the Victoria School Board passed a ruling that Chinese children would not be accepted in public school unless they were sufficiently acquainted with English.[73] As a result, all Chinese who could not enter public school went to Le Qun Free School instead, and it became overcrowded. In February 1908, the CCBA decided to build a new school to accommodate the increasing number of Chinese children;[74] within half a year, it had raised more than $7,000 to purchase a lot at 636 Fisgard Street.[75] The school, called Zhonghua Xuetang (the Imperial Chinese School), was officially opened on 7 August 1909 by Xu Bingzhen, Chinese Consul-General at San Francisco.[76] Unlike Le Qun Free School, the Imperial Chinese School had both English and Chinese classes and concentrated on improving the English standard of Chinese children so that they could be admitted to public schools.[77]

THE CHINESE CEMETERY

Another important function of the CCBA was to administer the Chinese Cemetery. In 1891, it purchased a piece of land covering an area of 3.54 hectares (8.75 acres) on the south slope of Christmas Hill facing Swan Lake.[78] Originally, it was to be used as a burial ground but Westerners in the neighbourhood objected to this use and the area remained empty.[79] The Chinese were buried in the low-lying southwest corner of Ross Bay Cemetery facing Ross Bay. During heavy storms, the wind and high waves eroded waterfront graves, sweeping away Chinese remains.[80] In view of this situation, the CCBA decided in May 1902 to sell its land on Christmas Hill and use the money to buy another piece of land for a cemetery.[81] The following year, the association purchased a lot covering an area of 1.4 hectares (3.5 acres) on a headland known as Foul Point between Foul Bay (Gonzales Bay) and Shoal Bay (McNeill Bay), establishing the Chinese Cemetery there.[82] In it, a brick house was built for the storage of bones sent to Victoria from Chinese communities across Canada, and once every seven years, these bones were shipped back to China.

DISCRIMINATION

Until 1947, Chinese were restricted in their commercial activities and employment opportunities in British Columbia because they were not on the provincial voters' list. For example, they could not sell liquor because licences were issued only to voters.[83] They could not be lawyers or pharmacists in British Columbia because anyone not on the voting list was ineligible to become a student or apprentice in those fields. Similarly, the Chinese were not entitled to a hand-logger's licence.[84] Thus the host society could use the voters' list as a way of discriminating against the Chinese.

Although British society in Victoria was proud of its sense of justice and fair play, this principle was denied to its Chinese residents; discrimination was rampant. In 1899, for example, City Council did not permit the Chinese community to hold an orchestra parade in the street, arguing that it was wrong to "patronize Chinese music when so many competent white musicians were to be found in the city."[85] In 1904, when Lee Mong Kow and his Chinese friends went to a show in the Victoria Opera House, they were told to sit in the gallery because no Chinese were allowed on the lower floor even though the tickets indicated no seating restrictions.[86] Another time, a store on Government Street put up a notice that "Chinese purchasers are strictly prohibited every Saturday night from 7 to 10 p.m." because the manager did not want Chinese men in the store while it was patronized by white women.[87]

The most common form of discrimination against the Chinese was employment. John Kurtz, a cigar manufacturer, vowed that he did not use slaves or Mongolians in production of his cigars, and he inscribed "No Chinese Labor" on his trademark.[88] Another cigar manufacturer, the T & L Company, advertised itself as a "white labor cigar factory" (Plate 33). Both companies were members of the White Cigar-Makers' Association of the Pacific Coast. In another example of employment discrimination, the City of Vancouver signed a contract with the British Sugar Refining Company in August 1890, giving the company a land grant, an exemption from municipal taxes, and an exemption from water rates if the company met several conditions, one of which was that "the company shall not at any time employ Chinese labour in and about the said works."[89]

After the 1900s, protests against the employment of Chinese workers grew stronger. In 1909, for example, the Victoria Labourer's Protective Association protested to City Council that Chinese labour had been employed in the construction of the Store Street pavement.[90] A year later, the Painter's Union of Victoria notified all contractors in the city that their members would not be permitted to work on buildings on which preliminary paintings and inside work were done by Chinese labourers.[91] Craft industry unions sent a deputation to provincial ministers requesting that carpentering and plastering not be done by "Celestials."[92] In those days many unions did not realize that they would have been more powerful if they had accepted Chinese workers as members. In July 1912, for example, when the woodworking trade union struck for shorter hours and better pay, the strike was more effective after a number of Chinese workers walked out of the mills to support their fellow white workers.[93] In this particular case, some white workers began to see the advantage of enlisting the support of Chinese workers, but most white workers thought otherwise.

Like Chinese labourers, Chinese merchants also faced protests from white merchants whenever they obtained government contracts. In 1913, three white-run laundries in Victoria protested strongly that washing for Rodd Hill and Work Point barracks was given to Chinese laundrymen simply because the white laundries wanted more money.[94] The same year, a local white caterer complained that a Chinese caterer provided the food supplies, and cooked and served the meals for workers at Albert Head Quarry.[95]

LAND OWNERSHIP PATTERNS

In the early 1880s, most Chinese merchants leased their business premises from Caucasian landlords. Only a few rich merchants could afford to own property in Chinatown. According to the Royal Commission Report of 1884, for example, only four Chinese people owned land in Chinatown, of whom Loo Chock Fan and Loo Chew Fan of the Kwong Lee & Company were the biggest landlords, with ten and a half lots.[96] In 1884, the Loo brothers disagreed about financial matters and brought their disputes to court.[97] Eventually they went bankrupt because of court expenses, and in 1887, Kwong Lee & Company was closed, with all the Loos' properties in Chinatown going up for auction.[98]

After the turn of the century, the pattern of land ownership underwent great changes. The landlord class, formerly made up of wealthy merchants from San Francisco, was replaced by the proprietors of small grocery stores, laundries, cafés, or restaurants who had either saved sufficient money themselves or pooled their resources to acquire land. By 1900, real estate owned by the Chinese in Victoria amounted to $296,000.[99] Some Chinese merchants even participated in land speculation when real estate prices rose in the early 1900s; most relied on heavy cash borrowings or carried large mortgages on their properties. By 1909, the Chinese people owned about half the land in four city blocks (see Figure 39). The major land investors were Chan Tong Ork, Lim Dat, Lee Dye, Lee Mong Kow, and Lee Fok Gay, who each owned three or more lots in Chinatown. In addition to these individual owners, various associations such as Chee Kung Tong, Gee Tuck Tong, Fung Toy Tong, Hook Sin Tong, the Chinese Empire Reform Association, the Hoy Sun Association, and Lee's Benevolent Association, also purchased land, erecting two- or three-storey brick buildings on their properties.

TABLE 36: Chinese business in Victoria, 1885, 1911

Business	Number of firms 1885	1911
Import and export	23	57
Tailoring	11	27
Groceries	6	16
Shoe-making and shoe-repairing	1	9
Laundries	9	8
Barbers	5	8
Employment agents and contractors	—	4
Restaurants	—	6
Bakery	1	3
Herbalists	4	1
Others	2	16
Total	62	155

Sources: compiled from *Victoria City Directory*, 1885, 1911

ECONOMIC GROWTH

During the blooming period of Victoria's Chinatown, a new class of small businessmen emerged. When a store employee or a street pedlar accumulated enough savings, he would invite his relatives or friends to start a new business together. Small cafés, stores, and other businesses thus multiplied in Chinatown. In 1885, for example, there were sixty-two business concerns in Chinatown; by 1911, that number more than doubled (Table 36). Many firms imported dry goods, opium, and other Chinese merchandise from China and redistributed them to other Chinatowns in the province. Tailors, grocery stores, cobblers, and laundries were numerous. Since Chinese merchants found it difficult, if not impossible, to obtain loans from Western banks, the alternative way to obtain investment capital was through partnership, usually of father and sons, uncles and nephews, brothers, in-laws, or close friends from the same native village from China. The business was usually run by a team in order to save the cost of employees. According to the Royal Commission Report of 1901, the average number of partners per firm was six in opium companies, five in wholesale import companies, and four in retail stores (Table 37).

In addition to Chinese business, many white-operated hotels and saloons were established in or near Chinatown as the result of the building of two railway terminals in the area. In 1887, the Esquimalt and Nanaimo Railway, which terminated on Russell Street in Esquimalt, was extended across the Inner Harbour of Victoria to end at the depot on Store Street facing Cormorant Street.[100] In 1902, another railway line, the Victoria and Sidney Railway,

TABLE 37: Patterns of partnership in Chinese business in Victoria, 1900

Business	No. of firms	No. of partners	Average no. of partners per firm
Wholesale import	14	64	4.6
Retailing	22	78	3.5
Opium importing and manufacturing	3	18	6.0
Dry goods retailing	12	19	1.6
Clothing manufacturing and import	5	14	2.8
Tailoring	14	17	1.2
Restaurants	9	15	1.7
Butchering and poultry	7	15	2.1
Drugs and general merchandise	5	12	2.4
Rice mills	4	4	1.0
Cannery contracting	4	15	3.8
Others	10	17	1.7
Total	109	288	2.6

Source: compiled from Report of the Royal Commission on Chinese and Japanese Immigration into British Columbia, 1901 (Ottawa: S.E. Dawson 1902), 212

which terminated on the north side of Hillside Avenue, was extended southward and ended at the site of the city's market building.[101] The original plan of the V and S Railway was to extend its line through Chinatown to join the E and N Railway terminus at Store Street, but after its request for the connection was turned down by the E and N Railway, the V and S Railway Company decided to change the location of its terminus in Chinatown. From the operational point of view, such a location was not ideal because the northbound train leaving the terminus had to climb a 2.3 per cent grade up Fisgard Street and then take a sharp fifteen-degree curve onto Blanshard Street.[102] Accordingly, in December 1910, the Market Building was vacated and the tracks in Fisgard Street were removed. A new station was built on the west side on Blanshard Street, just north of Fisgard Street.[103] After the railway terminus was removed, the hotel and saloon businesses in Chinatown declined.

OPIUM TRADE AND MANUFACTURE

The manufacture and sale of opium was by far the most profitable and important business in Victoria's Chinatown. Crude opium from India was exported to Hong Kong, where it was shipped to Canada and the United States. During the 1860s and 1870s, Tai Soong and Kwong Lee were the two large opium importers and manufacturers in Chinatown. Behind their stores on Cormorant Street, the manufacturing was carried out in small sheds. Balls of crude opium were cooked in boiling water for about twelve hours until they were converted into jelly form and then canned for sale. Many Chinese smoked opium, and the manufacturers had a large market in Canada and the United States.

The opium industry began to boom after February 1887 when the American government passed an act to prohibit the Chinese from importing opium into the United States and American citizens from trafficking in opium in China.[104] Opium which was sold for ten dollars a pound in Victoria was worth twelve dollars in London, Ontario, and would sell for twenty-five dollars in Chicago.[105] Thus smuggling opium into the United States became a profitable trade after 1887.

In British Columbia, opium was smuggled into the United States on vessels travelling between Victoria and American coastal ports such as Seattle, Tacoma, Port Townsend, Portland, and San Francisco. Another common route was to ship opium from Victoria to Hope, where it was taken to the Similkameen mines or to the Kootenay mines. From these mines, it was easy to smuggle the opium into the United States. Smuggling was also active in Chinatowns near the American border. For example, opium was delivered to Winnipeg, sent on to St. Paul and Minneapolis, or carried by rail to Windsor and then ferried across to Detroit. Sometimes opium was carried by rail to a location west of Winnipeg and then smuggled across the border into Montana and North Dakota.[106]

The demand for opium in the United States resulted in the prosperity of the opium industry in Victoria's Chinatown. In 1884, there were about six opium factories; by 1887, thirteen factories were in operation: Fook Yuen, Tai Soong, King Tye, Kwong On Lung, Bow Yune, Tai Yune, Hip Lung, Kwong Chung, Kwong On Tai, Tun Chung, Chu Ching, Lung Kei, and Sing Wo Chon (Figure 40). Together they produced an average of nearly 90,000 pounds of opium each year, and competition among them was keen.[107] For example, the Sing Wo Company, a large opium-manufacturing company in Macao, produced the finest opium, known as the "Tai Yuen brand." In 1886, it started a branch in Victoria under the name of the Sing Wo Chon Company. In an attempt to drive out its competitors, the Sing Wo Chon Company reduced the price of its opium from ten dollars to seven dollars per pound, suffering a loss of $40,000 within three years.[108] During the 1900s, several small factories could not compete with the strongly financed companies and had to close. By 1907, only four opium factories were still in operation: Tai Yune, Quong Man Fun, Kwong On Lung, and Shon Yuen.

The fully fledged international anti-opium movement in the 1900s also played an important part in the decline of the opium trade in Victoria. In 1906, the British parliament passed a motion condemning the opium trade as morally indefensible and requested the Indian government to bring it to a speedy close.[109] The same year, an imperial decree was issued by the Manchu

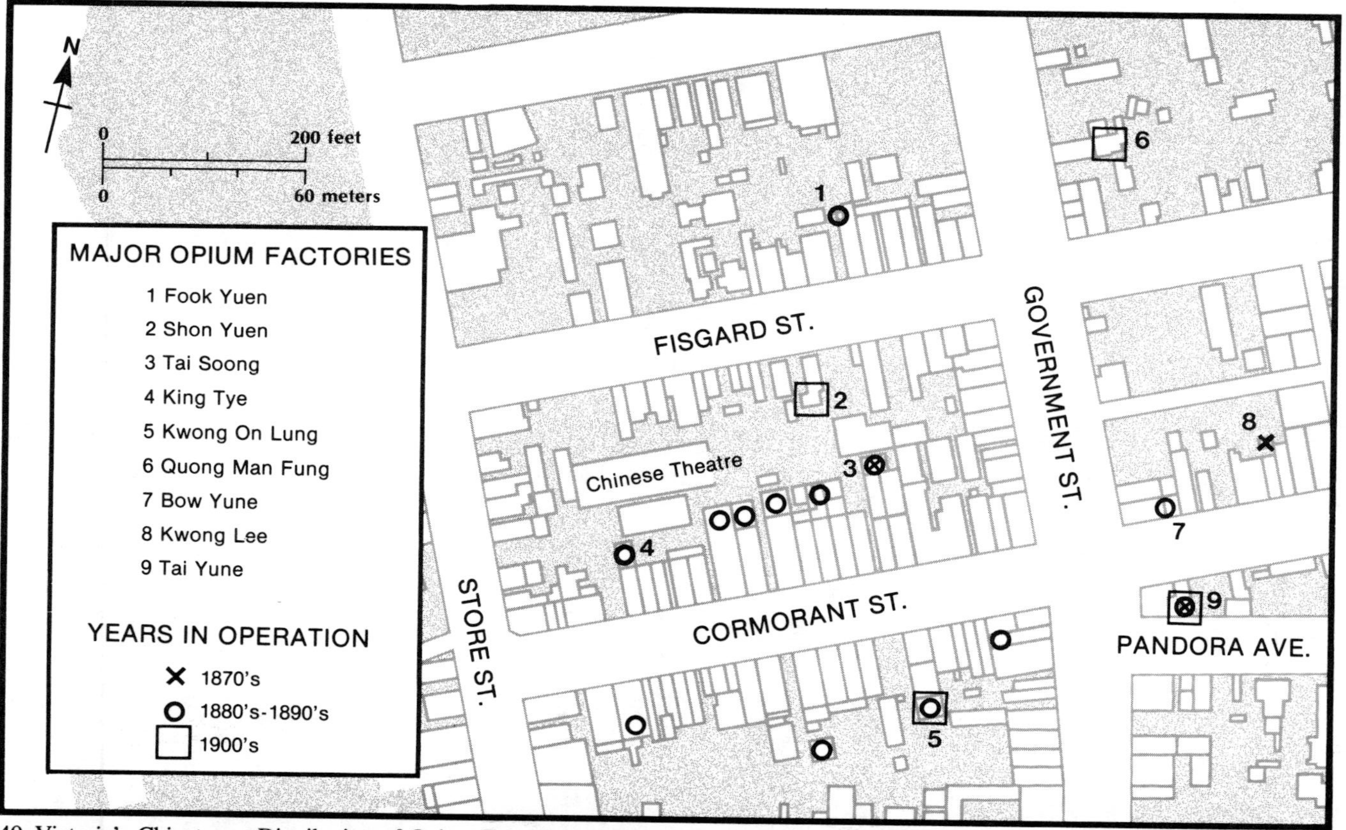

40 Victoria's Chinatown: Distribution of Opium Factories, 1870s–1900s

government commanding that the evils arising from opium be eradicated within ten years and the import of opium from India be ended by annual reduction. On 1 January 1909, an International Opium Commission met at Shanghai and was attended by representatives of thirteen governments (Austria-Hungary, China, Great Britain, France, Germany, Italy, Japan, The Netherlands, Persia, Portugal, Russia, Siam, and the United States).[110] At the conference, poppy cultivation and opium-smoking were condemned, and all countries were urged to adopt reasonable measures to stop the shipment of opium to "any country which prohibits the entry [of opium]" and to close opium dens in their concessions or settlements in China.

Until 1908, the opium business was legal in Canada. Neither the federal nor provincial governments banned the industry, mainly because it was an attractive source of income to them.[111] In 1906, for example, the value of crude and powdered opium import amounted to about $321,000; duty on it was nearly $47,500.[112] The annual licence fee on an opium firm in Victoria was $250 in 1886 and double that the following year.[113] In May 1908, W.L. Mackenzie King, Deputy Minister of Labour, was sent from Ottawa to Vancouver to investigate the Chinese and Japanese claims arising from riots the previous fall, and to his surprise, two Chinese opium merchants submitted claims for $600 each as compensation for business losses caused by the closure of their companies during the riots. In his report, he expressed dismay that the government permitted this industry, which was not only a source of human degradation but also a destructive factor in national life.[114] His report resulted in the passing of a federal act on 20 July 1908 to prohibit the importation, manufacture, and sale of opium in Canada.[115] Thus, after 1908, opium was no longer processed openly in Victoria's Chinatown.

LAND UTILIZATION

In the early 1880s, Cormorant Street between Store and Douglas streets was the commercial centre of Chinatown, where nearly half of Chinatown's business was concentrated. Behind the commercial façades of Cormorant Street were numerous wooden shacks or tenement buildings where most Chinese lived. At one time, Chinatown had three Chinese theatres, four temples, and a Chinese garden (Figure 41).

After the 1890s, the centre of Chinatown gradually expanded from Cormorant Street to Fisgard Street. The CCBA on Fisgard Street helped increase activity on the street as new arrivals from China came to register with the association, pupils came to the school, and worshippers came to its temple.[116] At the rear of the association building was the Chinese hospital, accessible from Fisgard Street through a narrow alley. The Chinese Methodist Church, the Imperial Chinese School, the Chee Kung Tong, and some other

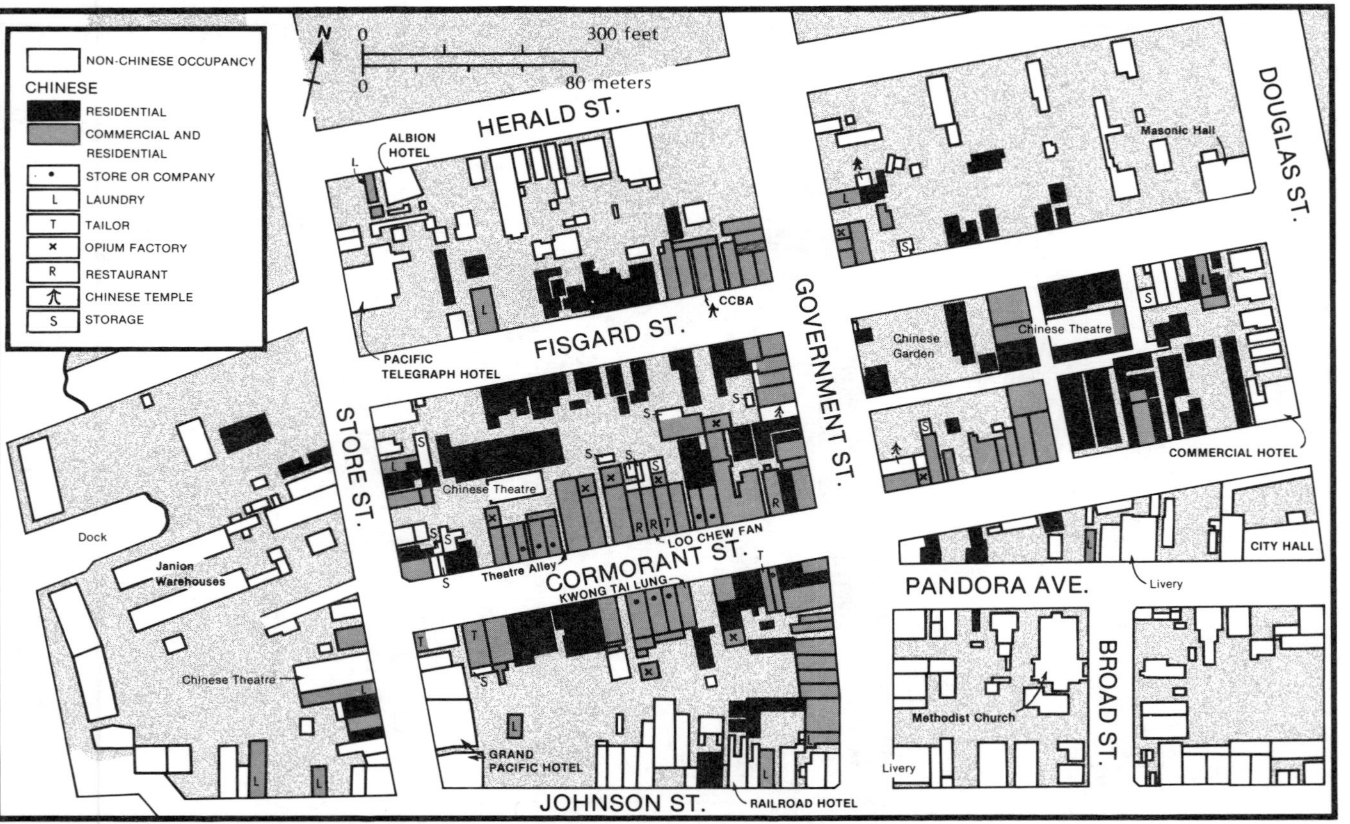

41 Victoria's Chinatown: Land Utilization, 1885

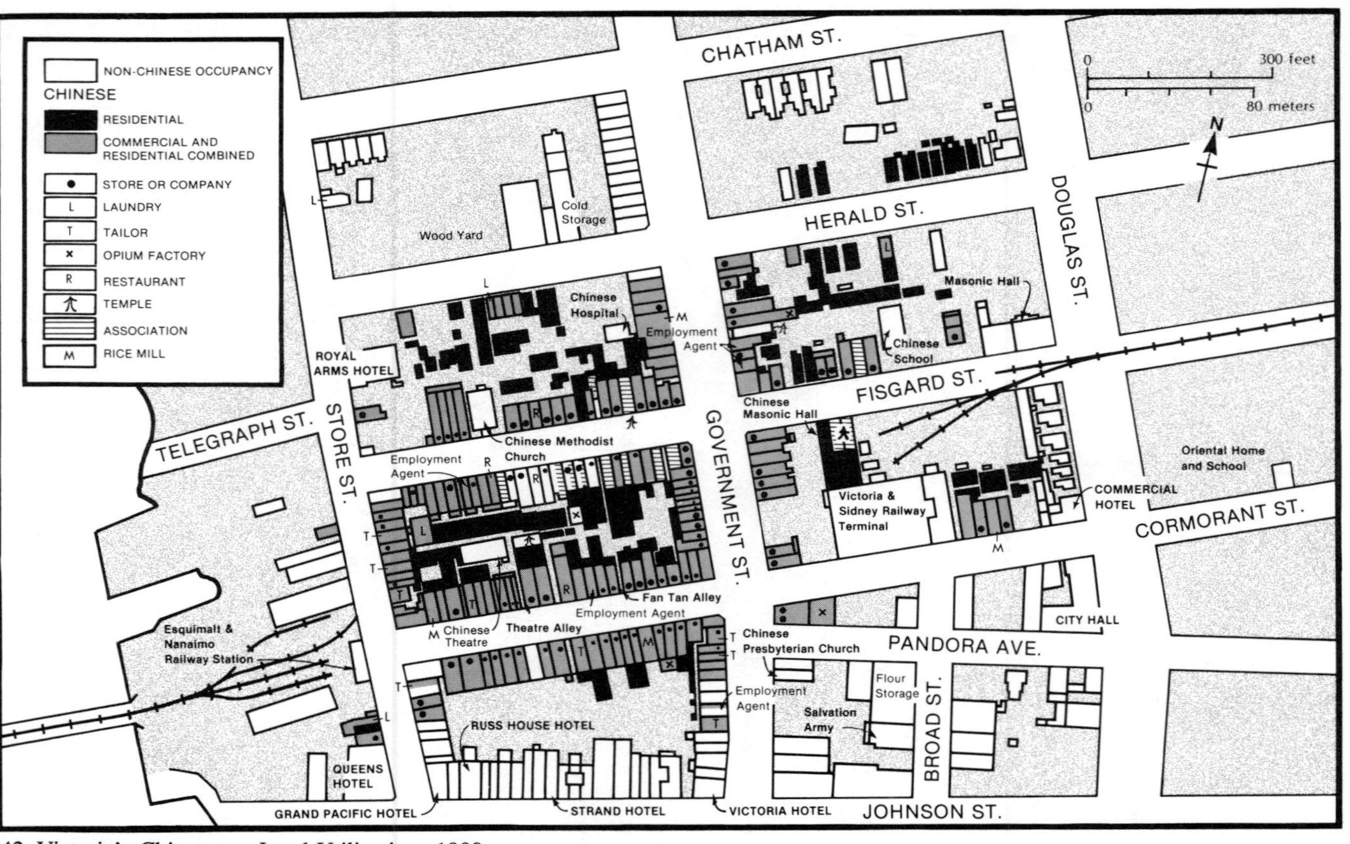

42 Victoria's Chinatown: Land Utilization, 1909

associations were also located on Fisgard Street (Figure 42). Because of all these developments, it attracted an increasing number of stores and residents. By 1911, of the 153 business concerns in Chinatown, about one-third were on Fisgard Street, another third on Cormorant Street, and the remaining third in other parts of Chinatown.[117] The old, established import and export companies and stores still remained on Cormorant Street, but new small grocery stores, fish markets, and restaurants were on Fisgard Street.

Although many upper floors of Chinatown buildings were used for residential accommodation, the major residential areas were at the rear of the commercial façade. The closely packed wooden cabins and tenement buildings, separated by narrow alleyways and tiny enclosed courtyards, provided accommodation for the poor. In September 1891, for example, a population survey revealed that the wooden cabins provided accommodation for 652 Chinese tenants.[118] All lodging cabins were partitioned into small rooms. Each room contained two-tiered bunks providing accommodation for four or more bunkmates. This violated the city's cubic air space bylaw, which required that each occupant should have a space of eleven cubic metres (384 cu. ft.). In other words, a room which measured about 2.4 metres square and two metres high (eight feet square and six feet high) should accommodate only one person. If several tenants shared a room this size and were caught by the sanitary or health inspectors, tenants would be fined.

In a passive attempt to stop the frequent raids, the CCBA finally passed a resolution in November 1893 that if any person were arrested during a raid on living quarters and sentenced to a fine for overcrowding, he should not pay the fine but should go to jail.[119] After his release from prison, the association would reward him with ten dollars. If anyone opted to pay the fine instead of going to prison, he would be fined ten dollars by the association. Unaware of this resolution, inspectors raided Chinatown and made arrests as usual. All those arrested took the jail sentence; this posed a problem for the police because the tiny city jail could not accommodate so many people at one time. Furthermore, if all of them were put into the small jail, the city would violate its own cubic air bylaw. Accordingly, all the arrested Chinese were released. After this incident, raids on Chinatown for overcrowding declined.

Poverty was a major cause of overcrowding in Chinatown. In 1900, the monthly rent for a room in a lodging house was about three dollars; most labourers could not afford this payment and had to share the room.[120] Usually five people shared it, each paying fifty cents a month. When a labourer was unemployed, he would stay with friends or relatives in their already congested rooms. The unheated living quarters were cold in the winter, so two or more people would share the same bed and cotton quilt in order to keep warm. During a severely cold winter, it was common for occupants to be crowded around the stove in one section of the room for warmth, leaving the

43 Victoria's Chinatown: Distribution of Brick Buildings, 1885

rest of the room empty. This overcrowding was exacerbated when labourers laid off until the summer had no work, money, or place to live.

TOWNSCAPE AND IMAGES

During the 1880s, the townscape of Chinatown was dominated by several three-storey brick buildings above a mass of wooden cabins (Figure 43). In May 1884, ten brick buildings were constructed in Cormorant Street, four owned by Kwong Lee & Company, three by Wang Foong & Company, and three by Carlo Bossi.[121] These buildings were constructed in the Italianate style and characterized by decorative cornices and windows and arcading on the ground floor. Their wooden balconies projected over the wooden sidewalk (Plate 34). On the opposite side of the Loo Chew Fan Building was an impressive brick building owned by the Kwong Tai Lung Company, a large import and export company in Chinatown. It was also fronted with a wooden balcony and had a heavy strong basement on the northern bank of the Johnson Street ravine (Plate 35). The most important brick building in Chinatown was the CCBA Building at 554 Fisgard Street (Plate 36). It was designed by John Teague, a famous local architect. Construction began in May 1885 and was completed two months later.[122] The front and rear façades of the brick building were wooden verandahs extending the full height of the structure. Small stores occupied the ground floor, the association office was on the second floor, and a free school and a temple filled the third floor. At one time, a room adjoining the temple was used by Reverend Gardiner for conducting the Chinese Mission School.[123]

Within the city block bounded by Cormorant, Fisgard, Store, and Government streets were densely packed wooden shacks, narrow alleys, and tiny enclosed courtyards. Many of the rickety wooden alleyways were held up by sticks and littered with garbage. In some partly covered passageways, holes were dug in the ground to permit waste water to pass through.[124] The enclosed courtyards were the only open space inside the block, where the residents dried their laundry, grew vegetables, and kept poultry.

The city government constantly condemned the overcrowding and unhygienic living conditions of Chinatown, but did nothing to improve them. Chinatown always had been viewed as a separate slum for only the Chinese, not as part of the urban fabric of the city. It was the threat of bubonic plague raging in Honolulu that prompted Victoria to improve its Chinatown. In December 1899, many buildings in Honolulu's Chinatown were burnt because of the plague.[125] When the news reached Victoria in January 1900, the public feared that the plague might gain a foothold in Victoria's unsanitary Chinatown.[126] In a clean-up of Chinatown, garbage was removed from alleyways and courtyards and some filthy wooden shacks were demolished. But as the

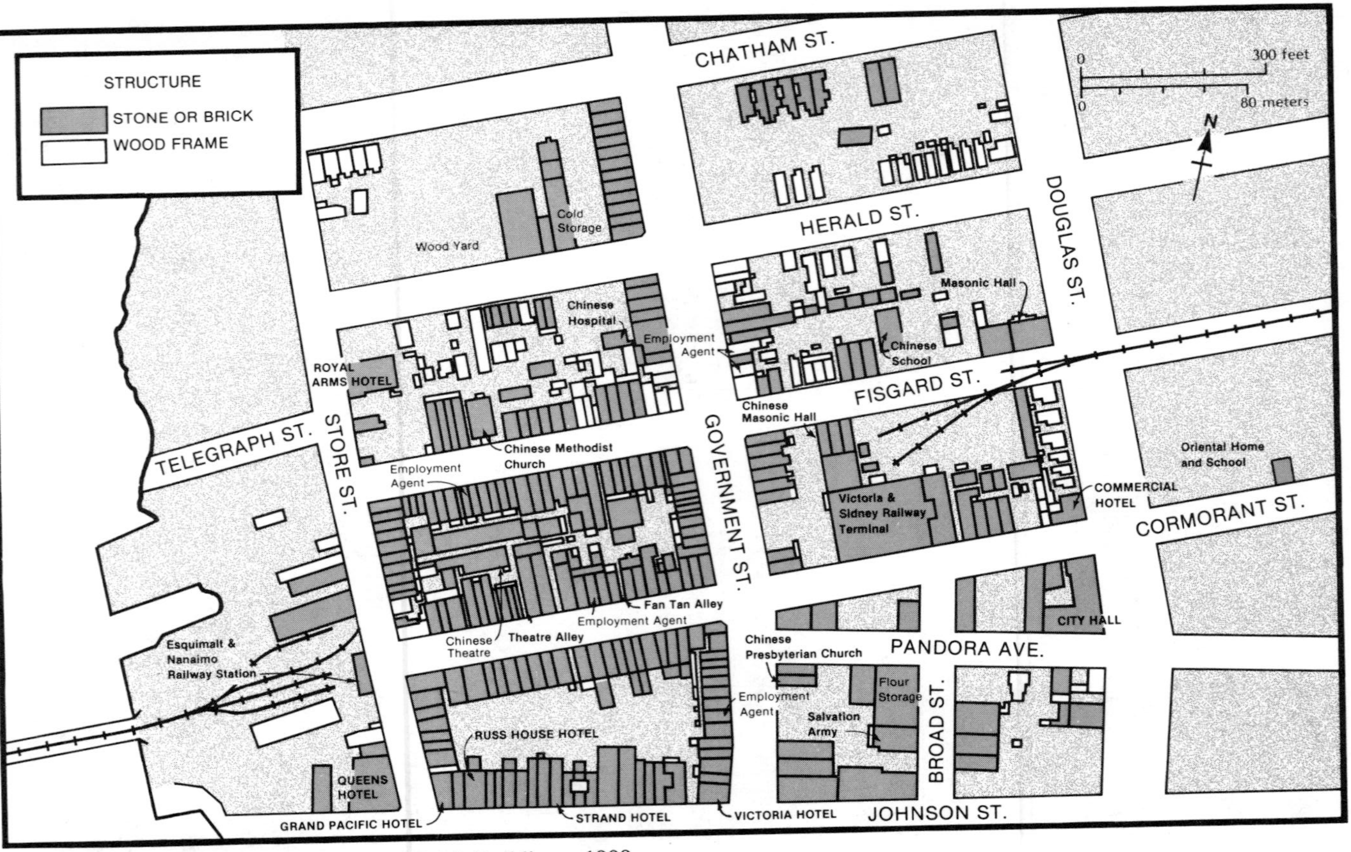

44 Victoria's Chinatown: Distribution of Brick Buildings, 1909

fear of plague subsided, Chinatown was again left alone.

The major transformation of the Chinatown townscape did not take place until October 1907, when two rats infected with bubonic plague were caught in Seattle.[127] Victoria City Council immediately carried out drastic improvement in Chinatown. Refuse in the filthy alleys and courtyards was collected and burned in a public incinerator. Fifty pounds of arsenic-poisoned oatmeal were spread under all wooden sidewalks and around cabins and other wooden structures throughout Chinatown. Unsanitary wooden sheds on platforms, dilapidated back verandahs, and rear fences which might be fire and health hazards were pulled down, and many outbuildings on the northern bank of the Johnson Street ravine were demolished as well.

In the following two years, the streetscape of Chinatown was improved further, as wooden sidewalks were replaced by concrete, hundreds of wooden shacks were demolished, the Johnson Street ravine was reclaimed, and 200 sewage connections were made. Many frame structures were replaced by brick buildings, which were constructed in rows, adjoining and backing onto one another and separated by covered or partially covered passageways or alleys (Figure 44). One of the most impressive new buildings in Chinatown was the school building at 636 Fisgard Street. It was designed by D.C. Frames, who formerly had worked with Francis Rattenbury, a famous Victoria architect.[128] This two-storey pagoda-like structure is set back twelve metres (forty feet) from the street and has a wrought-iron railing in front. The ground floor is reached from the main entrance by a staircase, and the building is decorated with Chinese architectural features such as tile-roofing, upturned eaves and roof corners, decorated wooden balconies, and an extended eave covering the main balcony. These features are incorporated with Western decorative elements such as Italianate cornices, Gothic trifoils, and Tudor arched windows.

Much of the vitality of the Chinatown streetscape was created by the activities and congestion of streets by people, and the noise and smell from restaurants, opium dens, stores, and association buildings. During the celebration of Chinese festivals, streets would be colourfully decorated. In the Chinese New Year, for example, red lanterns were hung outside most Chinese stores and colourful flags waved from the poles of various association buildings. Streets were littered with the red debris from firecrackers. Some buildings were freshly painted in red and other bright colours. In the evenings, Chinatown was lit up and celebrations went on in temples, theatres, association halls, and homes. After the Chinese New Year celebration, Chinatown was usually quiet and dull except for special occasions, as when the Marquis of Lorne visited Victoria in 1882.[129]

Gambling dens were usually housed at the rear of premises and accessible only from narrow alleys. Their operators installed various means to hinder police raids, providing escape routes for the gamblers. For example, some alleys were blocked by two or three doors, one after another, and had many exits. Gambling rooms had false partitions and barred windows with iron bars that could be removed so gamblers could escape to the outside alleys. Two doors were built to a washroom so that a fugitive could enter one door and disappear through the other. Thus, behind the street fronts, Chinatown was covered by an intricate network of alleys and courtyards, like a labyrinth. This hidden section, "the forbidden city" of Chinatown, open only to the Chinese people, was a unique section of the townscape in Victoria's Chinatown.

10

The Withering Period, 1920s–1970s

The population of Victoria's Chinatown began to decline after the passing of the Exclusion Act of 1923, and it continued to decrease until Chinese immigration was resumed after repeal of the act. But while the Chinese population in Victoria increased gradually, the depopulation of Chinatown did not stop. Unlike Chinatowns in big cities such as Vancouver and Toronto, Victoria's Chinatown did not receive a great share of the new Chinese immigrants in the 1960s and 1970s. As well, few Chinese entrepreneurs from Hong Kong and Taiwan invested in Victoria, mainly because the city, separated from the mainland, had no large local market nor did it have a large population base surrounding it. The economy of Chinatown continued to decline after the 1930s depression.

DEPOPULATION

In 1911, Victoria's Chinese population was surpassed by Vancouver's. Between the census years 1911 and 1921, the Chinese population in Victoria dropped slightly from 3,458 to 3,441, while that in Vancouver increased from 3,559 to 6,484. During the 1920s, many unemployed Chinese miners in Nanaimo and Cumberland went to Victoria or Vancouver to look for jobs because the coal industry was declining, so Victoria's Chinese population increased slightly to 3,702 in 1931. But as jobs were scarce in Victoria, many Chinese left for Vancouver.

Death, departure, and decentralization were the three major factors in the depopulation of Victoria's Chinatown. Deaths were caused primarily by old age or disease, of which tuberculosis was a leader. According to a report by Dr. Richard Felton, Victoria's medical health officer, the annual death rate from tuberculosis among the Chinese averaged about ten persons per thousand from 1923 to 1927 and rose to 12.6 from 1928 to 1932.[1] As the living environment in Chinatown improved, the average death rate dropped to six persons per thousand per year between 1934 and 1945, still triple that of the rest of the city's population.[2] Deaths could not be balanced by births because there were few adult Chinese women. Accordingly, there was little or no natural increase in the city's Chinese population.

Departure was another major cause of depopulation. After the 1930s, an increasing number of Chinese returned to China becuase of old age or financial difficulties. In the 1930s, many Chinese people accepted the government's offer of free passage back to China and remained there until employment conditions in Canada improved. In December 1934, for example, 150 Chinese left China at government expense.[3] This departure was interrupted by the outbreak of the Sino-Japanese War in 1937, but after it ended, homebound journeys of elderly people resumed. In 1947, for example, most of the 500 Chinese who left Victoria for China were in their sixties and over.[4] In 1959, the federal government permitted Canadians living outside Canada to continue receiving their old age pension, thus encouraging more elderly Chinese to return to Hong Kong or China. The aging Chinese population in Chinatown dropped sharply from 750 in 1959 to 350 in 1964.[5]

Decentralization was the last major factor in the reduction of the number of Chinese in Victoria and its Chinatown. Greater Victoria, including the City of Victoria and the three municipalities of Saanich, Oak Bay, and Esquimalt, had a Chinese population of 3,452 in 1941 (Table 38). Nearly 88 per cent of them lived in the city, and only 12 per cent in surrounding municipalities. After the Second World War, an increasing number of Chinese left Chinatown for better residential neighbourhoods, partly because they could afford better accommodations and partly because discrimination against them had subsided greatly. By 1961, for example, only 70 per cent of the Chinese in Greater Victoria lived in the city; the remaining 30 per cent lived in surrounding municipalities.

DEMOGRAPHIC CHARACTERISTICS

The registry of Chinese government bond sales is an important source for the study of the Chinese population in Victoria in 1938–9. The registry recorded each Chinese purchaser's full name, Victoria address, home county and village origins, sex, age, and occupation. The three volumes of the registry show that a total of 2,579 Chinese (2,447 males and 132 females) in Greater Victoria purchased Jiuguo Gongzhai (Bonds for Rescuing the Motherland) in 1938–9.[6] The purchasers represented about 75 per cent of the Chinese living in Greater Victoria. Analysis of the sex-age composition reveals that the aging Chinese population was predominantly male; 33 per cent of the men were between fifty and fifty-nine years old, and 15 per cent were sixty years of age or over (Table 39).

The Siyi people still outnumbered people from other counties, accounting

TABLE 38: Distribution of Chinese population in Greater Victoria, 1941, 1961, 1971

	1941		1961		1971	
Municipality	No. of persons	% of total	No. of persons	% of total	No. of persons	% of total
Victoria City	3,037	87.9	2,137	77.3	2,055	64.9
Saanich	323	9.4	491	17.8	910	28.8
Oak Bay	78	2.3	115	4.2	155	4.9
Esquimalt	14	0.4	20	0.7	45	1.4
Total	3,452	100.0	2,763	100.0	3,165	100.0

Sources: Census of Canada, 1941, 1961, 1971

TABLE 39: Age and sex composition of the Chinese population in Victoria, 1938–9

Age group	Male	Female	Total
10–14	3	0	3
15–19	22	8	30
20–24	33	13	46
25–29	65	12	77
30–34	109	12	121
35–39	114	15	129
40–44	246	17	263
45–49	392	13	405
50–54	436	8	444
55–59	378	5	383
60–64	233	1	234
65–69	108	1	109
70–74	18	2	20
75–79	2	1	3
Unspecified	288	24	312
Total	2,447	132	2,579

Source: Chinese government bonds sale records, Victoria, Jan. 1938–Apr. 1939

for more than half the Chinese population.[7] As well, over 80 per cent of the Ma, Guan, Xu, and Lei people came from Taishan, Kaiping, Panyu, and Zhongshan counties respectively (Figure 45). Occupational specialization was strongly related to the county and clan origins mainly because of nepotism. For example, half the cooks were from Taishan County and were surnamed Huang, Chen, Guan, or Ma; they were either relatives or fellow clansmen from the same county. This link between occupation and origins continued in Victoria's Chinatown until the late 1960s.

Although the 1961 census contained no specific information about Victoria's Chinatown, Enumeration Area 110 (EA 110), covering Chinatown and its adjoining areas, was distinguished by three main characteristics. First

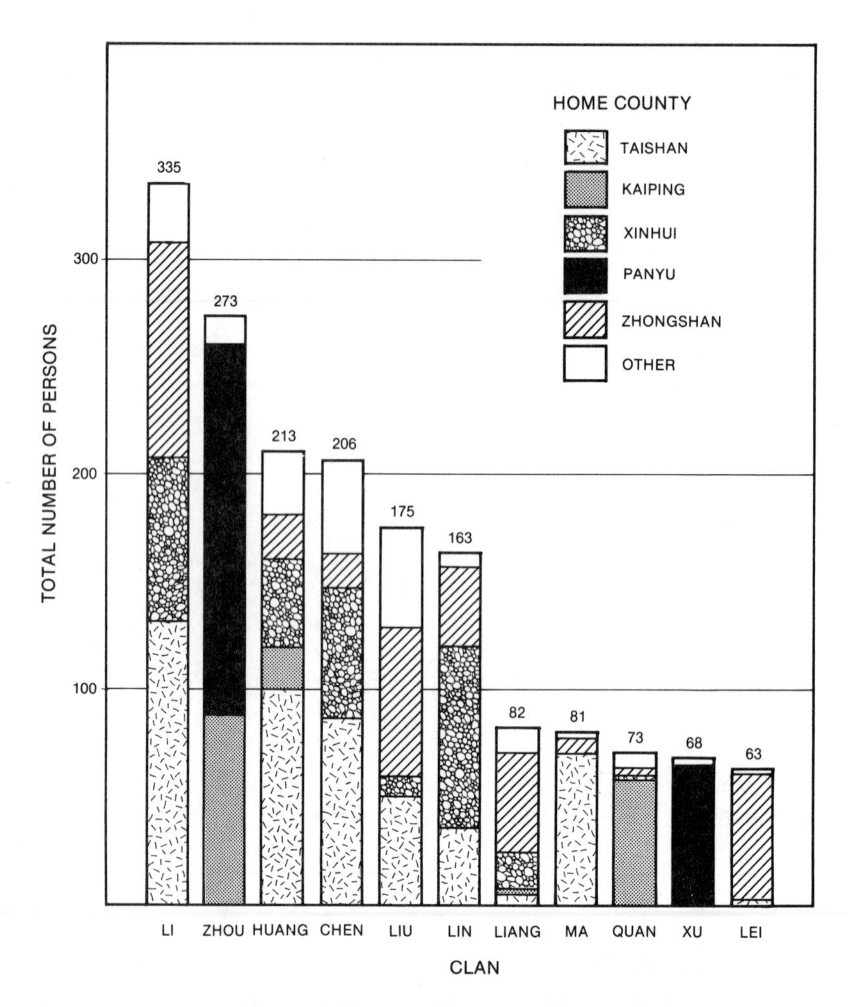

45 Victoria's Chinatown: Major Clans by Home County Origin, 1938–9

of all, EA 110 had the largest number of Asians of all other downtown enumeration areas; 659 of the 906 residents in EA 110, or 73 per cent of the area's total population, were Asians.[8] As Japanese, East Indians, and other Asian people in downtown Victoria were negligible in number, it was estimated that 99 per cent of the 659 Asians were of Chinese origin. Second, EA 110 had a ratio of nearly five men to one woman, four or five times higher than other downtown enumeration areas. Third, EA 110 was noted for the low educational level of its inhabitants. More than three-quarters of the population in the area over five years of age and not attending school had only elementary education. These three unique demographic phenomena were characteristic not only of Victoria's Chinatown but also of other Chinatowns in North America during the 1950s and 1960s.

The 1971 census reveals that these three demographic features still existed in Enumeration Area 308 (EA 308), covering the same area as EA 110 in 1961. About 260 of the 587 residents in EA 308 were sixty-five years of age or over.[9] As the ethnic origins of the residents were not identified in the census, I conducted a demographic survey of Victoria's Chinatown, covering three city blocks, in November 1971. The survey revealed that only 143 Chinese were still living in Chinatown, and that only thirty-five were over sixty-five.[10] The survey also revealed that since 1967 some low-income young couples with three or four children had moved into Chinatown, taking advantage of cheaper rents and proximity to work. The image that most residents in Victoria's Chinatown were elderly males was no longer accurate in the 1970s.

In May 1979, I conducted another survey which revealed that Victoria's Chinatown had fifty households with a total of 129 persons.[11] Unlike previous census returns, the sex-age ratio was balanced at nearly one male to one female. Only one-quarter of the population were pensioners; nearly half the populace was under forty years of age. Cheap rent, proximity to work, and the inability to communicate in English were three major reasons given by residents for living in Chinatown.[12] Ten households were composed of landed immigrants who had lived in Chinatown less than five years. The survey revealed that Chinatown was not only a place for elderly people but also a reception area for new immigrants who did not know English and needed time to adapt to Canadian society.

DISCRIMINATION AND ACCEPTANCE

From the 1920s through the 1940s, prejudice and discrimination against Asian people were strong in the host society. At first it was mainly the white working class which opposed Chinese labourers. After the Chinese merchant class began to expand in the 1910s and Chinese businesses, particularly laundries and grocery stores, spread throughout the city to cater to clientele outside Chinatown, white merchants in various cities began to complain about competition. Anti-Asiatic organizations were formed, and hate literature was published. After the First World War, for example, many returned soldiers who could not find jobs blamed cheap Oriental labour. In October 1921, they formed the Asiatic Exclusion League in Vancouver and engaged a number of canvassers to go to Victoria, Nanaimo, and other cities to canvass for members.[13] The objective of the league was obvious: to exclude Asians from Canada. The membership fee was only twenty-five cents, out of which no less than fifteen cents were paid to canvassers as commission. With a low membership fee and an attractive reward to canvassers, the league quickly got a membership of over 20,000. Another anti-Asiatic organization published *Danger: The Anti-Asiatic Weekly,* which consisted of inflammatory articles about the "yellow peril." Some writers, such as Hilda Glynn-Ward, a pseudonym for Hilda G. Howard, specialized in hate literature against Asians; her thinly disguised novel *The Writing on the Wall* was designed to convince eastern Canada of British Columbia's need for strict immigration regulations as protection against the "yellow peril."[14]

The most common form of discrimination against Chinese merchants and workers was in the awarding of government contracts. In 1922, for example, Charles Kent, a Chinese tailor, won the contract for making uniforms for the quarantine staff and other federal branches in Victoria. But organized labour protested, and the federal authorities had to confess that they had been misled by the name Kent and had not connected it with anyone Oriental.[15] Another incident occurred in 1939, when the Department of National Defence ordered 200 navy uniforms to three white tailors who, after six weeks' time, had finished only eighteen. In urgent need of these uniforms, the department turned to Chinese tailors, who completed the orders within two weeks.[16] As a result, the Victoria Assembly of the Native Sons of Canada protested against the use of Chinese tailors.[17] Mr. Herbert Anscomb, then MLA for Victoria, echoed this protest in the Legislature, saying that the work should be given to white workers.[18] The Department of National Defence later assured the Native Sons of Canada that in future "the practice [of employing Chinese] will be stamped out."[19]

As an increasing number of Chinese restaurants and grocery stores were established outside Chinatown, their owners opted to employ English-speaking waitresses to serve the public. In 1923, the provincial government proclaimed that Chinese people were prohibited from employing white women or permitting them to live or work in their place of business.[20] The CCBA hired a lawyer to fight this regulation, which was finally repealed.

Another form of discrimination was educational segregation, which was a complicated issue.[21] The demand to separate Chinese pupils from white pupils was first made in 1901 by some white parents at the Rock Bay Elementary School and the Trades and Labour Council on the grounds that Chinese children were unclean, untidy, and dangerous to white children.[22] The Victoria School Board placed the Chinese children in a separate class in the Rock Bay School. Six years later, some school trustees complained that many Chinese immigrants in their teens were not genuine students because they had spent only one year in a public school in order to obtain a certificate of attendance.[23]

Soon after they received the certificate, with which they could apply for the refund of their $500 head tax, they quit school. Thus the school board considered it a waste of public funds to educate them. Since the board could not prohibit Chinese pupils from attending public schools, it ruled on 29 August 1907 that no Chinese children be admitted to public schools until they understood English.[24] In January 1908 the school board passed another ruling that only native-born Chinese children could enter public schools. Under these two rulings, all Chinese children who arrived from China would be barred from attending public schools. Without a certificate of attendance, they could not collect their head tax refund.

The CCBA immediately engaged a lawyer to fight this discrimination.[25] In the meantime, the CCBA established the Chinese Imperial School (Chinese Public School) at 636 Fisgard Street for Chinese children who were not admitted to public schools. The issue was complicated further by the Chan Bun King incident. He was a boy of sixteen who was expelled from the Rock Bay Elementary School in March 1908 for drawing "obscene" pictures on younger white children's exercise books.[26] It was found that several Chinese teenagers in primary classes were overaged for their grades, but they had been put with the small children. Trustee Staneland worried that the Chinese youths would have immoral effects on the white children. This apprehension gave the school board an excuse to set up a segregated school in Chinatown. On 2 November 1908, the Fisgard Street Chinese School was opened for all Chinese pupils from First Primer (Grade 1) to Second Reader (Grade 4).[27] After they finished their Second Reader, they would be permitted to enter other graded schools and study with white children. Meanwhile in Vancouver the school board reversed the Victoria policy; it segregated older Asian students and left younger ones in integrated classrooms.[28]

The issue of educational segregation was dormant for over a decade, then erupted again in 1921. In that year, ninety of the 240 Chinese pupils in Victoria were studying in the segregated Rock Bay Elementary School, and the remaining 150 Chinese children were mixed with about 6,000 white children in public schools.[29] Some anti-Chinese agitators were still not pleased with this partial segregation and advocated that schools exclusively for the education of Orientals be set up because "white children sitting side by side with Orientals tended to develop the idea of social equality," which should be discouraged.[30] The following year, the school board decided to segregate all Chinese children from white children, because the Chinese were a "health menace" since they came from Chinatown, which was unsanitary.[31] As the "health menace" was unsubstantiated, the reason for segregation was changed: Chinese pupils were unable to speak "perfect" English, thereby retarding the satisfactory progress of the whole class.[32]

The CCBA again engaged a lawyer to fight the school board, deciding that if the school board carried out its segregation policy, all Chinese pupils should stay away from school to protest segregation.[33] When the school term began on 5 September 1922, all Chinese pupils in Boys' Central School and George Jay School were lined up and taken to King's Road School, a segregated school.[34] Before they reached the school, the Chinese pupils dispersed and returned home. This started a strike of the Chinese pupils against the school board, which lasted for a whole school year and only ended in the summer of 1923 after the school board readmitted most of the Chinese pupils to their former public schools.[35]

While fighting educational segregation, the CCBA was confronted with a greater issue: the prohibition of Chinese immigrants from entering Canada. In April 1923, the CCBA set up a Committee of Anti-Harsh Regulations and sent Joseph Hope (Liu Guangzu) to attend the mass meeting of Chinese in Toronto to discuss the federal government's "Forty-Three Harsh Regulations on Chinese Immigration."[36] To the Chinese, the Immigration Act of 1923 was humiliating. In May 1924, the CCBA declared that Dominion Day should be observed by all Chinese in Canada as "Humiliation Day."[37] It then published a circular relating the history of humiliation and discrimination against the Chinese in Canada, reminding the Chinese that they had never been accepted as Canadian citizens or treated with humanity and equality. They had been taxed upon landing in Canada and treated like imported animals. On Humiliation Day, the Chinese were to mourn their treatment by the host society by holding public meetings, distributing English-language news releases, refusing to fly Canadian flags, and wearing "Humiliation Mourning" lapel pins. On that day, the CCBA of Victoria would send to Chinese news agencies feature articles and news releases about Humiliation Day in Canada and would also appeal to all Chinese newspapers and print shops in Canada to insert the words, "July 1 Humiliation Day" in annual calendars. For the first few years after creation of Humiliation Day, many Chinese merchants in Victoria closed their shops on 1 July and attended mass meetings organized by the CCBA. As resentment against the 1923 act gradually diminished, however, observation of Humiliation Day lapsed.

Other forms of discrimination included exclusion of Chinese from certain places. For example, none were permitted to enter the Crystal Garden, a landscaped indoor swimming pool in Victoria. Soon after it opened, a Chinese man sought admittance, only to be rejected by whites who allegedly said, "What do you yellow dogs want to see our white women bathing for?"[38] As late as 1959, certain private clubs such as the Eagles' Club and the Tillicum Athletic Club were exclusively for whites.[39] An article by A.H. Fenwick in *Maclean's Magazine* on 15 January 1928 generally reflected the attitude of the host society towards the Chinese before the Second World War:

> Our acceptance of them as citizens was a mistake. Their ways are not our ways; their living standard is not our standard; they are of yellow races and we are of white; they are Oriental and we are Occidental; there is no thought on either side of the advisability of an attempt at assimilation; they and their children will always be alien to us.[40]

After Japan started a full military attack on China on 7 July 1937, the Japanese air force indiscriminately bombed hospitals, schools, and other non-military targets in Chinese towns and villages, causing enormous casualties among civilians. In occupied areas, Japanese soldiers robbed, raped, tortured, and murdered thousands of civilians; the world was horrified by these atrocities. The Canadian public began to feel sorry for China and was more sympathetic towards local Chinese anti-Japanese movements. In August 1939, for instance, a group of white sympathizers joined the Chinese to picket at the Yates Street wharf in Victoria to try to prevent the loading of scrap iron aboard a scow to be towed to a freighter bound for Japan.[41]

After the 1930s, white labourers considered the Chinese labourers their allies in opposing economic inequality under the capitalist system. The depression provided the catalyst for farmers, labourers, and socialists to unite and form the Cooperative Commonwealth Federation (CCF) in 1932. The CCF supported votes for Orientals during a debate in the House of Commons in 1934, and regularly promoted the franchisement of Oriental people in Canada.[42]

China became an ally of Canada after the United States and Britain declared war on Japan in 1941. The Canadian government had conscripted all Canadian youths of military age. However, Canadian-born Chinese youths were not conscripted because of the fear that if they were called for compulsory military service, they, like the Japanese veterans of the First World War, would later demand the right to vote.[43] Although they were not called up until August 1944, many Chinese youths had volunteered their service in Canada's fighting forces since the outbreak of the war; K.C. Lowe, a Victoria-born Chinese, was credited as the first Chinese to join the army.[44] Throughout the war years, the Chinese community in Victoria was active in services in the Air Raid Precautions and the Royal Jubilee Hospital and made generous contributions to Victory Bonds and Victory Loans.[45]

After the Second World War, discriminating regulations were repealed one after another, mainly because of the Chinese veterans' persistent fights for equality and the sympathetic support of the host society. In 1947, the Legislature of British Columbia permitted Chinese people to vote in a provincial election for the first time.[46] In the same year, Victoria City Council permitted Chinese property owners to vote on school and transportation bylaws

TABLE 40: Number of trade licences issued to the Chinese in Victoria, 1925

Trade	No. of licences
Vegetable and fruit pedlars	84
General and dry goods stores	80
Grocery stores	45
Laundries	22
Barbers	17
Tailors	14
Fish dealers and pedlars	13
Cafés or restaurants	12
Shippers or trucking	12
Shoemakers or shoe dealers	10
Butchers	9
Chinese herb stores	6
Greenhouses	3
Watchmakers and jewellers	3
Hardwares	3
Others	8
Total	341

Source: Compiled from *Report on Oriental Activities within the Province, Prepared for the Legislative Assembly of British Columbia* (Victoria 1927), 25

TABLE 41: Chinese business in Victoria's Chinatown, 1934–5, 1947, 1956–7, 1972

Business	1934–5	1947	1956–7	1972
Importing co. and grocery	26	22	15	7
Restaurant and café	8	8	9	7
Barber	6	7	3	1
Tailor	7	6	1	1
Laundry	5	2	2	2
Shoe-repair	3	1	1	1
Herbalist store	2	1	1	1
Other businesses	28	17	20	7
Total	85	63	52	27

Source: British Columbia Directory, 1935, 1947, and 1956, and field surveys in Nov. 1972

if they could produce a birth or citizenship certificate.[47] Two years later the city enfranchised the Chinese people.[48] Prejudice and discrimination against them disappeared rapidly as they became more integrated into the host society. Chinese children of the third generation, having grown up with white children, felt more Canadian than Chinese and entered the mainstream of Canadian life. Many had abandoned their mother tongue in favour of English and so could not maintain their Chinese heritage.

BUSINESS AND LAND OWNERSHIP

In 1925, the provincial government conducted a survey of Oriental activities in British Columbia. It revealed that 341 trading licences had been issued to Chinese in Victoria; about half the businesses were located in Chinatown (Table 40). After the 1930s, Chinese business establishments in Chinatown continued to decline in number, partly because of the depression and partly because of depopulation (Table 41). Service businesses such as barber shops, shoe repairs, tailors, and laundries disappeared rapidly as the demand for these services by bachelors declined. By 1956–7, Chinatown had both Chinese and non-Chinese business concerns. Of the fifty Chinese business concerns, the two largest companies were the Pioneer Fruit and Vegetable Wholesale Company on Chatham Street and the Lee Dye Sons Wholesale Company on Cormorant Street (Figure 46). During the 1960s, Chinatown businesses, small and large, were closing. In 1961, Lee Dye Sons Company sold its property to the city and moved out of Chinatown; seven years later, the Pioneer Company was sold to Canada Safeway Ltd. In 1963, a Chinatown Oriental Museum was set up by an investor, but it closed three years later because it failed to attract visitors.[49] By 1972, Chinatown's withered economy was represented by twenty-seven Chinese business concerns, of which about half were restaurants and grocery stores.[50] Seven years later, only half of the fifty-five businesses in Chinatown were owned by Chinese merchants.[51] The remaining establishments were non-Oriental: they carried on such activities as machine-welding, organ-manufacturing, and feed-distributing (Figure 47).

Many businesses in Chinatown were established after the 1960s. For example, nineteen business concerns which had been in operation for five years or less moved into Chinatown mainly because of cheap rent and the availability of space. Many firms in Chinatown left, claiming the area was dying;[52] others commented that unlike other downtown merchants, they could not depend on tourists.[53] Neither did Chinatown merchants rely on Chinese customers: fifteen merchants, particularly Chinese restaurant owners, claimed that 90 per cent of their patrons were non-Chinese.

During the withering period, Chinese land ownership patterns had undergone many changes. During the 1920s and 1930s, real estate in Chinatown offered a poor return on investment, and, owing to financial difficulties, many Chinese property owners defaulted in their mortgage payments: their property was either taken over by the mortagees or sold by the city tax collector at public auction. Several Chinese association buildings, such as those owned by the Chinese Empire Reform Association and Gee Tuck Tong were sold at a tax sale.[54] The Chinese Hospital was acquired by the city in 1929 for tax arrears.[55] By 1939, only fifteen lots in Chinatown were

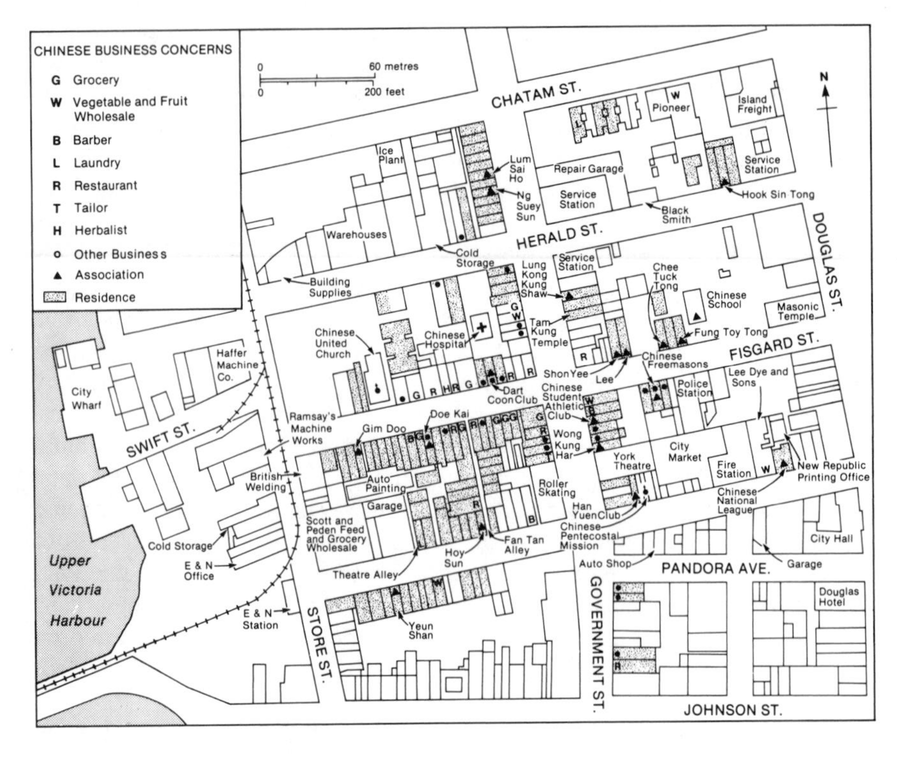

46 Victoria's Chinatown: Land Utilization, 1956–7

owned by Chinese associations or individuals. However, in the late 1940s, some Chinese merchants or investors began to purchase properties in Chinatown again because real estate prices were low. By 1979, Chinese landlords owned two-thirds of Chinatown's land (Figure 48).

INSTITUTIONAL ORGANIZATIONS

Throughout the 1920s, the unity of the Chinese community was weakened by rivalry between the Kuomintang and the Chee Kung Tong. In 1928, for example, a series of assaults resulted when the two factions disputed over funds needed to maintain the Chinese Hospital and the CCBA.[56] This discord subsided, however, after the Japanese attacked China. On 26 September 1937,

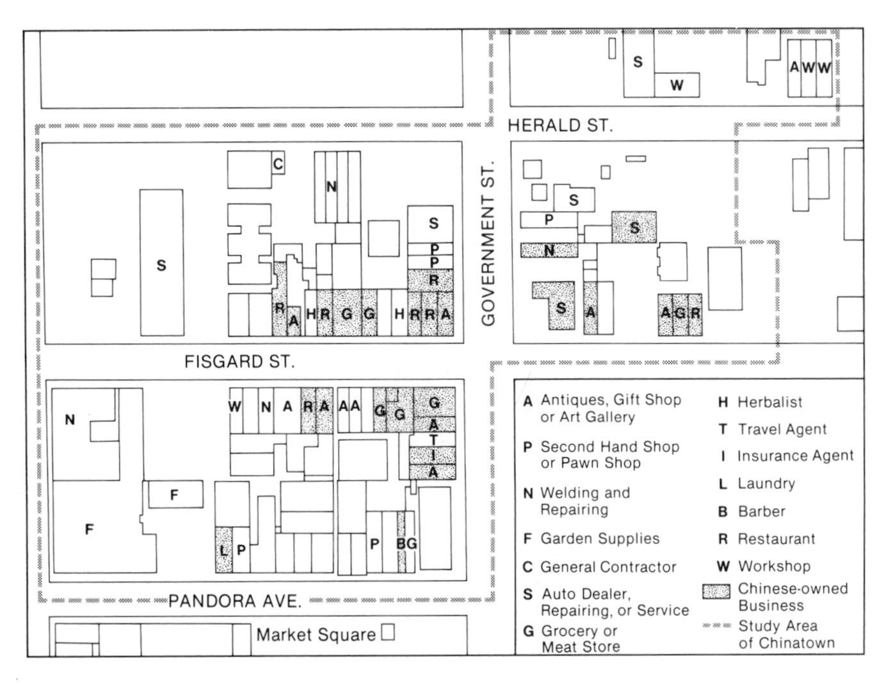

47 Victoria's Chinatown: Business Establishments, May 1979

the CCBA organized a public meeting in Chinatown to discuss Japanese aggression in Manchuria.[57] At the public meeting on 4 October, the CCBA established the first Chinese National Salvation Bureau in Canada, known as Huaqiao Kangri Jiuguohui (Victoria's Overseas Chinese Resisting Japan to Save the Nation Association). The bureau co-ordinated all forms of fund-raising campaigns in Victoria and helped unite the Chinese community in Victoria to fight against the Japanese. In March 1938, for example, the bureau, together with the CCBA and Chinese Freemasons, organized a lion dance to raise funds for Chinese war victims.[58] Throughout the war years, the Chinese community in Victoria was strongly united.

After the war, many clan and county associations were dissolved because they had few or no activities to attract new members, particularly young people. As old members were dying off or left for China, and few new members were recruited, many traditional associations died. Some, such as Gim Doo Tong (Ma Association), Oylin Kung Shaw (Zhou Association), Ng Suey Sun Tong (Wu Association), and Yen On Tong (Zengcheng County Association) did not own property in Chinatown and had no rental incomes.

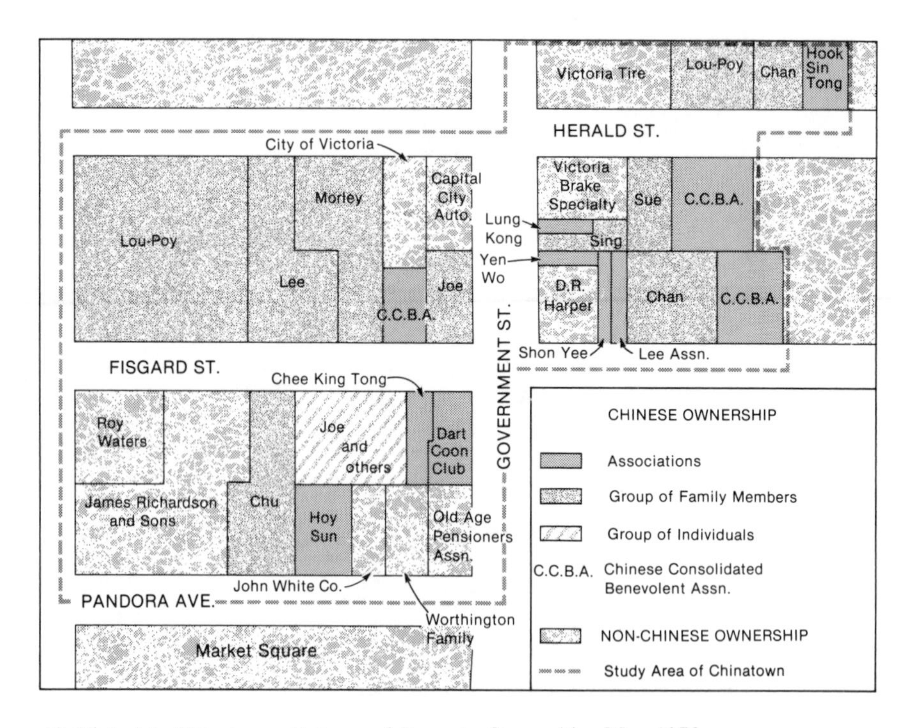

48 Victoria's Chinatown: Patterns of Property Ownership, May 1979

They became extinct in the 1960s as they did not have enough members to pay fees and make contributions. Some organizations fell apart because of discord among members; the closure of the Chinese United Church in 1962, for example, was caused by disunity among its members.[59]

In 1975, the provincial government passed a new Hospital Act under which the Chinese Hospital was disqualified as a licensed hospital, one reason being that it did not have a resident superintendent as required under the new legislation.[60] In November 1976, the Chinese Hospital housed nineteen elderly male patients, with two or three patients to a room.[61] The staff included two nurses, who each worked eight hours between 7:00 A.M. and 11:00 P.M.; one of the six orderlies was assigned for night duty. Under the new legislation, the Chinese Hospital needed at least three more graduate nurses and additional staff to do housekeeping and laundry.[62] In December, the Department of Health notified the CCBA that the Chinese Hospital should increase its staff to conform to the act's requirements, but as the CCBA had no money to employ more nurses, it decided to license the Chinese Hospital as a personal care home.[63] However, the hospital building did not meet regulations governing such homes. For example, one requirement was that each resident be provided with a single room, an ensuite toilet, and a handbasin.[64] The city owned the hospital building and did not wish to spend hundreds of thousands of dollars to convert it into a care facility.

In July 1977 the CCBA conducted an opinion survey which revealed that fifteen out of sixteen patients in the Chinese Hospital wanted to remain there, even if it were run as a private boarding home, cut off from the government subsidies.[65] As a result, the Chinese Hospital was licensed as a private boarding home after September. Without financial help from the government, it was run at a loss even though its staff had been reduced to only one nurse, four orderlies, and a cook. Eventually, the CCBA decided in June 1979 to close the "hospital" and transfer its remaining ten patients to James Bay Lodge.[66] The closure of the Chinese Hospital was another blow to the withering Chinatown.

The CCBA, once a powerful umbrella organization, ceased to be the formal mechanism of access to either the Canadian or the Chinese government. The younger generation, better educated and trained, was capable of communicating with Government officials. The CCBA also had lost many of its traditional activities, such as the shipment of bones to China, co-ordination of protests against discrimination, or mediation of interassociation disputes, although it continued to administer the Chinese Public School and the Chinese Cemetery. Nominally, the CCBA still represented the Chinese community in Victoria, but its leadership was not unchallenged.

In addition to traditional associations, new organizations, such as a Cub Pack, the Chinese Girls' Drill Team (Plate 37), and the Chinese Golf Club were formed to serve the needs of young people.[67] In 1956, a group of Chinese merchants and professional people established the Victoria (Chinatown) Lions Club, active not only in the Chinese community but also in the host society. Other new organizations, such as the University of Victoria Chinese Students Club and Victoria's Chinese Youth Association, were formed. Unlike the traditional associations, these new ones organized activities of interest to young people.

On 13 October 1970, Canada established diplomatic relations with China, and the following year a new organization known as the Chinese Canadian Friendship Association (CCFA) was established by thirteen Chinese merchants and workers in Victoria. For more than ten years, the CCFA, instead of the CCBA, sponsored touring troupes from China and organized welcoming parties when Chinese diplomats visited the Chinese community in Victoria. The CCBA regarded the CCFA as a "leftist association," whereas the CCFA did not recognize the CCBA's leadership and felt that it was "controlled" by Kuomintang members or pro-Taiwan people. There was no communication between the two associations until the revitalization programs of Chinatown were launched in 1979, when the two associations began to work together to improve Chinatown. Since the public had always confused the Chinese Canadian Friendship Association (CCFA) with the Victoria branch of the Canada-China Friendship Association (CCFA), the former association in 1982 changed its name to the Chinese Canadian Cultural Association (CCCA) and, two years later, joined the CCBA as a member association.

TOWNSCAPE AND IMAGES

After the Second World War, Chinatown was an aging district of patched-up brick buildings with peeling balconies and rusty fire escapes. Throughout the 1950s and 1960s, many dilapidated buildings in Chinatown were demolished and vacant sites used as parking lots.[68] Some old buildings were found unsafe and unsanitary and so were condemned. In June 1966, for example, after the 1800 block on Government Street was condemned, the landlord evicted his tenants, including several elderly Chinese.[69] All the upper floors were left vacant and only the ground floor was used commercially. Many other Chinatown buildings were the same: the upper floors were left vacant, with windows boarded up, and only ground floors were used as retail spaces. Street level floors were usually the only renovated sections of many old buildings in Chinatown, occasionally brightened by garish electric signs on Chinese restaurants or grocery stores.

Victoria's Chinatown retained much of its nineteenth-century townscape. Intricate networks of picturesque arcades, narrow alleys and enclosed courtyards were found behind the commercial façades of old buildings. This "hidden section" of Chinatown was no longer conceived by the host society to be dangerous, and many young people enjoyed exploring the "mystery" of historic Chinatown. A few cohesive groups of buildings with high heritage value were located on Fisgard and Pandora streets between Store and Government streets. Thirty heritage structures were identified in 1977 by the Ministry of Recreation and Conservation; sixteen had been built in the 1880s and 1890s.[70] Many of these buildings, constructed in Italianate and Edwardian styles, still dominate the streetscapes of 500 Block Fisgard and Pandora streets (Figure 49). For example, the Italianate Murray Building and Fan Building were distinguished by their fine, decorative brickwork, ornate cornices, and regular segmented or arched window patterns. The Joe Building, a typical Edwardian structure, was highlighted by a splayed corner

49 Victoria's Chinatown: Location of Heritage Structures, May 1979

entrance and balcony. The Joe and Fan buildings marked the entrance to Fan Tan Alley and its connecting narrow alleys and interior courtyards throughout the entire city block between Fisgard and Pandora streets.

Although all the Chinatown buildings were constructed in Western styles, they contained several Oriental features rarely found in other downtown commercial buildings. These features, best manifest in several Chinese association buildings, included recessed balconies, "cheater floors," upturned eaves and roof corners, tiled roofs, flagpoles, and inscriptions in Chinese characters.[71] The Gee Tuck Tong and Lee's Benevolent Association buildings were typical structures with recessed balconies and cheater floors. Of all association buildings, the Chinese Public School was the most outstanding in the Chinatown townscape. This two-storey, pagoda-like structure contained a host of Western and Chinese decorative elements and was considered one of the "gateways" to Chinatown.[72] These historic buildings in Chinatown were a unique feature of the historic downtown district, which the City Council of Victoria has always wanted to preserve.

FIRST REHABILITATION PROPOSAL, 1959

In 1959, the Capital Region Planning Board (CRPB) prepared an Urban Renewal Study for Victoria which recommended redevelopment or rehabilitation of four blighted downtown areas: Blanshard, Spring Ridge, Chinatown, and Victoria West.[73] The CRPB identified two major problems in Chinatown: poor housing conditions and a deteriorating physical environment. Its survey revealed that Chinatown had over half the derelict space in downtown Victoria, and most of Chinatown's 750 residents were elderly, single Chinese men who still lived in small communal groups and shared cooking and washing facilities in tenement-type buildings.[74]

The first blueprint for Chinatown rehabilitation called for two projects to be undertaken. The first was to convert the eastern half of 500 Block Fisgard Street into a decorative pedestrian mall and transform the shabby Chinatown into a tourist attraction. The scheme would eliminate thirty-one on-street parking spaces and might encourage property owners to demolish derelict buildings to meet the demand for off-street parking in Chinatown.[75] The other project was to rehabilitate two three-storey buildings and convert them into low-rental apartment buildings for elderly pensioners. One building was owned by the CCBA and the other one by the Chee Kung Tong. The cost of rehabilitation, including acquisition, was estimated at about $315,000. Since the federal government was prepared to pay 75 per cent of the cost of acquiring and rehabilitating existing housing in designated urban renewal areas, the remaining 25 per cent might be available from the provincial government or shared between the city and the Chinese community organizations.[76]

The proposal had little input from the Chinese community. Although the CCBA and Chee Kung Tong supported rehabilitation of their association buildings, owners of grocery stores in Chinatown were not pleased with the mall project, which would prevent them from loading and unloading goods in front of their stores. Restaurant owners also worried that their businesses might be affected by the removal of street parking on Fisgard Street because most patrons came to Chinatown by car. Some people also doubted whether Chinatown had much attraction for tourists.

At the same time as plans were being made to rehabilitate Chinatown, City Council was occupied with the development of Centennial Square for the city's centennial celebration in 1962. Around the landscaped square would be the refurbished old City Hall building, a new legislative wing on the Parliament Buildings, the McPherson Playhouse Theatre and restaurant, a Senior Citizens' Activity Centre, the Magistrate's Court Building, the renovated police headquarters, a multistorey parking building, and an arcade of specialty shops.[77] Pandora Avenue, on the south side of the square, would be realigned in order to link up with Cormorant Street between Government and Store streets, forming a major east-west crosstown artery for Esquimalt traffic.[78] The Centennial Square and street realignment projects were developed on the earliest site of Chinese settlement in Victoria, thus wiping out all trace of the oldest part of Victoria's Chinatown.

In August 1964, City Council decided to defer the Chinatown rehabilitation proposal. There were two reasons: first, the Blanshard Redevelopment Project, a low-rental housing project, had been given top priority for federal and provincial funding, and second, the demand for senior citizen housing in Chinatown seemed less imminent because, by 1964, only 350 elderly Chinese remained. It was expected that the aging Chinese population would continue to decline rapidly as more and more old men died or moved out of Chinatown.[79]

In 1964, the Council started a paint-up campaign in downtown Victoria; the cost was to be shared between the government and property owners. Chinatown landlords were also asked to participate, but most were not enthusiastic. As a result, Chinatown was left out of the downtown paint-up campaign. "Facelift Doomed in Chinatown," "City's Progress Hits Chinatown," "Will Success Spoil Chinatown's Face?," and other newspaper headings in the 1960s forecast a gloomy future for Chinatown.[80]

FAN TAN ALLEY REDEVELOPMENT PLAN, 1972

In January 1972, Milton Tisdelle, a design consultant in Victoria, conceived the idea of developing Fan Tan Alley into an historic commercial area like Vancouver's Gastown. The claustrophobic alley linking Fisgard Street and Pandora Avenue (formerly Cormorant Street), once lined by several gambling dens, cafés and stores, had been the busiest spot in Chinatown during the war. The vacant, dilapidated buildings on both sides of the alley had solid brick walls and were structurally sound. Tisdelle's plan was to renovate these buildings, allocating the ground floors for commercial use and the upper floors for cultural activities, such as museums or exhibition halls. The essence of his plan was to maintain the historic, social, and cultural continuity of Chinatown by rehabilitating its derelict buildings, bringing new life and excitement to Chinatown, as had been done in San Francisco.[81] In this way, Chinatown would be revitalized without drastic change in its historic townscape.

To carry out his plan, Tisdelle organized the Fan Tan Alley Development

Company Ltd. in the summer of 1972 and tried to get investors. With no confidence in the future of Chinatown, neither the alley's property owners nor developers were interested in the project. So, like the CRPB proposal, Tisdelle's plan was scrapped.

OTHER SCHEMES

Throughout the 1970s, Chinatown's future was uncertain and gloomy. On its south side was the proposed $25 million Reid Centre project, with its high-rise apartments and hotels. On the north was a CPR-owned city block which might be developed. Some people felt that sooner or later Chinatown would die after speculative developers bought all its properties at low prices.[82] Some Chinatown property owners also shared the same gloomy vision and would not spend money to restore or renovate their buildings. A few Chinese community leaders even regarded the survival of Chinatown as an anachronism: Chinatown was doomed. However, the city manager, William Hooson, remarked astutely that "Chinatown is very much in danger of being obliterated, not by the municipal government but perhaps by the—I wasn't going to say the indifference of the Chinese community, but perhaps their lack of awareness of the economic facts of life."[83] Council also felt that the initiative for rehabilitating Chinatown should come from the Chinese community and that the Inner Harbour development and other downtown renewal schemes should have priority in government funding.

The Chinese community did develop a redevelopment project in the early 1970s. Members of the Victoria Chinatown Lions Club thought that a Chinese cultural centre with a community hall would stimulate revitalization of socioeconomic activities in Chinatown. An ideal site for the proposed centre was the two pay-parking lots behind the Chinese Public Schools owned by the CCBA. Accordingly, the Chinatown Lions Club invited the CCBA to a joint meeting on 10 November 1971 to discuss the project.[84] At the meeting, it was proposed that a Chinese Cultural Centre (CCC) be established to raise funds for the community hall and organize social and cultural activities; the CCBA would donate the land and the Lions Club would contribute $10,000 to the centre, but the CCC would be financially and administratively independent of the CCBA and Lions Club.[85] Some CCBA directors objected to the proposal because without the pay-parking lots the CCBA would not have the rental income to support the school. Furthermore, some directors feared that in the future the CCC might challenge the CCBA's leadership in the Chinese community. As a result, no progress was made on the cultural centre and community hall project.

Four years later, the idea of building a community hall in Chinatown was raised again by the Chinatown Lions Club.[86] The club proposed to purchase the two parking lots from the CCBA at the nominal price of one dollar and raise the funds needed to construct the community hall.[87] Some CCBA directors objected because the community hall would then belong to the Lions Club and not to the CCBA. On the other hand, the Lions Club declined to sponsor the project if the land title was still under the CCBA's name. No agreement was reached, so the community hall project failed again.

In 1977, the city applied to Ottawa for Neighbourhood Improvement Program (NIP) grants for the Central Area of Victoria, which comprises Chinatown, Old Town, and their surrounding residential areas.[88] If an application were approved, home owners in a NIP area were entitled to borrow up to $10,000 from the Central Mortgage and Housing Corporation under the Residential Rehabilitation Assistance Program to renovate their premises. However, the NIP funding application for Chinatown was rejected because Ottawa did not consider Chinatown to be a NIP area. Thus, the scheme for Chinatown's rehabilitation under the NIP program failed.

Many people then felt that Chinatown's days were numbered. Their dismay about its future could be summarized by the headlines in local newspapers in the early 1970s: "Changed, Challenged, Obliterated . . . This is Chinatown's Epitaph," "Slow Death: Darkly-lit Chinatown begs for new life," "Disappearing Chinatown," and "Requiem for Chinatown."[89]

11

The Reviving Period, 1980s

All proposals for reviving Chinatown during the 1960s and 1970s were either scrapped or shelved, partly because they were planned without much input from the Chinese community and partly because the community itself was apathetic. City Council did not know what the Chinese community wanted to do about Chinatown or which Chinese associations represented Chinese views and opinions. For two decades, the future of the dying Chinatown was uncertain. It was the community-wide survey in 1979 and other surveys that resulted in its revival.

In September 1978, Victoria's Heritage Advisory Committee asked me to survey the views and opinions of the Chinese community about Chinatown. Two factors were important to the success of such a survey. The foremost factor was to inform the community well before the survey started and ascertain whether or not it would co-operate. The other important factor was for me to avoid being criticized as a "self-appointed co-ordinator or leader." Accordingly, I attended the meeting of the CCBA in March 1979 and informed the association of City Council's intention to conduct a survey. I was unanimously elected by twenty association representatives as the co-ordinator of the survey.[1] With this mandate, I undertook the community survey with the objective of making recommendations to Council for revival of Victoria's Chinatown.

A public involvement model of community planning was devised. The model consisted of three basic planning stages and aimed at citizen participation. The objectives of Stage 1 were to differentiate between the "Chinatown community" and the "Chinese community," to analyse the system of Chinese voluntary associations, and to identify the problems, needs, and concerns of Victoria's Chinatown (Figure 50). Stage 2 involved the organization of the survey and meetings with interest groups and individuals. These meetings would give the public the opportunity to express their feelings about the revival of Chinatown and control the future of their "town." Based on public input, a draft of the revitalization program was drawn up and

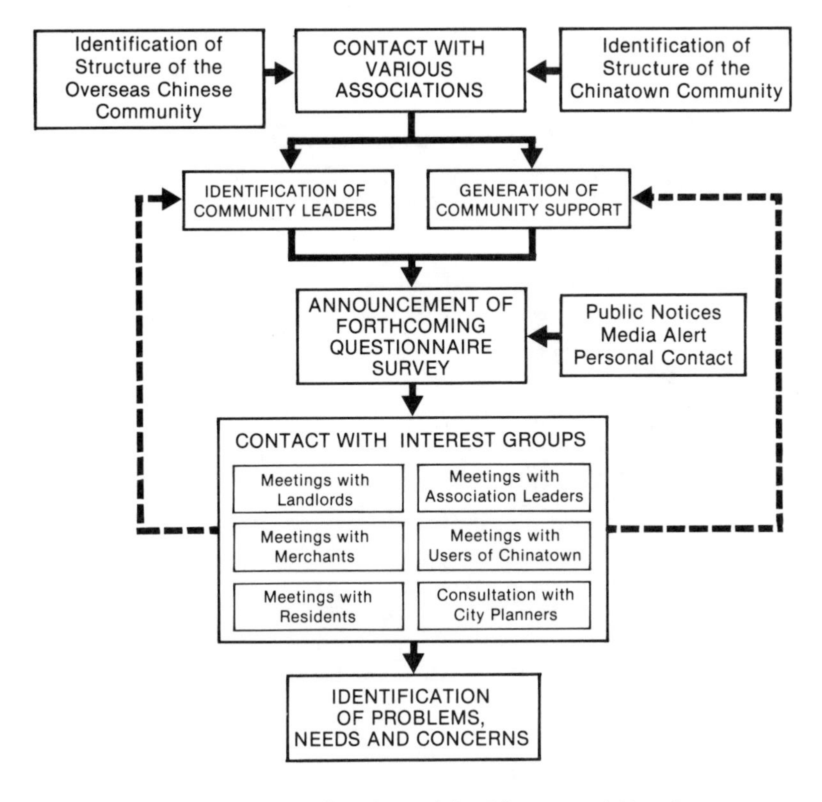

Stage 1: Identification of Problems and Needs

sented to City Council (Figure 51). Stage 3 aimed to mobilize the public to carry out the rehabilitation programs, and to apply for funding from three levels of government (Figure 52). Because experience has shown that different types of input are required at each stage, the model employed different kinds of strategy.

PUBLIC INVOLVEMENT MODEL: STAGE I

In much previous work on other Chinatowns in Canada, town planners tended to stress the need to obtain input from the Chinese community rather

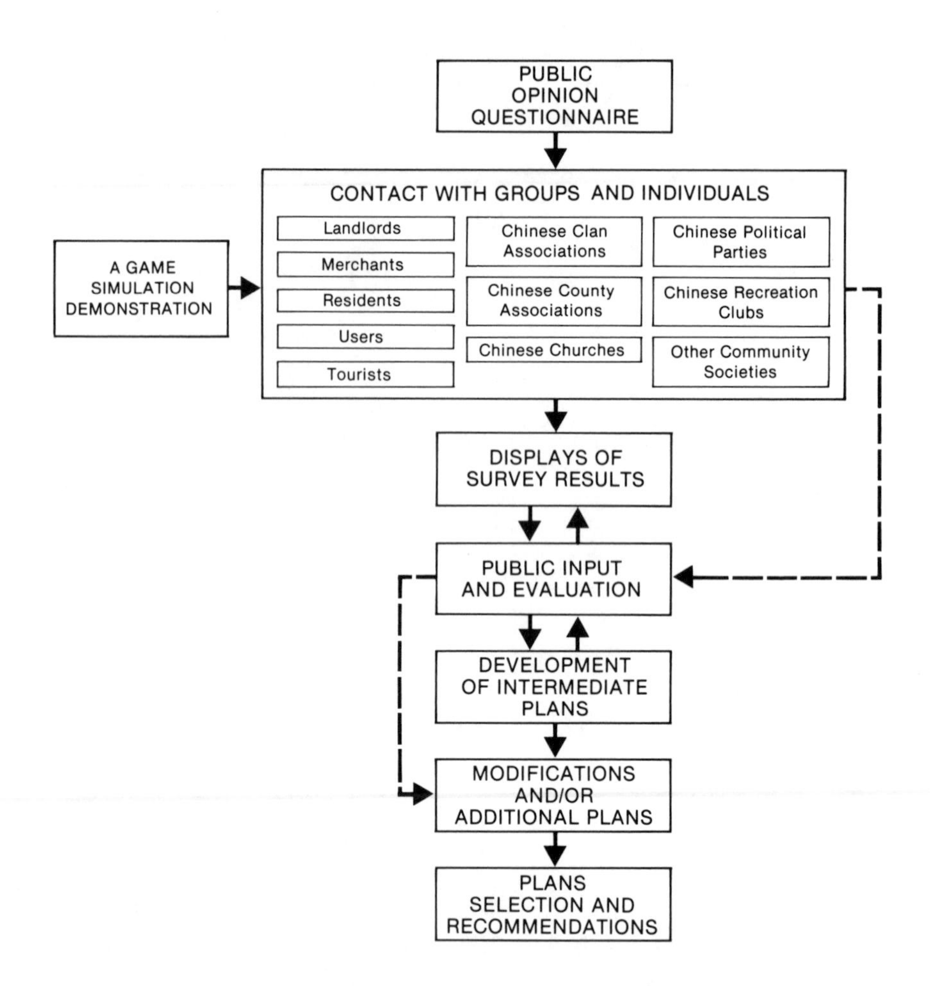

51 Public Involvement Model, Stage 2

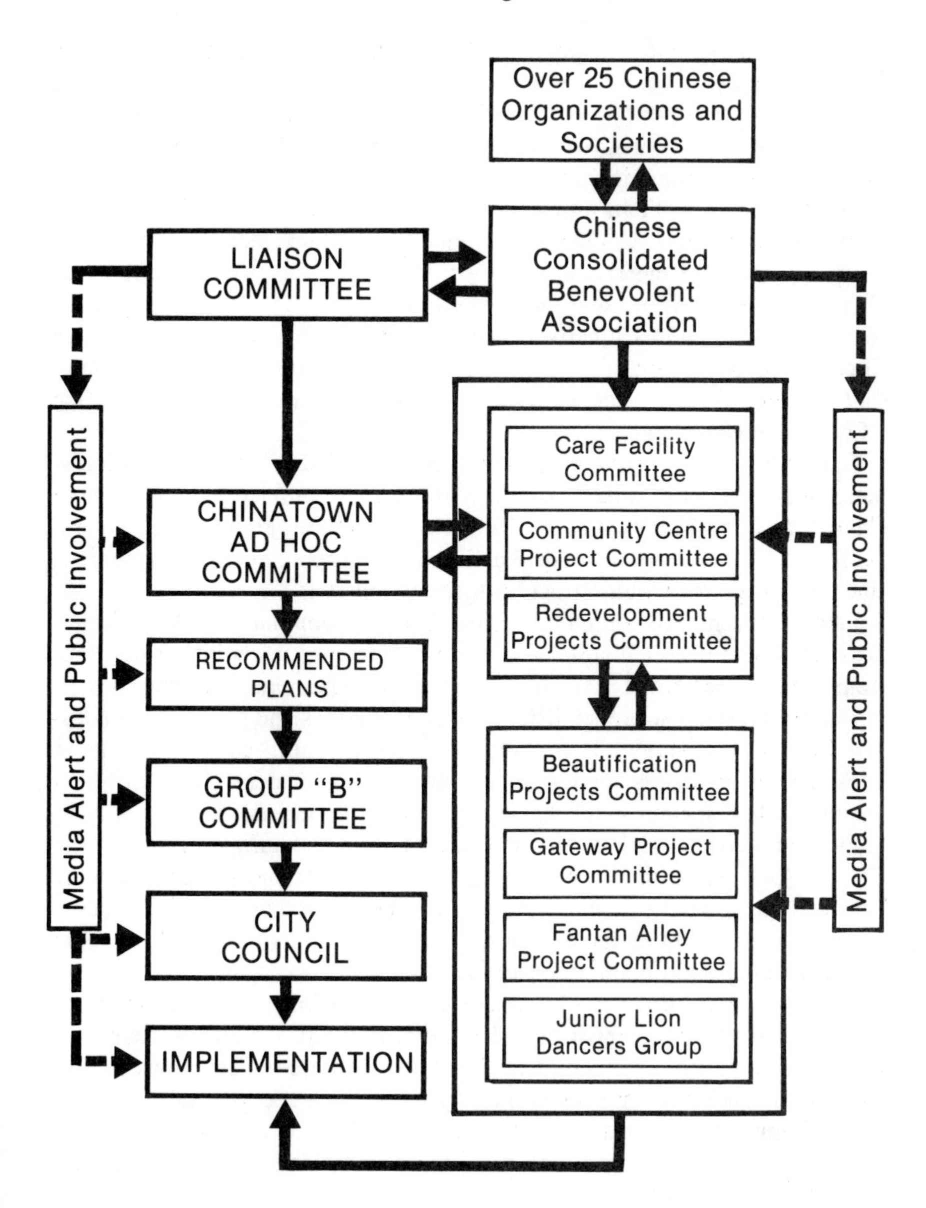

Stage 3: Development of Recommended Plans for Implementation

52 Public Involvement Model, Stage 3

than from non-Chinese residents. The "Chinese community" is loosely referred to as people of Chinese ethnic origins; most do not own land, operate businesses, or work or live in Chinatown. They are merely "users of Chinatown," coming to Chinatown to shop, dine, and meet friends. Some people in the Chinese community never come to Chinatown, do not participate in its activities, and are not concerned about its future. These people should be distinguished from the "Chinatown community," which includes Chinatown property owners, merchants, employees, and residents, regardless of Chinese, Caucasian, or other racial or ethnic origins. The latter are most affected by changes in Chinatown and, if united, are an important positive force in a revitalization program. A successful planning of Chinatown therefore should involve not only the Chinese community but also the "Chinatown community."

During Stage 1, five interest groups were identified: Chinatown landlords, merchants, residents, users, and tourists, each group consisting of both Chinese and non-Chinese people. The rehabilitation programs were to be regarded as joint projects of both the Chinese and non-Chinese communities since the programs would affect both communities in Chinatown.

Like other Chinese communities, the Chinese community in Victoria has factions. Local-born Chinese people have different viewpoints from those from China, Hong Kong, Taiwan, Singapore, or other places. Chinese immigrants from these places also differ in their aspirations and political attitudes. The Chinese community is further diversified by refugees from Vietnam, Laos, and Kampuchea, and visa-holding Chinese students from different countries. To understand the leadership pattern of Chinatown, three questions require investigation. What elements form the infrastructure of the Chinese society in Chinatown? How do these elements interact and function? How can the power structure, organization, and the decision-making process be understood? Without answering these questions, it is impossible to unite the community and obtain a consensus.

First, the basic structure of Chinese society in Chinatown is a system of voluntary associations which organize the community's political, religious, socioeconomic, and other activities. The thirty Chinese organizations in Victoria can be classified into seven major groups: clan, county, dialect, political, religious, recreational, and community-wide associations. Some have been inactive for many years, so it is important to identify those which are decision-makers. Second, some community leaders are active in both Chinese and non-Chinese organizations and support from them is vital to the rehabilitation of Chinatown. Third, the CCBA is still at the top of the power structure in Chinatown, although it exercises no power over its member associations. Unlike some Chinese Benevolent Associations in other Chinatowns, the CCBA in Victoria has been trying to stay away from China's politics and maintain a non-partisan status. As a result, it is accepted by most

Chinese organizations in Victoria as their representative and as a liaison between the Chinese community and government officials. Thus the CCBA took over leadership in rehabilitating Chinatown.

Before the questionnaire was prepared, concerned community leaders, representatives of various Chinese associations, and Chinatown landlords, merchants, and residents were invited to several meetings to express views and opinions. The questionnaire was based on their input. It was divided into five sets of questions, one for each interest group: landlords, merchants, residents, users, and tourists.[2] Questions were either fixed-alternative or open-ended, and respondents were encouraged to express their thoughts freely. Various means of communication, such as letters, posters, and newspaper reports, were used to alert the public about the forthcoming survey.

PUBLIC INVOLVEMENT MODEL: STAGE 2

From 15 April through 3 June 1979, research assistants were employed to deliver the questionnaire to Chinatown landlords, merchants, and heads of households, and to answer queries.[3] Representatives of various Chinese associations were also asked to help distribute the questionnaire to their members. A table was set up in Chinatown where pedestrians could complete and return the questionnaire. Because of the constraints of time and money, assistants were employed only for two Sundays to interview departing tourists at the downtown bus depot and at ferry terminals at Swartz Bay and the Inner Harbour.

At first, some Chinatown landlords and merchants were sceptical about the practicality of the survey. For the previous fifteen years or so, there had been several studies of Chinatown by the government, students, and other researchers which had not resulted in improvement of Chinatown, so some landlords and merchants did not respond as they felt that this questionnaire was yet another academic exercise. Other people also took a gloomy view of Chinatown's future and doubted whether the city government would spend money on a dying Chinatown. To gain the sceptics' confidence, I designed a simple model to demonstrate the importance of the questionnaire survey:

> If you place a coin on top of a strip of paper, the paper will sag immediately, because it cannot support the weight (Figure 53). This is an obvious fact. Similarly, it is an obvious fact that Chinatown is dying, and many previous proposals for its revival still remain on the shelf.
>
> Now, examine the model again. If you and I work and think together, we may eventually find a method which enables the piece of paper to support the coin. We have now achieved something which once seemed impossible.
>
> Similarly, if you complete the questionnaire and attend the meetings to

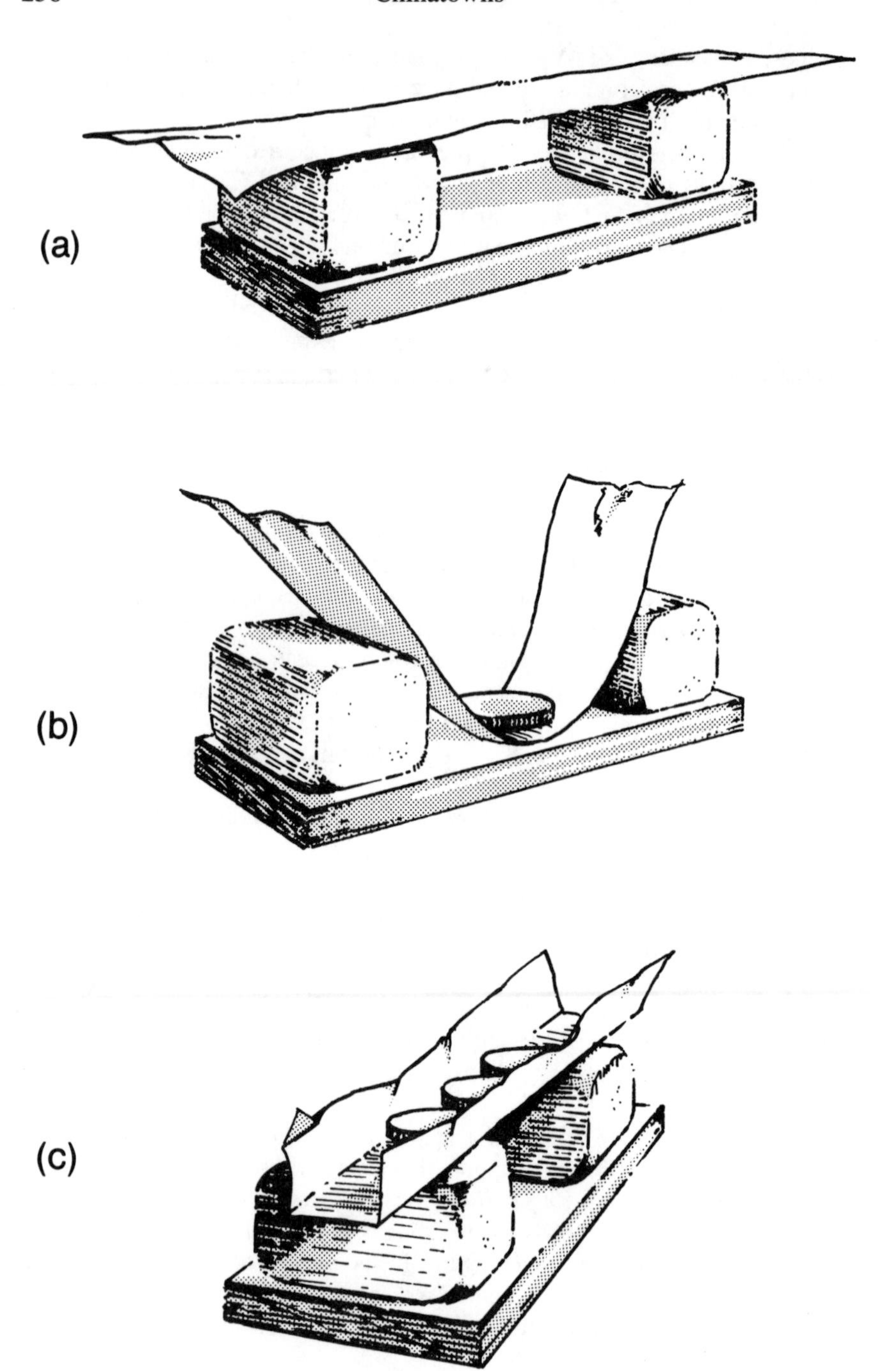

53 Demonstration Model on Public Input

express your views and opinions, we may find some ways to save our dying Chinatown. Your participation in the survey and involvement are essential to the Chinatown rehabilitation.

Demonstration of the model convinced many people to participate in the survey. A total of 1,116 questionnaires were distributed, and 71 per cent were completed and returned.[4]

The survey identified six major problems in Victoria's Chinatown: (1) no identifiable boundaries; (2) the physical deterioration and functional obsolescence of buildings; (3) the declining economy and lack of economic diversity; (4) the absence of a focal point for sociocultural activities; (5) a small population base; and (6) community indifference and lack of community confidence in itself as well as in city government.[5] Eighty per cent of respondents supported the idea of beautifying Chinatown and developing it into a tourist attraction.[6] Most people felt that a Chinese senior citizens' home, a Chinese hospital, a community hall, and a Chinese health clinic were essential facilities to be provided.[7] Two-thirds of the respondents felt safe walking at night in Chinatown, did not want Chinatown to be incorporated into the downtown area of the city, and wished to conserve and enhance Chinatown's appearance. Nearly 17 per cent of Chinese landlords did not want to sell their properties and wanted to preserve Chinatown's Oriental characteristics. The survey also revealed that 155 of 185 tourists had not been to Chinatown; sixty-seven tourists said that they did not know Victoria had a Chinatown.[8] Thirty tourists who visited Chinatown felt that it was "too small," "too dark," "somewhat run-down," or "full of drunkards."[9] Despite these comments, twenty-one tourists indicated that they would still recommend Chinatown because of its unique townscape.

The survey results were tabulated and displayed on the window of a restaurant in Chinatown. In consultation with interest groups, Chinese associations, and city planners, I ranked the Chinatown problems, listed solutions to them, and designed a draft plan for Chinatown rehabilitation. A final report, entitled *The Future of Victoria's Chinatown: A Survey of Views and Opinions,* was completed in June 1979 and presented on 17 July at a special meeting of Council members, the Advisory Planning Commission, and the Heritage Advisory Committee.[10] Seven major projects to revive Chinatown were recommended: the Street Beautification, Building Paint-Up, Chinese Gate, Care Facility, Subsidized Housing, Fan Tan Alley Rehabilitation, and Community Centre projects.

PUBLIC INVOLVEMENT MODEL: STAGE 3

In July 1979 City Council set up a Chinatown Ad Hoc Committee, chaired by Alderman Robert H. Wright, who was the driving force in Council to reha-

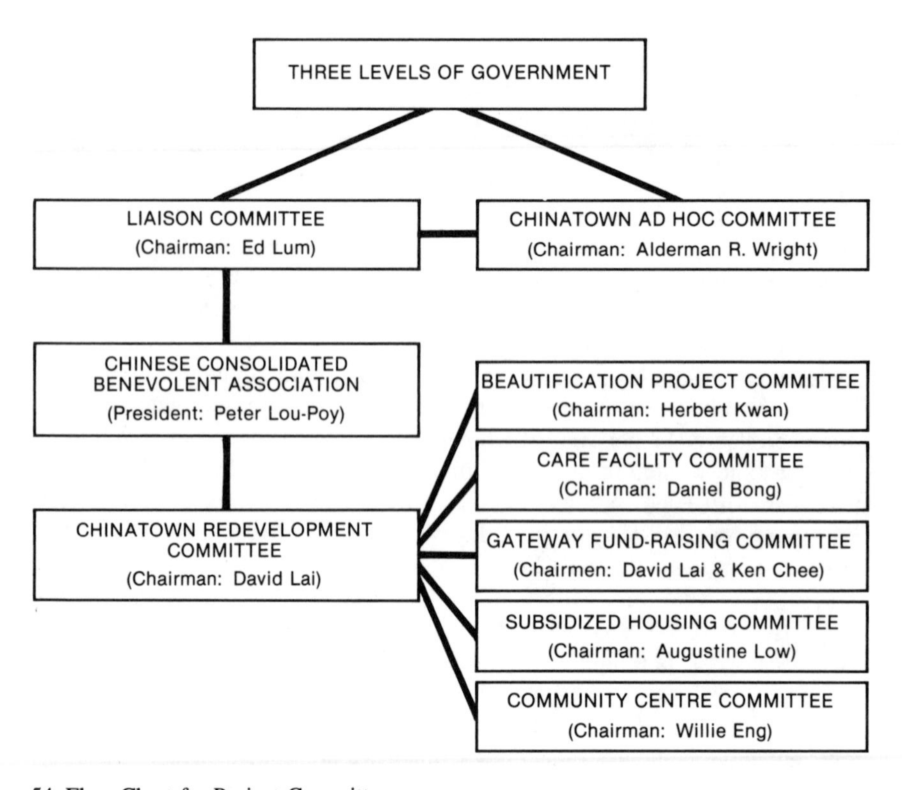

54 Flow Chart for Project Committees

bilitate Chinatown. The committee reviewed the report, consulted interest groups, worked on the budget for the projects, and made final recommendations. Meanwhile, the CCBA set up a Liaison Committee and a Chinatown Redevelopment Committee to co-ordinate the Chinatown projects (Figure 54).

In August, the Chinatown Ad Hoc Committee completed its report, *A Plan for the Rehabilitation of Chinatown,* which estimated that the total cost of the Chinatown facelift would amount to $985,000.[11] The most expensive item was the burial of overhead utility wiring. That cost, estimated at $700,000, would be shared by the city, the provincial government, and B.C. Hydro. The second expensive item, the Chinatown Gateway Project, would cost $96,000, which might be raised through public donations and contributions from the provincial government. The third item, estimated at $72,000, was the Street Beautification Project, which might be regarded as an extension of the established Old Town Beautification Program and thus paid for by the city, B.C. Hydro, and the B.C. Telephone Company. The Building Paint-Up Project, estimated at $59,000, would be paid by the Provincial Heritage Trust and the property owners on a 60-40 basis. Other projects, such as bilingual street signs, Oriental lamp posts, landscaping, and design guidelines, would cost $58,000, to be borne by the city and the Provincial Heritage Trust. A care facility might be built on the former Chinese Hospital site, which the city would then lease to a Chinese non-profit society at a nominal rent for ninety-nine years.[12] The project would rely on funding from the provincial Ministry of Health and the CMHC.

On 29 August 1979 the Chinatown Ad Hoc Committee submitted the Chinatown rehabilitation plan to the Group B Committee (Land Use and Zoning Committee).[13] The plan was approved by the committee and then by Council on 4 September 1979.[14] This date marked the beginning of the reviving period of Victoria's Chinatown: Phase 1 development (1979–81) included the Building Paint-Up, Street Beautification, Chinatown Gateway, Care Facility, and Junior Lion Dancers Group projects; Phase 2 development (1982–6) consisted of the Subsidized Housing, Fan Tan Alley Rehabilitation, and Community Hall projects.

The rehabilitation of Chinatown was heralded by a photographic exhibition on Chinese heritage in British Columbia 20–8 October 1979.[15] The exhibition had many purposes: to celebrate the ninety-fifth anniversary of the CCBA and the seventieth anniversary of the Chinese Public School; to coincide with the Heritage Canada Conference in Victoria; and to stimulate the public's interest in the heritage of Victoria's Chinatown and its rehabilitation program.

PAINT-UP AND STREET BEAUTIFICATION PROJECT

At first, Chinatown property owners wanted assurance from the city that they would be given a subsidy in the paint-up of their old buildings before they committed themselves to the facelift program. On the other hand, the Chinatown Ad Hoc Committee required that at least 60 per cent of Chinatown's property owners sign a financial commitment before the committee would apply for government subsidies. This was a Catch-22 situation. Accordingly, I spoke to the Chinatown landlords and convinced over 60 per cent of them to sign a letter agreeing to pay for half the expenses of exterior improvements of their buildings. By the end of 1980, over twenty Chinatown buildings had been cleaned or repainted according to the colour scheme for Chinatown: reds and greens were a main theme, and other supporting Oriental colours were carefully proportioned in the overall colour scheme for the building. The project resulted in an immediate visible improvement in Chinatown.

Street beautification was concentrated on the 500 Block Fisgard Street and the intersection of Fisgard and Government streets. All overhead wiring was placed underground. Sidewalks were resurfaced, patterned with large, stylish Chinese characters, and widened to accommodate rows of pines and flowering cherry trees. Old street lamp posts were replaced by red posts topped with pagoda-shaped lanterns (Plate 38). One public phone booth was painted red and gold, with a green Oriental-styled roof, and trash containers were decorated with red Chinese characters. New street signs for Fisgard and Government streets, with English letters and Chinese characters in red, were installed.

CHINATOWN GATEWAY PROJECT

Before the Second World War, it was customary to erect temporary arches to celebrate special events or welcome dignitaries on visits. Over 120 celebratory arches were built between 1869 and 1946 in various cities throughout British Columbia, including four in Victoria.[16] Therefore a permanent Chinese arch in Chinatown would be not only a way of preserving and recreating a historic past but also a monument to the Chinese heritage in Victoria. Furthermore, the arch would be a symbolic entrance to Chinatown and a special attraction. Tourists shopping in the Inner Harbour area would be attracted to the north part of the downtown commercial district, thus benefitting not only Chinatown but also other parts of the downtown area. The Chinatown Ad Hoc Committee decided that the Chinatown Gateway should be a community project, representing "the interest and active participation of all Victorians" and "the determination of all citizens to ensure its [Chinatown's] recognition and continued prosperity."[17] The first gateway fund-raising committee approached Chinatown property owners first, then Chinatown merchants, and the public last. The committee felt that property owners should lead the fund-raising drive because all the Chinatown projects would result in the increase of Chinatown's land value, benefitting the landlords most. If they did not play the leading role, the fund-raising committee would find it difficult to convince the public to support the gateway project. The committee also decided to install, on the pillar of the arch, a plaque with the names of individual donors contributing $100 or more and corporation donors contributing $500 or more; all donations would be tax-deductible. In less than two months, the first gateway fund-raising committee had raised nearly $30,000, presenting a cheque for this amount to the Council as the Chinese community's 25 per cent share of the construction cost.[18] The second fund-raising committee was immediately formed by Alderman Wright to seek donations from non-Chinese individuals, organizations, and corporations. In April 1980, Mel Cooper of CFAX Radio conceived the idea of selling share certificates valued at two dollars each, so that each buyer would own a piece of the gate.[19] This method of fund-raising generated public interest and involvement.

The gateway project received additional publicity when the city parade float, decorated with a model of the gateway, won seven first prizes in the Victoria Day parade, the Seafair Torchlight Parade in Seattle, and parades in other cities such as Vancouver and Port Angeles. On 19 June 1980, a special television program on Chinatown helped promote public interest in the gateway project and the fund-raising drive. On 30 November, a Chinatown Benefit Concert by Paul Mann was organized by Alderman Wright and Mel Cooper in co-operation with the McPherson Foundation. By the end of 1980, the second fund-raising committee had exceeded its target of $30,000.[20] Meanwhile, Alderman Wright was instrumental in obtaining a donation of $48,000 from the Provincial Capital Commission. By December 1981, the gateway fund had reached about $126,000, just meeting the cost of construction, estimated at $130,000.

The Chinatown gateway, spanning Fisgard Street, was designed by Mickey Lam, an urban design planner for the city, and built by the city (Figure 55). The materials were purchased from Taiwan, and the pair of stone lions was donated by the city of Suzhou, Victoria's sister city in China (Plate 39). Marti Wright chose the name "Gate of Harmonious Interest" (Tongji Men), which commemorated the co-operation of the entire community of Victoria to preserve the Chinatown heritage and revive its prosperity for the benefit and interest of all. The gate was officially dedicated on 15 November 1981 by Mayor William Tindal, Michael Young, chairman of the Provincial Capital Commission, and Peter Lou-poy, president of the CCBA.

CARE FACILITY PROJECT

Soon after closure of the Chinese Hospital, I immediately carried out a survey of the need for a care facility in Chinatown was conducted. A total of 277 Chinese (114 males and 163 females) fifty years old or over responded to the survey.[21] Ninety-nine per cent felt that if a care facility were located in Chinatown, Chinese patients would have more visitors because their children, relatives, or friends would drop in to visit them after shopping or dining in Chinatown. The survey also revealed that over 85 per cent of the pensioners could not speak English, that they preferred Chinese food to Western food,

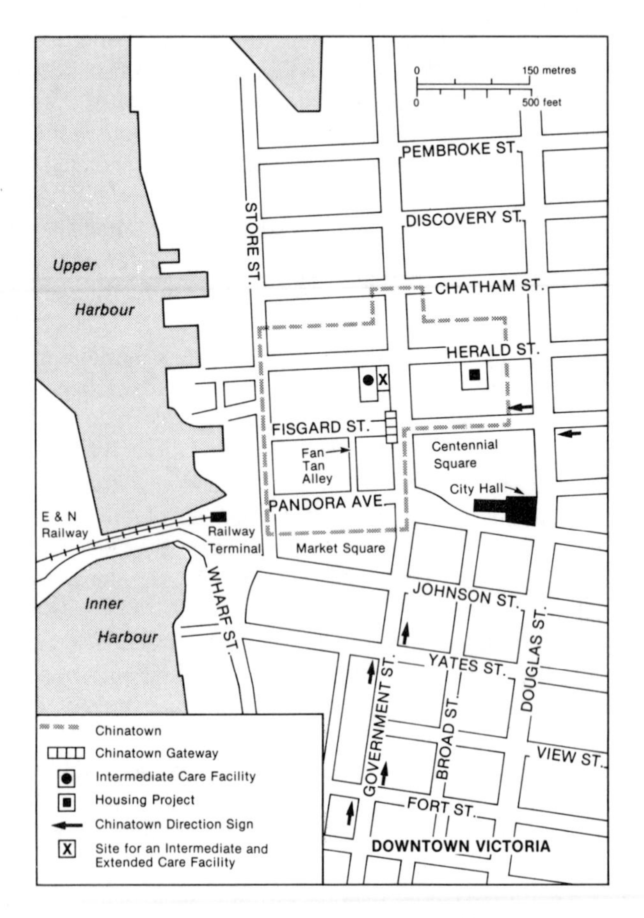

55 Locations of Chinatown Projects, 1979–87

and that over half the pensioners were widows.[22] Some Chinese and Caucasian patients in other care facilities indicated that they would consider moving to the proposed Chinatown care facility because they enjoyed Chinese cuisine. There was a need for an intermediate care facility in Chinatown, which would have Chinese-speaking staff, serve Chinese meals, and admit female patients.[23]

The Capital Regional Hospital and Health Planning Commission reviewed the report on *A Care Home in Victoria's Chinatown: A Survey of the Need* and supported the care facility project in Chinatown; it sent its recommendation to the provincial Ministry of Health on 29 November 1979.[24] For five months, senior officials in the health ministry had blocked the proposed facility on the grounds that the hospital site did not have enough green space for a care facility and that the ministry could not fund a care facility for only one ethnic group. In fact, the city had already waived the green space requirement for the facility and the officials had been assured repeatedly that the facility would be open to all patients, irrespective of ethnic origin. At a council meeting in April, Alderman Wright asked Mayor Tindall to write to the Honourable Rafe Mair, Health Minister, to expedite the approval of the project.[25] Meanwhile, the Liaison Committee also appealed to the minister. As a result, the project was approved on 21 May 1980.[26]

The three-storey brick building, designed by Daniel Bong, was an intermediate care facility with thirty beds, operated by the Victoria Chinatown Care Society, with Dr. David Liang as the first president. The site was leased for a dollar a year from the city, the construction cost of about $1.5 million was made available through CMHC, and the operation cost was subsidized by the Ministry of Health. When the facility was officially opened by the Honourable James Nielsen, Minister of Health, on 25 February 1982, over half its patients were Caucasians.[27]

REVIVAL OF SOCIAL ACTIVITIES

The Junior Lion Dancers Group, instructed by Master Wong Sheung, was formed in the summer of 1979 as a project to revitalize sociocultural activities in Chinatown.[28] Sponsored by the CCBA, the group, with Mrs. Lai Yin Bong as leader, was established with a grant of $3,160 from the provincial lottery fund.[29] Later that year, the Lotus Chinese Folk Dance Club, instructed by Mrs. Mow Ming Ho, was formed. Boys and girls between the age of seven and twelve were taught Kung Fu and Lion dancing, and girls of the same age group were taught Chinese folk dances. When the mothers took the children to Chinatown for training and performance, fathers often came along as well. The parents would shop in Chinatown while the children practised and often remained in Chinatown for dinner. Thus the children's groups not only revitalized social activities in Chinatown but also brought in new business.

CHINATOWN REDEVELOPMENT, PHASE 2, 1982–6

The second phase of Chinatown Redevelopment included a subsidized housing project, the Fan Tan Alley Rehabilitation Project, and the Direction Signs Project. In July 1982, a survey of the need for subsidized housing in Chinatown revealed that fifty-four respondents preferred living in Chinatown, seventeen preferred living outside Chinatown, and twenty had no preference.[30] Fifty per cent of the total respondents indicated that they preferred two-bedroom suites, and about 30 per cent preferred one-bedroom suites. The survey also revealed that there were about nineteen vacant floors in Chinatown, covering a total floor area of about 9,800 square metres (106,000 sq. ft.).[31] If rigid fire and building codes could be changed slightly, many of these vacant floors could be renovated for habitation.

In November 1982, the CCBA applied to CMHC for funding of a subsidized housing project, to be built on the parking lots behind the Chinese Public School.[32] The Chinese Consolidated Benevolent Association of Victoria Housing Society was incorporated, and John Di Castri was engaged to design the apartment building. According to the first design, the building was to consist of forty two-bedroom suites, a community hall on the ground floor, and underground parking. However, the design did not conform to CMHC's subsidizing guidelines and had to be revised several times in order to reduce the construction cost. One major change was that the CCBA had to pay for the community hall, which would cost over half a million dollars. As the CCBA could not raise this amount of money within a few months, it had only two options: to give up the housing project or to continue the project without the community hall. The subsidized housing committee decided to postpone the community hall scheme and proceed with the housing project; Mr. Ronald Lou-Poy, the CCBA lawyer, and Mr. Richard Ko, the CCBA accountant, have given valuable advice to the committee concerning various technical problems. Finally, CMHC approved funding of four million dollars to construct the five-storey apartment building, known as Chung Wah Mansion. It consisted of fifty-two residential units, four commercial units, and a covered parkade on the ground floor. Chung Wah Mansion was officially opened on 20 October 1984 by the Honourable Allan McKinnon, MP for Victoria, Mayor Peter Pollen, and Augustine Low, president of the CCBA.[33]

The second important project in Phase 2 was the Fan Tan Alley Rehabilitation. Fan Tan Alley, varying in width from one to three metres (four to eight feet), was inaccessible to a fire truck in the event of fire. Therefore premises on both sides of the alley were deemed uninhabitable unless fitted with sprinkler systems. As the installation cost was extremely high, property owners opted to leave their premises vacant (Plate 40). For many years, the alley had been a deserted lane, littered with broken beer bottles and frequented by drunks. I suggested an atelier model of planning which might be applied to the utilization of vacant premises on both sides of the alley: artisan or artist tenants would be responsible for renovating the vacant floors with their labour and materials and converting the floors into studios or workshops; in return, landlords would charge lower rents and ensure tenancy.[34] In this arrangement, landlords would have additional rental income without spending money on renovation; artisans and artists would pay low rents for workshops or studios. This project would help develop what I would call "alley art" in Chinatown. By the summer of 1983, several vacant premises on both sides of the alleys had been completely renovated and rented by fourteen artisans or artists. Fan Tan Alley, formerly a notorious gambling lane, is now the "Montmartre" of Victoria's Chinatown. In 1984, street signs for Fan Tan Alley were installed at each end of the alley.[35]

The third project, organized by Mr. Peter Lou-Poy, was to install a series of direction signs to direct tourists to Chinatown. The cost of manufacturing eight street signs pointing the way to Chinatown was shared between Chinatown merchants and the city.[36] In July 1984, these signs were installed at major street intersections between the Inner Harbour and Chinatown.[37]

IMPACT OF REHABILITATION

Today the physical environment of Chinatown is much improved. The upper floors of some buildings, vacant for many years, have been renovated by artisan or artist tenants and converted into studios or workshops. A few property owners have also upgraded their buildings to meet the requirements of new fire and building codes. In the early 1980s, new businesses, including five restaurants, two gift shops, a herbalist store, and some other small businesses, were established in Chinatown. Oil or watercolour paintings of the Chinatown gate or buildings are sold in various art galleries. Local tourist buses which formerly avoided Chinatown now include it on their routes. On 8 March 1983, Chinatown was honoured by the visit of Queen Elizabeth II and the Duke of Edinburgh—the first visit by a British monarch in its 125-year history.[38] In 1985, the Silver Thread Service organized the Week of Harmonious Interest, and Luis Ituarte organized the Chinatown Street Festival. Both events drew many local residents and tourists to Chinatown. During Expo 86, hundreds of tourists, brought by chartered buses from Vancouver, visited Chinatown and made it a vibrant shopping area again. As businesses and sociocultural activities revive in Chinatown, its residential population increases. Today, about 300 people live in Chinatown, two-thirds of whom are of Chinese ethnic origin. Many Chinatown residents are young couples with or without children who have great concern about the revitalization of Chinatown. Thus rehabilitation has restored not only prosperity but also community spirit and confidence in the Chinese community in the future of their Chinatown.

FUTURE PROJECTS

Although the rehabilitation of Chinatown has been completed, many other projects are still worth consideration. For example, the Chinese community has long dreamt of having a community hall in Chinatown. The ground floor of Chung Wah Mansion, originally designed for a 228-sq. m (3,200-sq. ft.) community hall with a stage and seats for 400 people, is temporarily used as a covered parkade. The Chinese community has to raise more than half a million dollars to convert the parkade into a hall and to lease or purchase a nearby parking lot for apartment tenants. Additional funds should be raised to maintain the hall since it will not generate enough rental income to meet its operating expenses.

Another important project is construction of a facility for extended care, required by patients unable to get in or out of bed or wheelchairs and who require constant medical supervision, twenty-four-hour-a-day nursing, and assistance in dressing, washing, and feeding. When patients in the intermediate care facility in Chinatown get to this stage, they would be transferred to an extended care facility. The ideal site is the property of Capital City Auto Ltd., adjoining the present care facility, because the property owner does not consider the Chinatown location an important factor for his automobile business.[39] On 1 April 1987, Mrs. Vivien Eng, President of the Victoria Chinatown Care Society, called an extraordinary general meeting at which it was decided that the society would purchase the property of Capital City Auto Ltd. at $250,000 for construction of an intermediate and extended care facility.[40] Having bought the property for $225,000, the society rents it back to the automobile company and uses the monthly rent to pay the interest on the $144,000 mortgage. After the mortgage is paid off, the society will start construction of a second care facility.

There is no senior citizen housing in Chinatown. As a result, many elderly Chinese now live in senior citizen homes in other parts of the city. Construction of a subsidized senior citizen home would bring elderly Chinese back to Chinatown and increase its population base. However, a study should be carried out to determine the demand for this type of housing.

By far the most important project is to preserve the historic nature of Chinatown. Behind the commercial façades of the old buildings on 500 Block Fisgard Street are intricate networks of picturesque arcades, narrow alleys, and enclosed courtyards. This hidden section of nineteenth-century Chinatown, formerly "the forbidden city" open only to the Chinese people, has not changed much. This area is unique in Canada since no other Canadian Chinatown has a similar intact and fascinating "lanescape." Victoria's lanescape is not only a physical reminder of the frontier spirit of the Chinese pioneers who built it but also a visual unity to Victoria's Old Town. Any new construction in Chinatown should conform to the overall nineteenth-century nature of the area. For example, new buildings should be designed to echo architectural details such as recessed balcony and semi-arched windows in the neighbouring heritage buildings. A comprehensive plan should be devised to retain the labyrinthine characteristics of "the forbidden city," enhance its viability as a heritage entity, and preserve it as "a socio-psychological well" to which the Chinese, old and young, can return to water their cultural roots and refresh themselves.

PART THREE

Conclusion

12

Retrospect and Prospect

Geographers are interested in urban ecology, which studies the total environment of a city as a human life-support system; the central problem is to understand how a population organizes itself in adapting to a constantly changing yet limiting environment.[1] One branch of urban ecology is the study of the spatial components of residential areas and the ecological and ethological ideas of community in space.[2] In each city, the structural analysis of an ecological organization entails examination of the spatial distribution of land-use types and population, such as the study of residential segregation among ethnic groups. Territorial segregation within a city increases the inter-relationship of territory and population activities. In North American cities, for instance, a Chinatown is an urban ethnic enclave which may be conceived of as an ecological niche in much the same way as different forms of plant life are in their ecosystems. This book has explored the growth and development of Canadian Chinatowns and used them to test the "stage hypothesis" that although Chinatowns in North America are different from one another, they tend to have followed a common growth cycle, experiencing a budding, blooming, withering, and dying or reviving stage, each stage having its own physical, demographic, and socioeconomic features. In spite of diverse origins and development, Chinatowns have a sort of unity, conforming at varying degrees to a stage-development model (see Figures 1, 2).

The evolution of Canadian Chinatowns was influenced not only by immigration acts and regulations but also by many other factors, such as changing political and socioeconomic conditions within Canada and China. For example, completion of the CPR, the depression of the 1930s, the outbreak of the Second World War, establishment of the People's Republic of China, the changing attitude of the host society, the arrival of the Chinese Vietnamese refugees, and the forthcoming return of Hong Kong to China in 1997 all have had repercussions on the demographic, socioeconomic, and political structures of Chinatowns in Canada. This concluding chapter attempts to recapitulate the major themes of the book.

CONCEPTS AND CLASSIFICATION OF CHINATOWNS

Ethnic groups are highly territorial in organization; Chinatown, Japanese towns, Italian towns, Jewish districts, and other ethnic enclaves are typical urban ethnic neighbourhoods in North American cities.[3] These neighbourhoods are seen to exist not only by their inhabitants but also by other city residents, but are not easy to define. For example, concepts about a Chinatown vary from person to person, place to place, and time to time. A Chinatown may be conceived of as Chinese living quarters in a particular section of a city or as an agglomeration of Chinese restaurants, grocery stores, and other businesses, or as a concentration of both Chinese people and businesses in one area. To some people, Chinatown is a slum or a skid row district, but to others it is a tourist attraction. Some people may regard a Chinatown as an identity and a root of Chinese heritage in the host society, but others may think Chinatown is a low-rent neighbourhood for poor and elderly people or a reception area for new immigrants who have problems assimilating into the host society. It is also impossible to formulate one acceptable definition for Chinatowns because they change in time and space; the definition depends on type and/or the stage of a Chinatown's growth according to the stage-development model.

This book has classified Chinatowns into four types: Old Chinatowns, New Chinatowns, Replaced Chinatowns, and Reconstructed Historic Chinatowns. The stage hypothesis and stage-development model apply to the evolution of Old Chinatowns established before the Second World War. They do not apply to the last three types of Chinatowns because they developed after the war and are still in a formative stage. Extensive observation of all major Chinatowns in Canada convinces me that the examination of Canadian immigration policies is essential to better understanding the growth and decline as well as demographic structures of the Chinese population and the evolution of Canadian Chinatowns. Based on historical and statistical evidence, Chinese migration to Canada can be divided into four periods, during which Old Chinatowns went through the four evolutionary stages of budding, blooming, withering, and revival or extinction.

PERIOD OF FREE ENTRY, 1858–84

Chinese immigrants came to Canada in response to the gold rushes of the late 1850s and 1860s in British Columbia. Another wave of Chinese immigration surged in the early 1880s with construction of the CPR. In both cases, a large volume of cheap Chinese labour was desperately needed to make up for the shortage of white workers in the province. These job opportunities brought to

Canada many Chinese immigrants who faced unemployment, poverty, and hunger in rural villages in South China. Like other international migration, the Chinese migration to Canada was shaped by "push and pull" factors.[4]

Most early Chinese immigrants had received little education in China. They did not speak English, nor did they understand Western culture. They were despised, intimidated, and even assaulted by the white community in British Columbia. Consequently these early Chinese immigrants, like their predecessors in San Francisco and other towns in California, became conscious of their own group identification and confined themselves to an area within a town in order to avoid hostility. However, the origin and growth of a Chinatown was shaped not only by prejudice and open discrimination from the white society but also by many other factors, such as the social nature of the Chinese, their inability to speak English, their desire to perpetuate their customs and way of life, their economic need to share room and board, and the influence of chain migration and social networks. Thus a Chinatown represents both involuntary and voluntary segregation; usually the involuntary choice of the Chinese was followed by the voluntary isolation as a defensive response to discrimination and insults.[5]

During the period of free entry, all Canadian Chinatowns were in British Columbia. Most conformed closely to the stage-development model, particularly in their budding stage. For example, all the Chinatowns were situated on the urban edge, mainly because the Chinese chose to live apart from white society instead of living in constant fear of being driven out by force. Demographically, all Chinatowns were similar during the budding period; nearly all their residents were young male labourers without families whose livelihood depended on a few merchants who had recruited them from China to work in Canada.

During the gold rushes, Chinatowns in Victoria and New Westminster were small and functioned as supply depots for gold-mining settlements on the mainland. Most new Chinese immigrants stayed briefly in Victoria and New Westminster and soon moved to mining districts. As a result, while Chinatowns in Victoria and New Westminster were still in the budding stage, Chinatowns in Yale, Barkerville, and other large gold-mining towns were booming. However, the duration of their blooming stage was short; for example, Barkerville's Chinatown was prosperous and busy in the early 1860s, but by the late 1860s it began to decline as gold returns decreased. In small gold-mining settlements, such as Leechtown, living quarters for Chinese miners were set up quickly and lasted for one or two years but were soon deserted after the gold was exhausted. Such short-lived quarters might be termed an "instant Chinatown," which completed the life-cycle of the budding, blooming, withering, and dying stages in a short time. These Chinatowns deviated from the growth patterns of the stage-development model because they did not last long enough to develop the physical and socioeconomic structures to qualify them as recognized Old Chinatowns.

PERIOD OF RESTRICTED ENTRY, 1885–1923

In 1885, the federal government attempted to restrict Chinese immigration by levying a $50 head tax on every Chinese immigrant; eight years later the tax was increased to $500. This discriminatory financial burden had a tremendous impact on the demographic and socioeconomic structures of the Chinese in Canada. Most Chinese immigrants, coming as labourers, had difficulty in raising enough money for their own passage and head tax, let alone for their families. As a result, the Chinese community in Canada remained predominantly male, and the shortage of Chinese women created social disruption. In the absence of a second generation, immigration remained the major source of population increase among the Chinese.

After 1886, Chinese immigrants began to move along the CPR from British Columbia to other parts of Canada. Budding Chinatowns gradually emerged in Moose Jaw, Calgary, and other railway towns on the prairies. A few adventurous immigrants reached Ontario and Quebec, where some Chinese people already had arrived from the United States, particularly New York and the eastern seaboard.

During the period of restricted entry, Chinatowns in Victoria and Vancouver were expanding, and their structures were increasingly diversified, ranging from crude wooden shacks and log cabins to two- to three-storey brick or wooden tenement buildings. Their morphology gradually changed from a cross-shaped to a reticulated pattern, conforming to the stage-development model. By the 1910s, Vancouver's Chinatown overtook Victoria's Chinatown as the largest Chinatown in Canada.

Outside British Columbia, various Chinatowns were budding or blooming as well. For example, while Moose Jaw's and Montreal's Chinatowns were blooming in the early 1910s, Chinatowns in Winnipeg and Toronto were still in the formative stage (see Figure 3). Chinatowns outside British Columbia were different from the stage-development model, which was more ideal for Chinatowns in Victoria, Vancouver, and other cities in British Columbia. The major differences occurred in the budding stage. For example, in the prairie provinces and central Canada, the first Chinese arrivals were mostly laundrymen from British Columbia or the United States. Since their businesses catered mainly to Western customers, the Chinese laundries were widely spaced so as not to compete with each other. Without an aggregation of Chinese people and businesses in one street, a Chinatown would not be formed. Although Chinatowns outside British Columbia were different from

British Columbian Chinatowns during the budding stage, the two groups had many striking similarities in later stages.

PERIOD OF EXCLUSION, 1924–47

During twenty-three years of exclusion, under the Exclusion Act of 1923, many married Chinese men, separated from their wives and children, could not lead a conjugal family life in Canada; they saw themselves as "sojourners" and eventually had to return to China if they wanted to be reunited with their families. Chinese women in Canada were rare, and intermarriage even rarer. After Chinese "bachelor sojourners" left Canada, they were not replaced by natural growth, and after the 1930s the Chinese population across Canada began to decline.

The Exclusion Act also resulted in the decline of Chinatowns across Canada. Without children or relatives to continue family businesses, many aging Chinese merchants had to close their companies or stores and retired to China. During the economic depression of the 1930s, increased unemployment in Chinatown forced many Chinese to return to China or to go to big cities where job opportunities were greater. Thus bigger Chinatowns depleted the population of nearby small Chinatowns: Victoria's Chinatown, for example, was depopulated by Vancouver's Chinatown, Hamilton's by Toronto's, and Quebec City's by Montreal's. From the 1940s on, Old Chinatowns, whether in small towns or big cities, were declining.

PERIOD OF SELECTIVE ENTRY, 1948 TO THE PRESENT

After the Exclusion Act was repealed in 1947, discriminatory regulations against Chinese immigrants were rescinded one by one. Wives, children, and other sponsored relatives of Chinese Canadians began to come to Canada. This immediate post-war migration helped lessen the imbalance in the sex ratio and social structure but could not help Chinatowns regain their populations. As discrimination and prejudice against the Chinese subdued, those with moderate incomes began to move out of Old Chinatowns to better residential neighbourhoods, leaving the poor and elderly behind. The post-war fighting between the Nationalists and the Communists in China, and establishment of the People's Republic of China, destroyed the dream of many elderly Chinese people of retiring to China. Thus the residential population in nearly every Old Chinatown was increasingly dominated by old men.

From the 1950s on, the suburban movement of the Chinese continued to deplete the population of the Old Chinatowns. Younger generations were better educated, entered various professional and non-professional occupations, and earned more money. Many began to acquire the sentiments and attitudes of the host society through education, intermarriage, the adoption of Christianity, and other means of integration. Many removed themselves literally and symbolically from their ghetto past by moving to suburbs. Therefore both behavioural and structural assimilation reduced their reliance on Chinatown, affecting its survival.

After the Second World War, all major Old Chinatowns had the characteristic features of the withering stage of the stage-development model although they did not manifest similar degrees of decay. They were mostly blighted inner-city neighbourhoods, considered eyesores by the host city. Many city planners felt that the only solution was to demolish Chinatown and redevelop it to fit in with other downtown urban renewal schemes. Many city councils also agreed that Old Chinatowns seemed to be approaching the end of their life-cycles and would disappear. The Chinese community was divided, as usual. Some people considered Chinatown a blot not only on the city but also on the Chinese community. Once a Chinese community leader was quoted as saying that "no progressive Chinese should think of having a Chinatown in the city, since people always link Chinatowns with either ghettos, slums, garbage-strewed streets, and crimes."[6] Others were fatalistic, believing that Chinatown would die sooner or later. Some were apathetic and had no concern about Chinatown, but there were also some community leaders who saw it as the root of Chinese culture and identity and wanted to protect it from redevelopment projects.

VIABILITY OF AN OLD CHINATOWN

Rose Lum Lee's hypothesis on the viability of an Old Chinatown tended to put too much emphasis on the size of population and was inapplicable to Canadian Chinatowns.[7] For example, Chinatowns in New Westminster and Moose Jaw in the 1940s met the three required demographic conditions of her hypothesis, but their Chinatowns did not survive. The provincial population in Canada was completely unrelated to the survival of a Chinatown, and the sizes of provinces are different from those of American states.

In contrast to the American experience, my study of Canadian Old Chinatowns revealed different requirements for their survival. First, the population base of an Old Chinatown should be at least ten or twenty Chinese residents. Second, the city where the Chinatown is located should have a Chinese population of about one hundred. Third, the Old Chinatown should have at least one active Chinese traditional association which owns its association building and which provides low-rental accommodation and has space for social activities for the Chinese in the city. An association would survive only when it has rental incomes or other sources of revenue and when some Chinese live near enough to participate frequently in its activities. Fourth, the Chinatown should have at least one Chinese restaurant and one grocery store whose operators own the properties. Fifth, natural hazards, such as fire, should not have occurred in the Chinatown. Last, the city government should not institute an urban renewal program to appropriate, eliminate, or relocate Chinatown.

The survival of Old Chinatowns in Canada was partly due to growing awareness of the social upheaval that resulted from postwar slum clearance and urban displacement. Numerous studies had pointed to the detrimental effects of gentrification.[8] Many town planners began to adopt other approaches to improving physical environment of Old Chinatowns. Victoria's Chinatown, for example, is the first Canadian Chinatown with a comprehensive plan for revitalization. In the past few years, council members or planners from Vancouver, Edmonton, Winnipeg, Toronto, and Portland, Oregon, have come to Victoria to collect information about its revitalization program.[9]

BIRTH OF NEW CHINATOWNS

After 1967, the Canadian government accepted immigrants of all racial groups on the basis of a universal point system which emphasized education, training, skills, and other criteria. Thus new Chinese immigrants came with more diversified occupational backgrounds than did early Chinese immigrants. The new immigrants, together with new generations of Canadian-born Chinese, have recast the entire structure of Chinese society in Canada. For example, many immigrants from Hong Kong, Taiwan, and other places were well educated and familiar with Western culture. They were urban dwellers and did not need Old Chinatown as a transitional place. They moved into the suburbs immediately on arrival in Canada, demonstrating great flexibility in adapting to the host society while maintaining their Chinese identification. To these new Chinese immigrants, the Old Chinatown was merely an area where they could taste genuine Chinese food and purchase Chinese commodities not available in other parts of the city. However, this situation changed rapidly when many Hong Kong and Taiwanese entrepreneurs established new restaurants and supermarkets outside the Old Chinatown. Chinese families of moderate to high means, particularly professionals, chose the suburbs. These "Chi-yuppies," or Chinese young upwardly mobile professional business people with disposable incomes, create a local market for Chinese restaurants, grocery stores, and other retail shops and services. Thus New Chinatowns began to emerge not only in suburbs of large metropolitan cities, but also in cities where no Chinatowns had been established before.

The growth of a New Chinatown follows a sort of "chain development" which plays a role like that in the creation of an Old Chinatown. A pioneer business is usually a Chinese restaurant or a grocery store which, if successful, will attract another Chinese restaurant or grocery store or other types of stores such as barbecue shops, Chinese bakeries, and seafood stores. Once begun, the establishment of other kinds of Chinese businesses tends to snowball; Chinese professional and financial service establishments such as doctors', lawyers', real estate, and insurance offices follow. Unlike the Old Chinatown, a New Chinatown does not have a residential Chinese population, nor traditional Chinese organizations such as clan and county associations. It is a commercial area, catering mainly to Chinese customers. To some people, such as Alderman Paul Cowell in Hamilton and Dennis Fong in Regina, a New Chinatown is an investment project to be financed mainly by Hong Kong investors and can be expanded to become an international village or a Replaced Chinatown. All New Chinatowns in Canada are still in the budding stage and their growth patterns will be different from those of the stage-development model of Old Chinatowns.

A New Chinatown may be linear or quadrangular. A linear New Chinatown consists of a group of Chinese businesses established in a row or interspersed with non-Chinese business establishments on one or two main avenues. New Chinatowns in Ottawa and Windsor are good examples. The other type is a quadrangular New Chinatown consisting of a square or rectangular shopping plaza where several or all the business establishments are owned and run by Chinese. Willowdale's Chinatown in the City of North York is this type. In some New Chinatowns, Chinese businesses spill over to nearby shopping plazas as they increase. For example, New Chinatowns in Richmond and Scarborough are commercial strips of small plazas in which Chinese businesses predominate.

Calvin Lee's study of American Chinatowns in the 1960s suggested that "the suburbanization of America has caused a recession of business for almost all of the Chinatowns in America."[10] In Houston, Monterey Park, and other cities, the new Chinese immigrants are interested in developing New Chinatowns and have no concern about Old Chinatowns. In 1986, for example, Calvin Leung, a Hong Kong developer, invested over U.S. $300 million to develop Tang City, twenty-three kilometres (fourteen miles) southwest of downtown Houston, Texas.[11] This New Chinatown, covering eighty-one hectares (200 acres), will include a trade convention centre, several shopping plazas, industrial parks, residential and office buildings, schools, churches, and recreational parks. Another example is Monterey Park's Chinatown in Los Angeles, which expands from the intersection of Atlantic and Garvey boulevards. It is commonly called "Little Taipei" because most of the Chinese businesses are run by the Taiwanese.[12] Little Taipei caters the Monterey Park Chi-yuppies, who show little interest in Los Angeles' Chinatown.

However, in surveys of about forty Chinatowns in North America, I found that Chi-yuppies also patronize Old Chinatowns in Philadelphia, Chicago, Boston, Toronto, Montreal, and other cities; suburbanization and the establishment of New Chinatowns in suburbs do not necessarily affect the prosperity of these Old Chinatowns. In Canada, a Little Taipei like that in Monterey Park has not been established because Taiwanese immigrants to Canada are fewer and less concentrated than those in the United States. Instead, many Hong Kong entrepreneurs and financiers who had immigrated to Canada with large amounts of capital have invested heavily in real estate and businesses in both Old and New Chinatowns. New businesses range from small stores to supermarkets, from bakeries to food-processing factories, and from rental agents to real estate investment corporations.

In recent years, some Chinese communities, such as those in Regina and Hamilton, have considered creating New Chinatowns. Based on the experience of Scarborough's Chinatown, I suggested that a New Chinatown should not be set up in an established neighbourhood because it makes local residents fear Chinese domination. If such a development is unavoidable, Chinese merchants should establish rapport with local residents and merchants by participating in their activities and contributing generously to local community events. But the best approach still is not to create a New Chinatown in an established neighbourhood. For example, Ottawa's New Chinatown, like Scarborough's Chinatown, is budding in an established neighbourhood where other ethnic groups have also set up their small businesses, such as La Roma Restaurant, Hoa Viet Groceries, Middle East Bakery, and the Indian Food Centre. Although there seems to be no great resentment against the budding Chinatown in Ottawa, it is advisable instead to develop a unique "HOME" (Harmonious Multi-Ethnic) Town, which is symbolic of the multi-ethnic nature of the street. Like Seattle's International District, a HOME town will operate like a multi-ethnic shopping centre and residential district.

CHINESE CULTURAL CENTRE

In almost every Chinese community in Canada, whether it has an Old, New, or Replaced Chinatown, or no Chinatown, a Chinese cultural centre is vital. If such a centre is built in an Old Chinatown, it provides many facilities such as a multipurpose community hall, Chinese library, Chinese museum, or a daycare facility, which are not available in traditional Chinese associations. The centre will also organize a great variety of activities, such as English or Chinese language classes, calligraphy, painting, and folk-dancing classes, and orientation programs for new immigrants, which are also rarely organized by traditional associations. These new activities and facilities will entice Chinese youth to return to Old Chinatown and learn more about its cultural heritage. An excellent example can be seen in the activities of the Chinese Cultural Centre in Vancouver's Old Chinatown and their impact on the city's Chinese community.

If a city does not have a Chinatown, a Chinese cultural centre will be a focal point for the Chinese community. For example, the 500 members of Nanaimo's Chinese community are widely scattered in residence and do not yet intend to re-establish their Chinatown. However, their Chinese Cultural Society actively promotes social and cultural activities, including the annual Chinese Cultural Festival, the Harvest Moon Festival, and "Rainbow Bridge," a monthly half-hour cable television show about the Chinese community.[13] Similarly, the Chinese community in the Halifax-Dartmouth area of Nova Scotia does not have a Chinatown, but some associations such as the Chinese Society of Nova Scotia and Chinese Community Development Centre organize various social and cultural activities.[14] All these communities want to have a Chinese cultural centre to provide facilities for their cultural and social programs. Thus a Chinese cultural centre may be considered a keeper of Chinese cultural tradition and provides a group identity which, no less than a political state, cannot exist without its own territory.

CONCLUSION

Chinatown is an ill-defined perceptual area because its characteristics, structures, images, and townscape have changed over time. Extensive observation of about forty major Chinatowns in North America has convinced me that every Old Chinatown established before the Second World War has passed through the stages of birth, growth, decay, and death or rejuvenation, and that each conforms somehow to the stage-development model. But the existence of a similar growth pattern does not imply uniformity in Chinatowns across Canada; numerical differences in population and business concerns bespeak a complex history of contrasting economic influences. Therefore local variations can be expected.

The development of Old Chinatowns in Canada has been shaped by immigration policies, white racism, the social nature of the Chinese, the economic advantages of business concentration, the arrival of new immigrants, land speculation, urban renewal programs, and many other socioeconomic and political factors over the past hundred years. Whatever similarities and dissimilarities exist, the common phenomena are that the public image of an Old Chinatown changed drastically after the 1970s. Traditionally, an Old Chinatown was a Chinese slum which existed side by side with other inner-city neighbourhoods, having little or no interaction with the latter. Today, an Old Chinatown is considered an historic district, an integral part of the downtown centre, and a tourist attraction. After revitalization, an Old Chinatown's economy changes; cheap bars, secondhand shops, chop suey-type restaurants, and dingy grocery stores are being replaced by elaborately decorated restaurants, well-lit supermarkets, gift shops, and other new businesses capitalizing on Chinese style and tourists' tastes. A revitalized Old Chinatown adds distinctive atmosphere to a city centre; it becomes a mecca for tourists and an asset to a city instead of a liability.

In general, the income of Old Chinatown residents is increasing, their living standards are rising, and demands for better housing, social services, and recreational facilities are growing. Affluent residents continue to move out of Old Chinatown, but if Chinese immigration continues, Old Chinatown will be replenished by new arrivals with lower incomes and limited knowledge of Western language and culture. Dilapidated tenement buildings accepted as normal a decade before become unacceptable. Thus provision of decent low-rental housing, care homes, daycare centres and the like in Old Chinatowns will continue to challenge town planners and city councils.

An Old Chinatown is still essential to many elderly Chinese, who cling to their cultural heritage despite many years' residence in Canada. To them, Chinatown is where they can live in continuity with the past. Some tend to live together in Old Chinatown for economic and social reasons. Thus, Old Chinatown is important to the "Chi-eppies," or Chinese elderly poor people, as well as those with moderate income. Many elderly Chinese who have moved to other parts of the city or its suburbs frequently drive or walk to Chinatown to meet friends, play mahjong, and read Chinese newspapers in the traditional association buildings. If they lived alone, they are isolated when ill. Accordingly, rehabilitation of an Old Chinatown ideally includes senior citizen homes and/or care facilities so that old people will not be isolated. When they live in Chinatown, they will still be in touch with Chinatown activities. Furthermore, their children or grandchildren will be able to drop in to see them whenever they go to Chinatown.

Between 1978 and 1984, about 23 per cent of the 44,414 immigrants from Hong Kong were fifty-five and over. (The retirement age for civil servants in Hong Kong is fifty-five).[15] In Montreal, Nancy Siew and Michael Maclean's survey of Hong Kong immigrants aged fifty-five and over revealed that 70 per cent of respondents preferred not to live with their married children, 88 per cent felt the need for a Chinese senior centre, and 96 per cent expressed the need for senior citizen housing.[16] Similar surveys in Victoria, Calgary, and other cities have also indicated a high demand for facilities for the elderly in Chinatown. Thus, an Old Chinatown will continue to function as a retirement centre for the Chinese in Canada.

Old Chinatowns in Vancouver, Victoria, Montreal, and other cities also have restored their former function as a warm and human place for new immigrants from China or Indochina. Unfamiliar with Western culture or unable to speak English, many refugees or recent immigrants from China need Chinatown as a transitional place to adjust to the new environment and to learn about Canadian culture. They use Chinatown as a springboard for acculturation and assimilation into Canadian society. Old Chinatown becomes important not only to "Chi-eppies," but also to "Chi-lippies," or Chinese low-income people.

Although Chinese heritage can be sustained by an Old Chinatown or a Replaced Chinatown, a New Chinatown manifests a Chinese identity in a Canadian city without an Old Chinatown. At present, most New Chinatowns are commercial areas but they may be zoned as residential, commercial, and institutional neighbourhoods. Many suburban New Chinatowns in Metropolitan Toronto and Vancouver have been developed by Hong Kong entrepreneurs. Since the beginning of the 1980s, many Hong Kong people have transferred their money to Canada and other countries because they have no confidence in Hong Kong once it becomes a Special Administrative Region of China in 1997. In the summers of 1985–7, several Hong Kong developers who declined to be identified, told me that they had invested or planned to invest several million dollars in real estate in Toronto, Vancouver, and other metropolitan cities. Two investors said that they were interested in buying or building shopping plazas, and developing them into New Chinatowns. As the 1997 deadline approaches, it is expected that more money and immigrants from Hong Kong, particularly "yacht people" (wealthy refugees), will enter Canada, and more New Chinatowns will emerge.[17]

Suburban New Chinatowns in Toronto and Vancouver will continue to take away customers from the two cities' Old Chinatowns. Some Chi-yuppies hardly ever go to these Old Chinatowns, which are among the most heavily travelled downtown areas. Instead, they go to the quadrangular New Chinatown designed for the automobile; entrance to the plaza and access to the restaurants and stores is geared to the convenience of car-drivers. Since a quadrangular New Chinatown cannot expand, a new plaza will be built nearby, containing similar Chinese business concerns. Thus a suburban New Chinatown which originally consisted of one quadrangular plaza gradually becomes a strip of quadrangular plazas, each competing with the other. This growth pattern is exemplified by the development of the New Chinatowns in Scarborough and Richmond. The linear New Chinatown, of which Toronto's Chinatown West is a good example, is expected to continue to grow rapidly, expanding and consuming nearby streets. Like an Old Chinatown, it will change from linear to cross-shaped and eventually reticulated. Today many New Chinatowns also have severe traffic congestion and parking problems, particularly on weekends and holidays.

A Chinatown whether old, new, or replaced is a unique component of the urban fabric of Canadian cities and part of Canada's multicultural mosaic. Its charm is created not only by its Chinese mercantile structures and ornately decorated commercial façades but also by the noise, smells, and congestion on its streets. Many Chinese Canadians and some non-Chinese Canadians still have deep personal attachment to Old Chinatowns. Transformation and adaption are the heart of Old Chinatowns; they are always evolving. I hope that this book will be useful for its insight into what Canadian Chinatowns were and what they are becoming.

Appendices

1: Chinese Canadians Elected to Public Office, 1957–June 1988

Name	Official position	Period
Gordon Chong	alderman, Toronto	1980–2
Harry Chow	alderman, Colwood, British Columbia	1985–7
	mayor, Colwood	1988–
George Ho Lem	alderman, Calgary	1959–65
Ying Hope	alderman, Toronto	1969–85
John Hums	councillor, McGarry, Ontario	1978–82
	reeve, McGarry	1982–
Douglas Jung	MP, Vancouver Centre	1957–62
David See-chai Lam	lieutenant-governor of British Columbia	1988–
Art Lee	MP, Vancouver East	1974–8
	leader, provincial Liberal party, British Columbia	1985–7
Ed Lum	alderman, Saanich, British Columbia	1965–73
	mayor, Saanich	1974–7
	alderman, Saanich	1979–80
Wayne Y.S. Mah	alderman, Eston, Saskatchewan	1982–4
	mayor, Eston	1984–5
Jack Mar	alderman, Central Saanich, British Columbia	1988–
Peter Wing	mayor, Kamloops	1966–71
Bob Wong	MLA, Fort York, Toronto,	1987–
	Minister of Energy, Ontario	1987–
Ken Wong	Alderman, Winnipeg	1972–7
Peter Wong	mayor, Sudbury	1983–
William Yee	alderman, Vancouver	1982–6

Sources: personal interviews with Alderman Ying Hope, Mayor Peter Wong of Sudbury, and other elected politicians; Dr. K.C. Li, Mr. Lee Tung Hai, and other community leaders; city hall managers or clerks, Ministry of Municipal Affairs in Victoria, and Union of British Columbia Municipalities.

2: Chinese Canadians Appointed to the Order of Canada, 1976–July 1988

Name	Place of residence	Year of appointment
Ernest C.F. Chan	Saskatoon	1984
Harry Con	Vancouver	1982
Joseph N.H. Du	Winnipeg	1985
Lori Fung	Vancouver	1985
Shiu Loon Kong	Toronto	1979
David Chuenyan Lai	Victoria	1983
David See-chai Lam	Vancouver	1988
Wah Leung	Vancouver	1988
Clara Yee Lim	Richmond	1979
David T.W. Lin	Montreal	1986
Jean B. Lumb	Toronto	1976
William Wen	Toronto	1988
Peter Wing	Kamloops	1976
Peter Bowah Wong	Victoria	1977

Source: *Order of Canada: Companions, Officers, Members,* February 1988 (Ottawa: Advisory Council Members of the Order of Canada 1988). All appointees are MC (Members of the Order of Canada, except Wah Leung, who was appointed OC (Officer of the Order of Canada)

3: Chinese Canadians Awarded Honorary Citizenship, City of Victoria, 1971–June 1988

Name	Year of appointment
Philip Chan	1977
David Chuenyan Lai	1980
Bang Lim	1973
Augustine Low	1973
Rose Lum	1986
Jack Tang	1971
Bessie Tang	1986
Peter Wong	1977

Source: *A List of Honorary Citizens of the City of Victoria* (Victoria: City of Victoria)

Notes

ABBREVIATIONS

BCD	*British Columbia Directory*
CCBA	Chinese Consolidated Benevolent Association of Victoria
Cen.	*Census of Canada*
CH	*Calgary Herald*
Col.	*Colonist*, Victoria
EJ	*Edmonton Journal*
GM	*Globe and Mail*, Toronto
JLABA	*Journals of Legislative Assembly of British Columbia*
NAC	National Archives of Canada
NFP	*Nanaimo Free Press*
PABC	Provincial Archives of British Columbia
PAC	Public Archives of Canada (now National Archives of Canada)
Pro.	*The Province*
SBC	Statutes of British Columbia
St.	*Standard*, Victoria
TST	*Toronto Star*
TV	*Times*, Victoria
VG	*Victoria Gazette*
VN	*Vancouver Daily News-Advertiser* (1886–1917), also known as *Vancouver News* or *Daily News-Advertiser*. After 1917, it was absorbed by the *Vancouver Sun*.
VS	*Vancouver Sun*
WFP	*Winnipeg Free Press*

NOTES TO CHAPTER ONE

1 For example, the umbrella organization is called Zhonghua Huiguan (Chinese Consolidated Benevolent Association) in Victoria, Wengehua Zhonghua Huiguan (Chinese Benevolent Association of Vancouver) in Vancouver, and Dangjin Zhonghua Gongsuo (Chinese Benevolent Association of Duncan) in Duncan.

2 John Meares, "Voyages Made in the Years 1788 and 1789, from China to the North West Coast of America" (London 1790), 2; and "Authentic Copy of the Memorial to the Right William Wyndham Grenville, 30 April 1790" (London 1790), 11 (unpublished manuscript presented to the House of Commons, 13 May 1790).

NOTES TO CHAPTER TWO

1 *Daily Globe*, San Francisco, 15 May 1858.

2 An agreement signed between Hop Kee & Co. and Allan Lowe & Co., 24 June 1858 (PABC)
3 J.D. Pemberton, *Facts and Figures Relating to Vancouver Island and British Columbia* (London: Longman 1860), 170; and VG, 1 Mar. 1858.
4 J.S. Macdonald, and L.D. Macdonald, "Chain Migration, Ethnic Neighborhood Formation and Social Networks," *The Milbank Memorial Fund Quarterly*, 42 (1964), 82–91.
5 VG, 22 Oct. 1859. In those days, Chinese immigrants were commonly called "Johns," "John Chinamen," "Celestials," "Heathen Chinee," and other derogatory names.
6 *Col.*, 26 April 1860.
7 Ibid.
8 *Col.*, 9 June 1860.
9 Chinese Labour (Vertical File, PABC).
10 *Col.*, 19 June 1865.
11 Chuen-yan David Lai, "Home County and Clan Origins of Overseas Chinese in Canada in the Early 1880's," *BC Studies*, 27 (1975), 17.
12 H.F. MacNair, "The Relation of China to Her Nationals Abroad," *The Chinese Social and Political Science Review*, VII (1923) 543–4.
13 G.T. Staunton and N.B. Edmondstone, "A Letter to the Governor at Prince of Wales Island," *Journal of the Indian Archipelago and Eastern Asia*, 6 (1852), 147.
14 Tsen-ming Huang, *The Legal Status of the Chinese Abroad* (Taipei: China Cultural Service 1954), 2.
15 T. Chen, *Chinese Migrations with Special Reference to Labour Conditions* (Taipei: Ch'eng Wen 1967), 18–19.
16 F.W. Howay, *British Columbia from the Earliest Times to the Present* (Vancouver: S.J. Clarke 1914), II, 41.
17 Ibid., 72.
18 United Kingdom, *Blue Books of Statistics, Colony of British Columbia* (London: Colonial Office 1861), 148–9.
19 Gordon R. Elliott, *Barkerville, Quesnel & the Cariboo Gold Rush* (Vancouver: Douglas & McIntyre 1978), 24–5.
20 Howay, *British Columbia*, 190.
21 Chuen-yan David Lai, "Ethnic Groups," in *Vancouver Island: Land of Contrasts*, ed. by Charles C.N. Forward, Western Geographical Series, No. 17 (Victoria: University of Victoria, Department of Geography 1979), 33.
22 M.A. Ormsby, *British Columbia: A History* (Vancouver: Macmillan 1971), 221.
23 J.C. Goodfellow, *The Story of Similkameen* (n.p., n.d.), 36.
24 R.E. Wynne, "Reaction to the Chinese in the Pacific Northwest and British Columbia, 1850–1910" (unpublished PH.D. thesis, University of Washington 1964), 140.
25 United Kingdom, *Blue Books of Statistics, Colony of Vancouver Island* (London: Colonial Office 1865), 224–5; and *Blue Books, British Columbia*, 1867, 140.
26 *Col.*, 28 Mar. 1861.
27 VG, 31 Mar. 1859.
28 *Col.*, 6 Mar. 1860.
29 Ibid., 8 Mar. 1860.
30 Ibid., 10 Mar. 1860.
31 Ibid., 8 Mar. 1860.
32 L. Kidder, *The Psychology of Intergroup Relations* (New York: McGraw-Hill 1975), 20; and G. Allport, *The Nature of Prejudice* (Cambridge, MA: Addison-Wesley 1954), 15.
33 Ormsby, *British Columbia*, 212.
34 Ibid., 221.
35 The *British Colonist* was first published in December 1858 and later had various names such as the *Daily British Colonist*, the *Victoria Daily Chronicle*, the *Daily Colonist*, and now the *Times-Colonist*. For brevity, the newspaper is referred to as the *Colonist*. Similarly, the *Daily Standard*, also known as the *Evening Standard*, is referred to as the *Standard*.
36 *St.*, 22 July and 16 Aug. 1875, and *Col.*, 23 July 1875.
37 *Col.*, 9 Aug. 1878.

38 SBC, 1872, 121.
39 SBC, 1876, 3.
40 Ibid., 13.
41 Consolidated SBC, 1877, 567.
42 Walter Sage, "Federal Parties and Provincial Groups, 1871–1903," *B.C. Historical Quarterly*, 12 (1948), 153.
43 *A Constitution of the Workingmen's Protective Association* (PABA).
44 BCD, 1892, 1, 188.
45 *Col.,* 1 Oct. 1878.
46 Ibid., 5 Nov. 1878.
47 Ibid., 11 Nov. 1879.
48 Ibid., 12 Oct. 1879.
49 Ibid., 18 Oct. 1879.
50 Ibid., 11 July 1885.
51 Huang, *Legal Status*, 35.
52 SBC, 1878, 30.
53 *Col.,* 2 Oct. 1878; and G.V. La Forest, *Disallowance and Reservation of Provincial Legislation* (Ottawa: Queen's Printer 1955), 68.
54 JLABA, Vol. 8, 1879, appendix XXV
55 Pierre Berton, *The Last Spike* (Toronto: McClelland & Stewart 1971), 206.
56 Johan Gibbon, *Steel of Empire* (New York: Bobbs-Merrill 1935), 241.
57 Ibid.
58 J.M. Gibbon, *The Romantic History of the Canadian Pacific* (New York: Tudor Publishing 1937), 241.
59 *Col.,* 8 June 1883.
60 Canada, *Royal Commission on Chinese Immigration: Report and Evidence* (Ottawa: Printed by Order of the Commission 1885), 398.
61 Chinatown News, 18 Aug. 1982, 14.
62 BCD, 1882–3, 362.
63 Patricia E. Roy, "A Choice between Evils," in *The CPR West*, ed. by Hugh Dempsey (Vancouver: Douglas & McIntyre 1984), 16.
64 Canada, House of Commons, *Debates*, 12 May 1882, 1476.
65 Ibid., 1477.
66 *Cen.,* 1880–1.

NOTES TO CHAPTER THREE

1 Robert McClellan, *The Heathen Chinee: A Study of American Attitudes toward China, 1890–1905* (Columbus: Ohio State University Press 1971), 6.
2 W. Peter Ward, *White Canada Forever* (Montreal: McGill-Queen's University Press 1978), 24.
3 T.R. Balakrishnan, "Ethnic Residential Segregation in the Metropolitan Areas of Canada," *Canadian Journal of Sociology*, 1 (1976), 482.
4 Edgar Wickberg, ed., *From China to Canada* (Toronto: McClelland & Stewart 1982), 42–3.
5 Ward, *White Canada Forever*, 64.
6 *Vancouver Daily World*, 1 Oct. 1898.
7 *Pro.,* 15 Jan. 1901.
8 D.Y. Yuan, "Voluntary Segregation: A Study of New York Chinatown," *Phylon: The Atlanta University Review of Race and Culture*,XXIV (1963), 258.
9 George E. Hartwell, "Our Work among the Japanese and Chinese in British Columbia," *Missionary Bulletin*, 9 (1913), 518.
10 Ibid., 519.
11 Frances Macnab, *British Columbia for Settlers* (London: Chapman and Hall 1898), 88.
12 United Kingdom, *Blue Books of Statistics, Colony of Vancouver Island* (London: Colonial Office 1867), 140.
13 *Col.,* 24 July 1867. A white labourer was paid two dollars a day.

14 *Blue Books, British Columbia*, 1871, 134.
15 *Col.*, 8 May 1867, and 7 May 1871.
16 *Guide to the Province of British Columbia* (Victoria: T.N. Hibbon 1877), 345.
17 *First Victoria Directory and British Columbia Guide* (Victoria: E. Malladaine 1874), 57.
18 *Guide to the Province of British Columbia* (Victoria: T.N. Hibbon 1877), 36.
19 Ibid., 101, 106, 108.
20 Lynne Bowen, *Boss Whistle: The Coal Miners of Vancouver Island Remember* (Lantzville: Oolichan Books 1982), 49.
21 A. Woodland, *New Westminster: The Early Years, 1858–1898* (New Westminster: Nunaga Publishing 1973), 14.
22 *Blue Books, British Columbia*, 1867, 140.
23 A.B. Ramsey, *Barkerville: A Guide in Word and Picture to the Fabulous Gold Camp of the Cariboo* (Vancouver: Mitchell Press 1961), 25.
24 British Columbia, Minister of Mines, *Annual Reports*, 1893, 211, and *Cariboo Sentinel*, 22 Sept. 1869.
25 S.M. Lyman, W.E. Willmott, and B. Ho, "Rules of a Chinese Secret Society in British Columbia," *Bulletin of the School of Oriental and African Studies*, XXVII (1964), 531.
26 Ibid., 534.
27 *Cariboo Sentinel*, 7 Aug. 1869.
28 British Columbia, Report of Chinese in Barkerville, Lightning Creek and Quesnellmouth Polling Divisions, submitted by J. Lindsay to the provincial secretary of British Columbia, 30 November 1879.
29 *Blue Books, British Columbia*, 1861, 148–9; 1862, 226–7; and 1865, 66–7.
30 Chuen-yan David Lai, "Chinese Imprints in British Columbia," *BC Studies*, 39 (1978), 27.
31 *Royal Commission, 1885*, 363–5.
32 Chuen-yan David Lai, "Home County and Clan Origins of Overseas Chinese in Canada in the Early 1880's," *BC Studies*, 27 (1975), 17.
33 Gregory Tong, Director of Sing Tao Newspapers, private interviews held in Chicago, April 1983; and community leaders of Seattle's Chinatown, private interviews, Seattle, May 1982.
34 Mark Lai, former Director of the Chinese Historical Society of America, San Francisco, private interviews held in San Francisco, April 1984.
35 Gustavo Da Rosa, *A Feasibility Study for the Development of Chinatown in Winnipeg* (Winnipeg: Winnipeg Chinese Development Corporation Ltd. 1974), 62–4.
36 *Royal Commission, 1885*, 363–5.
37 BCD, 1882–3, 169.
38 Ibid., 156.
39 *Royal Commission, 1885*, 156.
40 NFP, 5 Nov. 1884.
41 Ibid., 4, 5 Aug., 29 Sept., 10 Oct. 1883.
42 Ibid., 7, 10 Jan. 1885.
43 Gordon R. Elliott, *Barkerville, Quesnel & the Cariboo Gold Rush* (Vancouver: Douglas & McIntyre 1978), 52.
44 BCD, 1882–3, 363.
45 Linda J. Eversole, "Chinese Community in Yale," Yale, Jan. 1984, 4 (mimeographed).
46 *Royal Commission, 1885*, 363.
47 Ibid., 365.
48 W.E. Willmott, "Some Aspects of Chinese Communities in British Columbia Towns," *B.C. Studies*, 1 (1968–9), 29.
49 *Royal Commission, 1885*, 365.
50 BCD, 1884–5, 174–5.
51 BCD, 1882–3, 219.
52 *Royal Commission, 1885*, 363.
53 *Cen.*, 1880–1.

NOTES TO CHAPTER FOUR

1 B.W. Tuchman, *The Proud Tower* (New York: Bantam Books 1966), 177.
2 Huang, *Legal Status,* 82.
3 SBC, 2nd Session, 4th Parliament, 1884, Chap. 2, 3; Chap. 3, 5–6; and Chap. 4, 7–12.
4 Canada, Statues of Canada, "An Act to Restrict and Regulate Chinese Immigration into Canada," 1885. Ottawa, 48–9 Victoria, Chap. 71, 207–12. The head tax did not apply to consular officials, tourists, merchants, students, and returning Chinese residents who had paid a dollar for a certificate of leave before they left Canada and who returned within twelve months.
5 *Royal Commission, 1885*, 366.
6 *Col.*, 22, 27 Jan. 1886.
7 P.A. Phillips, *No Power Greater: A Century of Labour in British Columbia* (Vancouver: BC Federation of Labour, 1967), 11–12.
8 Ibid., 19.
9 *Col.,* 10 Apr. 1892.
10 Ibid., 29 Aug. 1896.
11 British Columbia, Legislative Assembly, *Sessional Papers*, 1899, 1, 385.
12 SBC, 1900, Chap. 11, 36.
13 David Chuen-yan Lai, "Chinese Attempts to Discourage Emigration to Canada: Some Findings form the Chinese Archives in Victoria," *BC Studies* 18 (1973), 35–7.
14 Canada, Report of the Royal Commission on Chinese and Japanese Immigration, 1902, *Sessional Papers* No. 54, Vol. 13, Ottawa, 1902, 2.
15 Canada, Statutes of Canada, An Act Respecting and Restricting Chinese Immigration, 1900. Ottawa, 63–4 Victoria, Chap. 32, 215–21.
16 Canada, Statutes of Canada, An Act Respecting and Restricting Chinese Immigration, 1903. Ottawa, 3 Edward VII, Chap. 8, 105–11.
17 Laurier Papers Series. Correspondence, Vol. 638–41, 174002 (NAC). The restrictive measures and the exclusion policy both helped enrich the provincial and federal governments. For sixty-two years, from 1886 to 1947, all head taxes, pecuniary penalties, departure fees, and revenues from other sources under various Chinese Immigration Acts amounted to over $23 million. (See British Columbia, Department of Mines and Resources, Annual Report, 1947, 245.)
18 Chuen-yan Lai, "Chinese Attempts to Discourage Emigration," 38–40.
19 In March 1906, for example, ten Chinese labourers who were brought to Penticton by a contractor to clear land were aroused from sleep one evening by a mob, told to pack their belongings, and were driven out of town (*Pro.*, 23, 28 Mar. 1906).
20 Canada, Department of External Affairs, *Documents on Canadian External Affairs*, Vol. 1, 1908–18, Despatch 565, 598.
21 Ibid., 602–4.
22 Canada, Department of the Interior, Superintendent of Immigration, Annual Reports, 1904–17; and Department of Immigration and Colonization, Annual Reports, 1918–25.
23 Chuen-yan Lai, "Chinese Attempts to Discourage Immigration," 45.
24 Ibid.
25 Edgar Wickberg, ed., *From China to Canada* (Toronto: McClelland & Stewart 1982), 137.
26 Canada, Department of Immigration and Colonization, Annual Reports, 1918, 1919.
27 Canada, Order in Council, PC 1202, "Landing of Immigrants at Certain B.C. Port of Entry, Prohibited," 9 June 1919, *Canada Gazette*, 14 July 1919, 3824.
28 Canada, Statutes of Canada, 1920, 10–11 George V, Chap. 46, *An Act Respecting the Election of Members of the House of Commons and the Electoral Franchise*, 182–3.
29 Patricia E. Roy, "Educating the 'East': British Columbia and the Oriental Question in the Interwar Years," *BC Studies*, 18 (1973), 52.
30 Canada, House of Commons, *Debates,* 8 May 1922, 1509.
31 Canada, Statutes of Canada, 1923, 13–14 George V, Chap. 38. *An Act Respecting Chinese Immigration*, 301–15.
32 Ibid.

33 Stanislaw Andracki, *Immigration of Orientals into Canada with Special Reference to Chinese* (New York: Arno Press 1978), 102.
34 *Col.*, 6 Oct. 1883. This estimated number would be higher if it included those who had left Victoria by steamers and been smuggled into Portland and San Francisco.
35 James Morton, *In the Sea of Sterile Mountains: The Chinese in British Columbia* (Vancouver: J.J. Douglas 1974), 227.
36 The three Chinese non-census sources are: 1892–1915 donor receipts from Victoria's Chinese Hospital; a register of boxes of bones shipped to China in 1937; and a register of boxes of bones stored in Victoria's Chinese Cemetery in the late 1930s.
37 Canada, Department of Mines and Resources, Annual Report, 1947, 245.
38 British Columbia, Legislative Assembly, Report on Oriental Activities within the Province, 1927, 5.
39 *Cen.*, 1931, Vol. I, Table 3, 236–7.
40 Canada, Department of Immigration and Colonisation, Annual Report, 1931–2, Vol. II, 65.
41 VS, 28 Dec. 1934; TV, 11 Feb. 1935; *Col.*, 3 May 1935.
42 Canada, Department of Immigration and Colonisation, Annual Report, 1934–5, Vol. III, p. 79.
43 W.E. Willmott, "Some Aspects of Chinese Communities in British Columbia Towns," *BC Studies*, 1 (1968–9), 29.
44 Walter Isard, ed. *Methods of Regional Analysis: An Introduction to Regional Science* (Cambridge, MA: MIT Press 1960), 259.
45 *Cen.*, 1941, Table 32, 500–1.
46 Ibid., 442–3.

NOTES TO CHAPTER FIVE

1 Paul Yee, *A Walking Tour of Vancouver's Chinatown* (Vancouver: Weller Cartographic Services Ltd. 1983), photo 2.
2 Clyde Gilmour, "What, No Opium Dens?" *Maclean's Magazine*, 15 Jan. 1949, 16.
3 Bruce Ramsey, *Ghost Towns of British Columbia* (Vancouver: Mitchell Press Ltd. 1970), 48.
4 *Similkameen Star*, 10 Sept. 1915.
5 W.M. Hong, *And so that's How It Happened: Recollections of Stanley Barkerville, 1900–1975* (Quesnel: published by the author, 1978), 185.
6 Ibid., 189–90.
7 Lynne Bowen, *Boss Whistle: The Coal Miners of Vancouver Island Remember* (Lantzville: Oolichan Books 1982), 49.
8 T.W. Paterson, *Ghost Town Trails of Vancouver Island* (Langley: Stage Coach Pub. 1975), 66, 88.
9 *Cen.*, 1901.
10 Canada, *Royal Commission on Coal Mining Disputes on Vancouver Island* (Ottawa: King's Printer 1913), 5.
11 A.J. Wargo, "The Great Coal Strike: The Vancouver Island Coal Miners' Strike, 1912–1914" (unpublished BA graduating essay, 1962), UBC, 86.
12 Ibid., 92.
13 Ibid., 119.
14 Paterson, *Ghost Town Trails of Vancouver Island*, 72.
15 Commox District *Free Press,* 18 May 1966.
16 Ibid.
17 Letter, A.F. Buckham of Cumberland to W.E. Ireland, Provincial Archivist of British Columbia, 6 Apr. 1960.
18 Lynne Bowen, *Boss Whistle: The Coal Miners of Vancouver Island Remember* (Lantzville: Oolichan Books 1982), 89. Bevan's Chinatown was levelled by a fire in 1922.
19 Ibid.
20 Bill Johnstone, *Coal Dust in My Blood: Heritage Record No. 9* (Victoria: Provincial Museum 1980), 109.
21 Ibid.

22 Kamloops *Daily Sentinel*, 22 Apr. 1965.
23 TV, 12 Mar. 1965.
24 Susan Mayse, "Coal town, Boomtown, Ghost Town?," *Canadian Heritage* (Oct.–Nov. 1985), 18.
25 *Col.*, 14 Oct. 1979, and NFP, 9 Mar. 1908.
26 *Pro.*, 5 May 1910.
27 NFP, 22 June 1908. Pine Street outside Chinatown was paved.
28 T.A. Rickard, "A History of Coal Mining in British Columbia," *The Miner* (June 1942), 33.
29 British Columbia, Land Registry Office, *Certificate of Title No. 12864-N*, 1929: ABCDE of Lot 4 in Block N and ABCDEFG of Lot 1 and 2 in Block O and all of section G of Lot 3 and Section G of Lot 4 in Block N, 9 May 1929.
30 *Vancouver Sun Magazine Supplement*, 29 Jan. 1955.
31 Cen., 1921, 1941.
32 A. Woodland, *New Westminster: The Early Years, 1858–1898* (New Westminster: Nunaga Publishing 1973), 16–17.
33 BCD, 1897–8, 68, and 1901, 463.
34 VS, 26 Mar. 1913.
35 *Pro.*, 23 Nov. 1906.
36 BCD, 1921, 428.
37 Ibid., 1938, 583.
38 *Pro.*, 19 Apr. 1940.
39 Mary Balf, *Kamloops: A History of the District up to 1914* (Kamloops: History Committee, Kamloops Museum 1969), 49.
40 Ibid., 89.
41 Ruth Balf, *Kamloops: 1914–1945* (Kamloops: History Committee, Kamloops Museum 1975), 120.
42 Kamloops *Daily Sentinel*, 1 Feb. 1927.
43 Ormsby, *British Columbia*, 297.
44 VN, 2 June 1886.
45 Eric Nicol, *Vancouver* (Toronto: Doubleday 1970), 56.
46 *Col.*, 24 May 1887.
47 Alan Morley, *Vancouver: From Milltown to Metropolis* (Vancouver: Mitchell Press 1961), 97.
48 VN, 27 July 1886.
49 *Col.*, 13 Jan. 1887.
50 Ibid., 13, 14 Jan. 1887.
51 VN, 15 Jan. 1887.
52 *Col.*, 15 Jan. 1887.
53 VN, 4 Feb. 1887.
54 Ibid., 9 Feb. 1887.
55 Ibid., 25 Feb. 1887.
56 *Col.*, 26 Feb. 1887.
57 Ibid., 26 Jan. 1887.
58 Ibid., 27 Jan. 1887.
59 Ibid., 2 Mar. 1887.
60 Patricia E. Roy, "The Preservationof the Peace in Vancouver: The Aftermath of the Anti-Chinese Riot of 1887," *BC Studies*, 31 (1976), 58.
61 BCD, 1889, 535–6.
62 Vancouver *Daily World*, 28 Apr. 1899.
63 Ibid., 7 Apr. 1899.
64 Ibid., 23 Mar. 1893.
65 S.S. Osterhout, *Orientals in Canada* (Toronto: United Church of Canada 1929), 80.
66 Vancouver *Daily World*, 11 July 1889.
67 Ibid., 8, 9 Sept. 1892.
68 Ibid., 23 July 1898, and 24 Oct. 1899.
69 *Cen.*, 1891, 1901.

70 *British Columbia Gazetteer and Directory*, 1901, 588.
71 Raymond Edger Young, "Street of Tongs: Planning in Vancouver Chinatown" (unpublished MA thesis, UBC 1975), 62.
72 Paul Richard Yee, "Chinese Business in Vancouver, 1886–1914" (unpublished MA thesis, UBC, 1983), 46.
73 *British Columbia Gazetteer and Directory*, 1904, 911.
74 Ibid., 1905, 50, 140.
75 Paul Yee, *A Walking Tour of Vancouver's Chinatown* (Vancouver: Weller Cartographic Services Ltd. 1983), photo 1.
76 R.E. Wynne, "American Labor Leaders and the Vancouver Anti–Oriental Riot," *Pacific Northwest Quarterly*, 57 (1966), 176.
77 T. Ferguson, *A White Man's Country: An Exercise in Canadian Prejudice* (Toronto: Doubleday Canada 1975), 4.
78 Ken Adachi, *The Enemy that Never Was* (Toronto: McClelland & Stewart 1976), 72–6.
79 Canada, *Report on Losses Sustained by the Chinese Population of Vancouver B.C. on the Occasion of the Riots in that City in September, 1907* (Royal Commission: W.L. Mackenzie King) Sessional Papers, No. 74f, 1908, 18.
80 *Cen.*, 1921, 1931.
81 *Pro.*, 15 Oct. 1941.
82 VS, 11 Aug. 1944.
83 VN, 4 May 1945.
84 VN, 4Feb. 1941, and VS, 4 Feb. 1941.
85 VS, 18 July 1945.
86 Alan F.J. Artibise, "The Urban West: The Evolution of Prairie Towns and Cities to 1930," *Prairie Forum*, 4 (1979), 245.
87 Ibid., 247.
88 J. Brian Dawson, "The Chinese Experience in Frontier Calgary: 1885–1910," in *Frontier Calgary: Town, City and Region*, ed. by A.W. Rasporich and H.C. Klassen (Calgary: McClelland & Stewart West 1975), 126.
89 Ibid., 127.
90 Gunter Baureiss, "The Chinese Community in Calgary," *Alberta Historical Review*, 22 (1974), 3.
91 *Cen.*, 1901.
92 P. Voisey, "Chinatown on the Prairies: The Emergence of an Ethnic Community," *Selected Papers from the Society for the Study of Architecture in Canada Annual Meetings in 1975 and 1976* (Ottawa: Society for the Study of Architecture in Canada 1981), 35.
93 *Calgary City Directory*, 1910, 576.
94 Dawson, "The Chinese Experience in Frontier Calgary," 137.
95 Gunter Baureiss, "The Chinese Community of Calgary," *Canadian Ethnic Studies*, 2 (1971), 51–2.
96 Osterhout, *Orientals in Canada*, 99–100.
97 *Cen.*, 1941.
98 Lethbridge *News*, 7 May 1890.
99 *Pro.*, 22 Apr. 1904.
100 *Lethbridge City Directory*, 1909, 43–63.
101 A. Johnson, and Andy A. den Otter, *Lethbridge: A Centennial History* (Lethbridge: City of Lethbridge and Whoop-Up Country Chapter, Historical Society of Alberta 1985), 57, 86.
102 L.G. Ellis, "Love Thy Neighbour: Social and Racial Attitudes in Lethbridge, 1885–1945" (unpublished ms), 9.
103 J.B. Joyner, "Lethbridge Chinatowns: An Analysis of the Kwong On Lung Co. Building, the Bow On Tong Co. Building, and the Chinese Free Masons Building, 1985" (unpublished ms submitted to Historic Sites Service, Edmonton, Alberta), 14.
104 Editorial, "Edmonton's Growth," *Chinese-Canadian Bulletin,* No. 122 (1981), 42.
105 J.P. Day, "Wong Sing Fuen and the Sing Lee Laundry," Historical and National Science Services, Edmonton Parks and Recreation Department, Nov. 1978, 6, 9.

106 City of Edmonton, Planning Department, "Downtown Plan Working Paper No. 1: The Future of Chinatown, 1978," 24–5.
107 Osterhout, *Orientals in Canada*, 102.
108 City of Edmonton Planning Department," Downtown Plan Working Paper No. 1," 25; and Sing Tao Jih Pao, 28 Jan. 1984.
109 Ben Seng Hoe, *Structural Changes of Two Chinese Communities in Alberta* (Ottawa: Canadian Centre for Folk Culture Studies 1976), 113.
110 Anon., "A Brief History," *Century Regina, 1882–1982* (Regina: Southland Mall 1982), 12.
111 Tim Yee et al., "An Ethnic Study of the Chinese Community of Moose Jaw, Moose Jaw, 1973" (unpublished report on Opportunities for Youth Project, May–Aug. 1973), 8.
112 Ibid., 18.
113 Peter S. Li, "Chinese Immigrants on the Canadian Prairie, 1910–1947," *Canadian Review of Sociology and Anthropology,* 19 (1982), 533.
114 Osterhout, *Orientals in Canada*, 104.
115 Statutes of the Province of Saskatchewan, 4th Session, Second Legislature, 1912: Chap. 17 An Act to Prevent the Employment of Female Labour in Certain Capacities (assent given 15 Mar. 1912).
116 Yee, "An Ethnic Study of the Chinese Community of Moose Jaw," 10.
117 Ibid., 12.
118 *Cen.*, 1931.
119 Ernest Chan, retired school-teacher in Saskatoon, interview, July 1986.
120 *Regina City Directory*, 1907.
121 Ibid., 1914.
122 Ibid., 1940.
123 WFP, 19 Nov. 1877.
124 Y.L. Chan, "Planning for Change: The Winnipeg Chinese Community and Its Responsiveness to Government Services" (unpublished Master of Community Planning thesis, University of Manitoba 1962), 80.
125 Julia Kwong, "Transformation of an Ethnic Community: From a National to a Cultural Community," in *Asian Canadians and Multiculturalism: Selections from the Proceedings of the Asian Canadian Symposium IV, May 1980,* ed. by Victor Ujimoto and Gordon Hirabayashi, 90.
126 Gan Low, Chinese old-timer in Winnipeg, private interview, June 1986.
127 Kwong, "Transformation of an Ethnic Community," 90.
128 Gan Low, Chinese old-timer in Winnipeg, private interview, June 1986.
129 Gunter Baureiss and Julia Swong, "The History of the Chinese Community of Winnipeg" (unpublished report, Chinese Community Committee 1979), 29.
130 *Cen.*, 1901, 1911.
131 A. Pan, "History of Winnipeg Chinese United Church," *Chinatown News*, 18 October 1983, 16.
132 Osterhout, *Orientals in Canada*, 106–7.
133 Gunter Baureiss and L. Driedger, "Winnipeg Chinatown: Demographic, Ecological and Organizational Change, 1900–1980," *Urban History Review*, x (1982), 15.
134 Kwong, "Transformation of an Ethnic Community," 88.
135 Ibid., 92.
136 Winnipeg *Tribune*, 2 Mar. 1942.
137 Rebecca B. Aiken, "Montreal Chinese Property Ownership and Occupation Change: 1881–1981" (unpublished PH.D. thesis, McGill University 1984), 60–3.
138 Valerie A. Mah, "The 'Bachelor' Society: A Look at Toronto's Early Chinese Community from 1878 to 1924" (unpublished BA essay, University of Toronto 1978), 6–9.
139 Ibid., 18.
140 Ibid., 20–2.
141 *Toronto City Directory*, 1910.
142 Lao Bo, "Hostages in Canada: Toronto's Chinese (1880–1947)," *The Asianadian*, 1 (1978), 11.

143 K. Paupst, "A Note on Anti-Chinese Sentiment in Toronto Before the First World War," *Canadian Ethnic Studies,* 9 (1977), 57.
144 Ibid.
145 Dora Nipp, "The Chinese in Toronto," in *Gathering Place: Peoples and Neighbourhoods of Toronto 1834–1945*, ed. Robert R. Harney (Toronto: Multicultural History Society of Ontario 1985), 156–7, 167–8.
146 *Cen.,* 1911.
147 Ibid., 1931.
148 *Ottawa City Directory*, 1931.
149 Ibid., 1941.
150 Wenxiong Gao, "Hamilton: The Chinatown that Died," *The Asianadian*, 1 (1978), 15.
151 Ibid., 16.
152 May Wong et al., "A Report on the Development of the Chinese Community in Hamilton" (unpublished ms, Chinese Cultural Association of Hamilton 1984), 17.
153 Rebecca B. Aiken, "Montreal Chinese Property Ownership and Occupation Change: 1881–1981" (unpublished PH.D. thesis, McGill University 1984), 52.
154 Ban Seng Hoe, "Chinese Community and Cultural Traditions in Quebec City," in *Chinese Consolidated Benevolent Association 1985 Tri-Celebration Special Issue* (Victoria: Chinese Consolidated Benevolent Association, 1986), 131.
155 Percy A. Robert, "Dufferin District: An Area in Transition" (unpublished MA thesis, McGill University 1928), 14.
156 *Montreal City Directory*, 1910–11.
157 Ibid., 1921–3.
158 Jane Hong et al., *Chinese Community in Newfoundland* (unpublished ms, Chinese Student Society of Memorial University of Newfoundland 1976), 24.

NOTES TO CHAPTER SIX

1 Canada, House of Commons, Debates, Ottawa, Vol. III, May 1947, 2644–6.
2 B.L. Sung, *The Story of the Chinese in America* (New York: Collier 1967), 80–1.
4 Canada, Statutes of Canada, An Act to Amend the Immigration Act and to Repeal the Chinese Immigration Act, 1947. Ottawa, 11 George VI, Chap. 19, 107–9.
5 Canada, Order in Council, PC 2115, 16 Sept. 1930.
6 Canada, Order in Council, PC 1378, 17 June 1931.
7 *Cen.,* 1941. That year there were only 6,860 locally born Chinese, constituting about 20 per cent of the Chinese population in Canada.
8 Canada, Dominion Bureau of Statistics, *Canada Year Book*, Ottawa, 1948–9, 174.
9 Canada, Department of Citizenship and Immigration, Immigration Statistics, 1957.
10 Foon Sien, "Yesterday, Today, Tomorrow," *Chinatown News*, 2 Nov. 1956, 16.
11 EJ, 25 May 1960. Many Chinese came to Canada as "paper sons," immigrants who used real or falsified identification papers of deceased people with living relatives in Canada.
12 Ibid., 24 June 1966.
13 Edgar Wickberg, ed., *From China to Canada* (Toronto: McClelland & Stewart 1982), 137.
14 EJ, 23 Oct., 22 Nov. 1962.
15 Canada, Order in Council, PC 1962–86, 18 Jan. 1962.
16 *Pro.,* 12 Jan. 1965.
17 Ibid., 8 Oct. 1968.
18 Canada, Dominion Bureau of Statistics, *Canada Year Book,* Ottawa, 1968, 227.
19 Canada, Statutes of Canada, An Act Respecting Immigration to Canada, 21–2 Elizabeth II, Chap. 27, 1973.
20 Canada, Department of Manpower and Immigration, Highlights from the Green Paper on Immigration and Population, Ottawa 1975.
21 Ching Ma, *Chinese Pioneers* (Vancouver: Versatile Publishing 1979), 75–82.
22 Canada, Statutes of Canada, An Act Respecting Immigration to Canada, 25–6 Elizabeth II, Chap. 52, 1976–7.

23 Canada, Ministry of Industry, Trade and Commerce, *Canada Year Book* (Ottawa, 1980–1), 125.
24 Canada, Department of Employment and Immigration, *Kaleidoscope: The Quarterly Review of Immigration and Ethnic Affairs* (1982–3), 3.
25 Canada, Minister of Supply and Services, *Canada Year Book* (Ottawa 1985), 48.
26 *Times-Colonist,* 26 Oct. 1985.
27 Canada, Employment and Immigration Commission, *Immigration*, 1 (1986), 2.
28 *Financial Post*, Toronto, 17–23 Nov. 1986.
29 *Chinese Times,* 31 Oct. 1987; Canada, Department of Employment and Immigration, *Annual Report to Parliament on Future Immigration Levels* (1987), i–ii. In 1987, nearly 134,000 immigrants came to Canada, exceeding the 125,000 level (*Times-Colonist*, Victoria, 2 Mar. 1988).
30 *Times-Colonist*, 4, 23 Apr. 1987.
31 *Financial Post*, Toronto, 11–17 May 1987.
32 Canada, Minister of Employment and Immigration, news releases, 27 Apr. 1988 and 27 May 1988.
33 Emily Lau, "On to Greener Pastures," *Far Eastern Economic Review*, 18 June 1987, 23.
34 Philip Bowring, "Directing the Elections," *Far Eastern Economic Review*, 7 July 1987, 30.
35 Canada, Department of Manpower and Immigration, *Immigration Statistics*, Table 3, 1975 and 1976, 6.
36 *Chinatown News*, 18 May 1985, 8.

NOTES TO CHAPTER SEVEN

1 Hope *Standard*, 29 Sept. 1982.
2 Canada, Historic Sites and Monuments Board of Canada, "Program of the Unveiling Ceremony of the Plaque at Yale, 25 Sept. 1982, 1.
3 Cumberlalnd and District Historical Society, Schedule of Events on Workers' Memorial Day at Cumberland, 21 June 1986, 3.
4 *Chinatown News*, 3 Feb. 1987, 16.
5 VS, *Magazine Supplement*, 29 Jan. 1955.
6 NFP, 1 Oct. 1960.
7 Ibid., 30 Sept. 1970.
8 Ibid., 27 July 1962.
9 *Col.,* 17 Sept. 1967.
10 Kamloops *Daily Sentinel*, 22 Apr. 1965.
11 Gustavo Da Roza, *Winnipeg Chinatown: A Proposal* (Winnipeg: Winnipeg Chinese Development Corporation 1971), 1–3.
12 *Chinatown News*, 3 Feb. 1982, 33.
13 Wei-man Lee, Manager of Cathay House Restaurant, private interview, Ottawa, Apr. 1983.
14 CH, 12 July 1973.
15 Kamloops *Daily Sentinel*, 6 Mar. 1946.
16 Jaime Jung, owner of Paradise Café, interview, July 1987.
17 Sien Lok Society, Calgary, "National Conference on Urban Renewal As It Affects Chinatown," 6–9 Apr. 1969, 2.
18 Leonard Marsh, *Rebuilding a Neighbourhood: Report on a Demonstration Slum Clearance and Urban Rehabilitation Project in a Key Central Area in Vancouver, Research Publication No. 1* (Vancouver: University of British Columbia 1950), 3, 71.
19 City of Vancouver, Planning Department for the Housing Research Committee, Dec. 1957, 48.
20 Ibid., 12, 15, 31, 43.
21 Ibid., 37.
22 City of Vancouver, Planning Department, Urban Renewal in Vancouver, Progress Report, No. 7, 1966, 8.
23 VS, 5 Oct. 1960.
24 Ibid.
25 City of Vancouver, Planning Department, Urban Renewal in Vancouver, Progress Report, No. 7, 1966, 3–4.

26 vs, 23 Jan. 1963.
27 Ibid., 7 Jan. 1961.
28 Ibid., 16 Mar. 1961.
29 Ibid., 9 Jan. 1962.
30 Ibid., 19 Jan. 1963.
31 *Pro.*, 23 Jan. 1963.
32 Ibid., 19 Jan. 1963.
33 vs, 23 Jan. 1963.
34 *Pro.*, 30 Mar. 1963.
35 vs, 30 Mar. 1963.
36 Ibid., 29 Apr. 1963.
37 Ibid., 24 July 1963.
38 Ibid., 14 Dec. 1963.
39 Ibid., 13 Mar. 1964.
40 City of Vancouver, Planning Department, Urban Renewal Proposed Study under Part v of the National Housing Act, Aug. 1966, 21.
41 *Pro.*, 26 Aug. 1966.
42 United Community Services of the Greater Vancouver Area, *Urban Renewal Scheme III, Strathcona*, July 1966, Appendix vi, page iii.
43 United Community Services of the Greater Vancouver Area, Draft Report of Committee on Redevelopment and Relocation (n.d.), 1.
44 City of Vancouver, Urban Renewal Proposed Study under Part v of the National Housing Act, 22.
45 vs, 17 Oct. 1967.
46 Ibid., 19 Oct. 1967.
47 *Pro.*, 23 Oct. 1967.
48 vs, 18 Jan. 1968.
49 City of Vancouver, Urban Renewal Program Scheme 3: Sub-area 1—Strathcona, Appendices, n.d., "Brief to City Council from a Citizens Group," 1.
50 Submission of the Chinatown Property Owners Association to Mr. J.V. Clyne, Royal Commission, Property Expropriation, Regarding Vancouver Redevelopment in Area A, n.d.; and A Brief of the Chinatown Property Owners Association Presented to the Royal Commission on Expropriation Laws and Procedures, 17 July 1961.
51 Richard Nann, "Urban Renewal and Relocation of Chinese Community Families" (unpublished PH.D. thesis, University of California, Berkeley 1970), 70.
52 Ibid., 77.
53 Government of Canada, Report of the Task Force on Housing and Urban Development, Ottawa, Queen's Printer 1969, 12.
54 Larry I. Bell, and Richard Moore, *The Strathcona Rehabilitation Project: Documentation and Analysis* (Vancouver: Social Policy and Research Department of United Way of Greater Vancouver, December 1975), 12.
55 vs, 14 Aug. 1969.
56 David N. Spearing, W.W. Wood, and W.H. Birmingham, *Draft of a Report of Rehabilitation through Cooperation in Strathcona* (Dec. 1970), 3.
57 City of Vancouver, Strathcona Rehabilitation Committee, *Strathcona Rehabilitation Program: Recommendations of the Strathcona Rehabilitation Committee* (July 1971), 1.
58 *City of Vancouver, Strathcona Rehabilitation Project, Stage II Evaluation* (1977), 61, 63.
59 Ibid., 73.
60 Vancouver, Chinese Cultural Centre, *Chinese Cultural Centre* (leaflet, n.d.).
61 *Vancouver Chinese Cultural Centre Reports*, Vol. I (Nov. 1976), 1.
62 University of British Columbia, Alma Mater Society, *The Ubyssey*, 28 Sept. 1978.
63 Committee to Democratize the CBA, *CBA Issue* (Jan. 1978), 2.
64 *Chinese Times* (Dahan Gongbao), 13 May 1978.
65 *Cen.*, 1971, 1981.
66 British Columbia, Gazette, Vol. CXI, No. 7, 18 Feb. 1971.

67 City of Vancouver, Planning Department, Chinatown Planning Newsletter, Nov. 1976, unfolioed.
68 The $6.6 million Dr. Sun Yat-Sen Park and Garden project has two components. The Classical Chinese Garden cost $5 million, of which the provincial government contributed $1.5 million, the City of Vancouver $500,000 (donation of the land), China $500,000 (donation of materials and labour), and individual and corporate donors $2.7 million. The Garden, administered by the Dr. Sun Yat-Sen Garden Society of Vancouver, was officially opened on 24 April 1986. An admission fee is charged. The other component of the project is the Dr. Sun Yat-Sen Park, which cost $1.6 million; the federal government contributed $1.5 million, and individual and corporate donors $100,000. The park, administered by the Vancouver Parks Board, was officially opened on 16 September 1986. No admission fee is charged.
69 *Chinese Times* (Dahan Gongbao), 14 Feb. 1986.
70 Paul Richard Yee, "Chinese Business in Vancouver, 1886–1914" (unpublished MA thesis, University of British Columbia 1983), 31.
71 *Chinatown News*, 18 Feb. 1988, p. 18.
72 City of Vancouver, Planning Department, *Reports to Council: North Park Development Concept*, 12 Mar. 1986, 1.
73 Ibid., 12.
74 CH, 19 June 1945.
75 City of Calgary, Planning and Building Department, Chinatown Area Redevelopment Plan, 1986, 6.
76 City of Calgary, Planning Department, Calgary Chinatown Design Brief, September 1976, 9.
77 City of Calgary, Planning Department, Chinatown-Calgary, Nov. 1972, 15.
78 City of Calgary, Chinatown Design Brief, 1.
79 CH, 16 Sept. 1976.
80 City of Calgary, Chinatown Design Brief, 32.
81 Ibid., 2–4.
82 CH, 24 Nov. 1978.
83 Ibid., 12 Mar. 1979.
84 Ibid., 30 Aug. 1977, and CT, 11 July 1979.
85 CH, 28 April 1982.
86 Chinatown Ratepayers Association of Calgary, *Chinatown Development Proposal*, Oct. 1982, 7–9.
87 *Canadian Chinese Times*, Calgary, No. 68, 8 Oct. 1982 (in Chinese script).
88 CH, 23 Oct. 1982, and Chinatown Workshop Proposal Call, City of Calgary, 16 Nov. 1982.
89 City of Calgary, Chinatown Area Redevelopment Plan, 7.
90 Ibid., 2.
91 City of Calgary, Planning and Building Department, *Chinatown Handbook of Public Improvements*, 1985, 18.
92 City of Edmonton, Planning Department, Downtown Plan Working Paper No. 1: The Future of Chinatown, 1978, 2.
93 Ibid., 35.
94 EJ, 28 July 1973.
95 Ibid., 14 Sept. 1974.
96 City of Edmonton, Future of Chinatown, 36–7.
97 EJ, 14 Apr. 1977.
98 City of Edmonton, Future of Chinatown, 31–2.
99 City of Lethbridge, Planning and Development, Downtown Phase II Area Redevelopment Plan, 1979, 3.
100 Letter, Dr. Michael Carley, field representative of Historical Resources, Historic Sites Service, to Mr. Felix Michna, Director of Planning Department, 1 Mar. 1983.
101 City of Winnipeg, Department of Housing and Urban Renewal, Final General Report of Urban Renewal Plan No. 2, Winnipeg, Jan. 1968, 163.
102 Sien Lok Society, Calgary, National Conference on Urban Renewal, 24.
103 WFP, 9 Aug. 1975.

104 Da Roza, Gustavo, *Winnipeg Chinatown: A Proposal* (Winnipeg: Winnipeg Chinese Development Corp. Ltd 1971), 30–2.
105 Hung Yuen Lee, President of Chinese Benevolent Association, Winnipeg, interview, June 1986.
106 WFP, 9 Aug. 1975.
107 Ibid., 25 Jan. 1980.
108 A. Pan, "History of the Winnipeg Chinese United Church," *Chinatown News,* 18 October 1983, 35.
109 Winnipeg, Winnipeg Chinatown Development (1981) Corporation, *Chinatown Redevelopment Project, 1984*, B1.
110 City of Winnipeg, Winnipeg Core Area Initiative Status Report of Program Activities, 30 Sept. 1985, 56.
111 Winnipeg, Chinatown Redevelopment Project, B4.
112 WFP, 4 May 1984.
113 City of Winnipeg, *Core Area Initiative Status Report*, 56.
114 WFP, 2 Nov. 1984.
115 Ken Wong, Chairman of the Housing Corporation, interview, June 1986; and *Manitoba Chinese Post*, 1 Dec. 1985, 3.
116 Joseph Du, Chairman of Winnipeg Chinatown Development (1981) Corporation, and Hung Yuen Lee, President of Chinese Benevolent Association, interview, June 1986.
117 Martin Cash, "Hong Kong Dollars Head for the Prairies," *MicroSense: Business in the Information Age*, 2 (May 1986), 27.
118 Ken Wong, "A Letter to the Publisher of the Manitoba Chinese Post," *Manitoba Chinese Post,* 1 Jan. 1986, 18.
119 Letter, from Dr. Joseph N.H. Du to Dr. David C.Y. Lai, 10 Sept. 1986.
120 *Manitoba Chinese Post*, 1 Nov. 1986.
121 *Downtowner*, Winnipeg, No. 42, 22 Oct. 1986.
122 Letter, from Dr. Joseph N.H. Du, Chairman of the Winnipeg Chinatown Development (1981) Corporation, to Dr. David C.Y. Lai, 17 June 1987.
123 Edgar Taam, City Planner, interview, June 1986.
124 WFP, 21 Feb. 1986.
125 GM 25 Nov. 1965.
126 TST, 16 May 1967.
127 W.R. Winnicki, "Chinatown in Transition: The Impact of the New City Hall and Court House on Toronto's Chinatown" (unpublished MA thesis, University of Waterloo 1969), 40.
128 TST, 17 March 1967.
129 Ibid., 30 Aug. 1967.
130 Ibid., 24 Feb. 1969.
131 Sien Lok Society, National Conference on Urban Renewal, 27.
132 TST, 15 Apr. 1969.
133 GM, 17 June 1969.
134 TST, 15 Oct. 1969.
135 Ibid., 23 May 1970.
136 Ibid., 7 June 1971.
137 Ibid., 18 Aug. 1971.
138 Letter, from City of Toronto Planning Board, Commissioner of Planning, to Committee on Buildings and Development, 6 Feb. 1979, 1.
139 TST, 30 May 1975.
140 Ibid., 21 June 1975.
141 Letter, from City of Toronto Planning Board, Commissioner of Planning, to Committee on Buildings and Development, 6 Feb. 1979, 3.
142 Letter, from City of Toronto Planning Board, Commissioner of Planning, to Land Use Committee, 23 Mar. 1982.
143 Sien Lok Society, National Conference on Urban Renewal, 29.
144 Chinatown News, 3 Sept. 1985, 29.

145 Sien Lok Society National Conference on Urban Renewal, 29.
146 *Chinatown News*, 3 Feb. 1982, 3.
147 Zhenqiao Tan, "A Preliminary Suggestion of the Development of Montreal's Chinatown," *Quartier Chinois de Montréal* (2 Mar. 1982), 2 (in Chinese script).
148 Abraham Cohen, "City Councillor Has Views on the Planning and Development of Montreal Chinatown," *Quartier Chinois de Montréal* (2 Mar. 1982), 3.
149 *South China Morning Post*, Hong Kong, 3 July 1977.
150 Henry K.C. Ng, "Preliminary Study/Report on Redevelopment and Revitalization of Montreal Chinatown," Urban Planning Department, City of Montreal (unpublished ms), 14 Jan. 1982, 20.
151 *Chinatown News*, 3 Mar. 1983, 31.
152 Rongxu Tan, "The Sod-Turning Ceremony of Montreal's Chinese United Building," *Quartier Chinois de Montréal* (Dec. 1983), 1.
153 Kwok B. Chan, "Ethnic Urban Space, Urban Displacement and Forced Relocation: The Case of Chinatown in Montreal," *Canadian Ethnic Studies*, XVIII (1986), 70.
154 City of Montreal, By-law 6513: Zoning of the Downtown Area Located South of Sherbrooke Street between Bleury and Sanguinet streets, 22 Oct. 1984.
155 *Chinatown News*, 3 May 1985, 23.
156 *The Gazette*, Montreal, 13 Mar. 1985; *Globe and Mail*, Toronto, 14 Mar. 1985.
157 Montreal Chinese Professional and Businessmen's Association, "Report of the Public Hearings on the Impacts of Municipal Zoning By-law No. 6513 upon Chinatown in Montreal" (unpublished ms, n.d., unfolioed).
158 Ibid., 18 Aug. 1985.
159 City of Montreal, By-law 6748: By-law to amend By-laws 6513 and 6612, 26 Aug. 1985.
160 *Asian Leader*, Montreal (Dec. 1985–Jan. 1986), 3.
161 *Chinatown News*, 3 Mar. 1986, 16.
162 *Chinese Press* (Huaqiao Shibao), Montreal, 10 May 1976, 8.
163 *Chinatown News*, 3 Aug. 1986, 3.
164 *Chinese Press*, Montreal, 7 Nov. 1987.
165 Montreal Chinese Community United Centre. "First Public Opinion Survey," 13–16, 18–19 July 1987.
166 Ibid., "Second Public Opinion Survey," 14–15 Nov. 1987.
167 City of Montreal, Urban Development Department. *Quartiers du coeur de Montréal* (Mar. 1988), 21.
168 Father Thomas Tou, Chinese Catholic Church, interview, Apr. 1988.
169 Pierre Boucher, Executive Director, Montreal Chinese Hospital, interview, Apr. 1988.
170 Montreal Chinese Community United Centre, *Great Wall Campaign*, undated (probably published in 1987), 6–9.
171 L. Tom, Chairman of Montreal Chinese Community United Centre, interview, Apr. 1988.
172 EJ, 21 June 1979.
173 City of Edmonton, Planning Department, Revised Chinatown Plan, 1980, Appendix A, Enclosure III.
174 EJ, 20 Mar. 1980.
175 Ibid., 12 June, 24 July 1980.
176 City of Edmonton, Planning Department, Boyle Street/McCauley Area Redevelopment Plan Bylaw, 1981, 78.
177 City of Edmonton, Planning Department, Revised Chinatown Plan, 1980, 12.
178 City of Edmonton, Final Report of Mayor's Task Force on the Heart of the City, 1984, 51.
179 Letters, from Armin Preiksaitis, Manager, Area Planning Branch, City of Edmonton, to David C.Y. Lai, 18 Mar., 7 Apr. 1986.
180 *Edmonton Chinese News*, 4 Apr. 1986; *Chinatown News*, 18 May 1986, 19.
181 *Chinatown News*, 18 Feb. 1987, 36.
182 "Edmonton's New Chinatown Chinese Entrance Gate Plan," Shenzhen Gardens Design & Decorative Engineering Co., 5 Apr. 1987.
183 EJ, 25 Oct. 1987.

184 Kulbir Singh, Senior Planner, Planning and Building Department, City of Edmonton, interview, July 1987.
185 *Cen.*, 1951, 1961.
186 Letter, from Derek Chao, student at University of Saskatchewan, to Dr. David C.Y. Lai, 26 Nov. 1986.
187 *Ottawa City Directory, Street Guide* (1961), 5.
188 Wei-Man Lee, Manager of Cathay House, interview, Apr. 1984.
189 City of Ottawa, Planning Branch, *Dalhousie Study: Existing Conditions Report* (Nov. 1975), 16.
190 Ibid., 23.
191 City of Ottawa, Planning Branch, *Dalhousie North Redevelopment Plan* (July 1980), 1.
192 Ibid., 20.
193 *Chinese Canadian Community News*, Ottawa (May 1982), 4, 9.
194 Ottawa, Chinatown Development, "Minutes of Second Meeting," 2 February 1986, 1.
195 *Chinese Canadian Community News*, Ottawa (Aug. 1986), 14.
196 *Centretown News, Ottawa, 11 Apr. 1986.*
197 *Citizen*, Ottawa, 14 June 1986.
198 Ibid., 24 Apr. 1986.
199 *Chinese Canadian Community News*, Ottawa (July 1986), 18.
200 *Citizen*, Ottawa, 9 July 1986.
201 Ibid., 10 July 1986.
202 Ibid., 7 Aug. 1986.
203 Ibid., 16 Aug. 1986.
204 *Centretown News*, Ottawa, 25 Sept. 1987.
205 *Citizen*, Ottawa, 28 Oct. 1987.
206 Ibid., 11 Nov. 1987.
207 *Centretown News*, Ottawa, 6 Nov. 1987.
208 Letter, from L.O. Spencer, A/Director, Community Planning Branch, City of Ottawa, to David C.Y. Lai, 8 Jan. 1988.
209 Letter, from David C.Y. Lai, to Rose Kung, Project Co-ordinator, Community Development Department, City of Ottawa, 20 Jan. 1988.
210 *Ottawa Herald*, 8 Feb. 1988.
211 *Centretown News*, Ottawa, 5 Feb. 1988.
212 *Citizen*, Ottawa, 30 Jan. 1988.
213 *Cen.*, 1981.
214 C. Lau, "Proposal For a Chinese Cultural Center and Development of a 'Chinatown' Project in Essex County, 2 February 1982," 1.
215 *Sing Tao Daily*, Vancouver, 18 Oct. 1983.
216 *Chinese Times*, 16, 17 July 1987.
217 City of Toronto, Planning Board, *Official Plan Proposals: South-East Spadina*, Jan. 1978, 30.
218 Dora Nipp, "The Chinese in Toronto," in *Gathering Place: Peoples and Neighbourhoods of Toronto, 1834–1945*, ed. Robert F. Harney (Toronto: Multicultural History Society of Ontario 1985), 171.
219 *Telegram*, Toronto, 21 June 1971.
220 Valerie A. Mah, "An Indepth Look at Toronto's Early Chinatown, 1913–1933" (unpublished ms, 1977), 20.
221 City of Toronto, Planning Board, Official Plan for the City of Toronto Planning Area, Part I, (Oct. 1969), 36.
222 GM, 10 Nov. 1972.
223 Ibid., 12 June 1975.
224 Ibid., 19 Mar. 1973.
225 Ibid., 10 Nov. 1970.
226 Forbes Brown, "Mon Sheong Home for Elderly Chinese," *Living Places*, 11 (1975) 2.
227 City of Toronto, Planning Board, *South-East Spadina Appraisal* (1971), 5, and *Towards a Part II Plan for South-East Spadina* (1972), 7–9.

228 A Brief on the Chinese Community in Toronto, prepared by the Chinese Canadian Association and presented to the Special Committee on the Chinese Community, 18 June 1970, 3–4.
229 City of Toronto, Planning Board, *Planning in South-East Spadina Status Report* (June 1973), 3.
230 Edgar Wickberg, ed., *From China to Canada* (Toronto: McClelland & Stewart 1982), 264.
231 City of Toronto, Planning Board, *Towards a Part II Plan for South-East Spadina* (1972), 90.
232 Richard Yao, Managing Editor, Sing Tao Newspapers (Canada) Ltd., Toronto, interview, June 1987.
233 Anthony B. Chan, *Gold Mountain* (Vancouver: New Star Books 1982), 155.
234 Wickberg, *From China to Canada*, 265–6.
235 City of Toronto, Planning Board, *Final Recommendations–South-East Spadina* (Mar. 1979), 70.
236 TST, 28 Apr. 1986.
237 John Tam, 17 Egan Avenue, Toronto, interview, Apr. 1984.
238 City of Toronto, Planning Board, *Towards a Neighbourhood Plan: South Riverdale* (1977), 85.
239 TST, 19 Jan. 1980.
240 *Chinatown News*, 3 Mar. 1984, 35.
241 *Chinese Canadian Community News*, Ottawa (Jan. 1986), 1.
242 Woodgreen Community Centre of Riverdale, *Riverdale: The Changing Community* (n.d.), 12, 18.
242 TST, 30 July 1977.
244 Chinese restaurant owner, Glen Watford Plaza, interview, Apr. 1984.
245 *Chinatown News*, 3 June 1984, 6.
246 TST, 14 May 1984.
247 Ibid.
248 TST, 8 Aug. 1984.
249 *Mirror*, Scarborough, 9 May 1984.
250 Ibid., 23 May 1984.
251 Scarborough, CBL TV, transcript by MediaReach, No. 1509, "For Your Information," 29 May 1984, 1–2.
252 Toronto *Sun*, 29 May 1984.
253 Ibid.
254 Scarborough, CBL TV, transcript by MediaReach, No. 1504, "Metro Morning Program," 28 May 1984, 2.
255 Toronto *Sun*, 29 May 1984.
256 *Mirror*, Scarborough, 30 May 1984.
257 Margaret Hunter, "The Danger of Canada's 'Open Door' Immigration Policy," 2.
258 GM, 21 Aug. 1984.
259 TST, 16 Aug. 1984, 10 Jan. 1985.
260 GM, 9 Jan. 1985.
261 Scarborough, Planning Department, Glen Watford Commercial Area Study, 20 Feb. 1985, 19.
262 TST, 7 Feb. 1985.
263 Ibid., 3 Feb. 1986.
264 City of Regina, Economic Development Department, *Discussion Paper Regarding the Possibility of a Chinatown in Regina* (Oct. 1985), 5–7.
265 Letter, from William McKim, City Manager, City of Regina, to Mayor and Council, 2 Feb. 1987.
266 Regina *Leader-Post*, 2 Feb. 1987.
267 Letter, from William McKim, City Manager, City of Regina, to Mayor and Council, 22 Dec. 1986.
268 Regina *Leader-Post*, 21 Nov. 1986.
269 Bill Lim, "Plan to Create a Chinatown Fraught with Problems," newspaper clipping sent by Dennis S. Fong, Chairman, Regina Chinatown Steering Committee, to David C.Y. Lai, 19 May 1987.
270 Regina *Leader-Post*, 13 Jan. 1987.
271 Hamilton *Spectator*, 1 Apr. 1985.

272 *Chinatown News*, 2 Apr. 1987.
273 Paul Cowell, Hamilton alderman, interview, June 1987.
274 Alastair W. Kerr, "Barkerville in the Eighties: The Future of Barkerville's Past," *Datum*, 6 (1981), 17.
275 British Columbia, Heritage Conservation Branch and Parks and Outdoor Recreation Division, Barkerville Historic Park Concept Plan (Victoria 1981), 28.
276 Ken Mather, "Barkerville: Then and Now," *Datum*, 7 (1982), 7.
277 British Columbia, Ministry of Lands, Parks and Housing, Barkerville Provincial Historic Park, Victoria: Queen's Printer 1986, unfolioed).
278 British Columbia, Barkerville Historic Park, *Chi Kung Tong Research Contract, Schedule A* (May 1987).
279 TV, 20 Sept. 1962.
280 Ibid., 23 Nov. 1962.
281 *Chinatown News*, 3 June 1984, 28.
282 Ibid., 18 June 1985, 24.
283 Village of Cumberland, *Application for Expo '86 Legacy Funding to Build and Develop a Restoration of Cumberland's Historic Chinatown* June 1986 (pages unfolioed); and letter, William Moncrief, Jr., Mayor of Cumberland, to David C.Y. Lai, 12 Jan. 1987.
284 Cumberland Heritage Conservation Society, *A Legacy of the Comox Valley* (1987), 15.
285 Ibid., 7–8.

NOTES TO CHAPTER EIGHT

1 Herbert P. Plasterer, *Fort Victoria: From Fur Trading Post to Capital City of British Columbia, Canada* (Victoria: Colonist Printer n.d.), 9.
2 C.N. Forward, *Land Use of the Victoria Area, British Columbia* (Ottawa: Geographical Branch, Department of Energy, Mines and Resources 1969), 9.
3 Harry Gregson, *A History of Victoria, 1842–1970* (Victoria: Victoria Observer 1970), 6.
4 Mathew Macfie, *Vancouver Island and British Columbia* (London: Longman, Roberts and Green 1865), 65.
5 Ibid., 66.
6 Ibid., 86.
7 The Victoria Incorporation Act, 1863, British Columbia Directory and Victoria Directory (Victoria: Howard and Barnett 1863), 94.
8 M.A. Ormsby, *British Columbia: A History* (Vancouver: Macmillan 1971), 167.
9 VG, 1 Mar. 1858.
10 The Victoria Incorporation Act, 1863, British Columbia Directory and Victoria Directory (Victoria: Howard and Barnett 1863), 94.
11 *Col.*, 16 May 1861.
12 Ibid., 19 May 1865.
13 Ibid., 20 May 1865.
14 Ibid., 19 June 1865.
15 BCD, 1871, 95.
16 First Victoria Directory and British Columbia Guide, (Victoria: E. Malladaine 1874), 1.
17 *St.*, 22 July 1875.
18 *Col.*, 13 Apr. 1878.
19 Harry Gregson, *A History of Victoria, 1842–1970* (Victoria: Victoria Observer 1970), 124.
20 VG, 14 July 1858.
21 British Columbia, Land Registry office, *Absolute Fees Book*, Vol. 2, Fol. 619, No. 2567 and No. 14877.
22 Ibid., Vol. 5, Fol. 324, No. 1861a.
23 Ibid., Vol. 5, Fol. 776, No. 2371.
24 *Col.*, 13 Sept. 1878.
25 "Trades' Licence Act, 1860,' *Supplement to the Daily British Colonist*, 8 May 1861, 1.

26 VG, 30 June 1858.
27 Commemorative Issue of the Opening of the Yue Shan Society Building in Vancouver, Vancouver (1949, Chinese script), 2–3.
28 In the late 1860s, Tai Soong & Company was moved from Johnson Street to Cormorant Street and operated by Chan Tan.
29 *British Columbia Directory and Victoria Directory* (1868), 77.
30 *Col.,* 26 Apr. 1860.
31 *Col.,* 12 Oct. 1860.
32 British Columbia, Land Registry Office, *Absolute Fees Book*, Vol. 2, Fol. 619, No. 2567, and No. 14877.
33 Ibid., Certificate No. 14877–3.
34 *Col.,* 23 Apr. 1866.
35 British Columbia, Land Registry Office, Certificate No. 14877–15.
36 *Col.,* 24 Mar. 1865.
37 *Col.,* 10 Aug. 1865.
38 JLABA, 1871, "Goods in Bond at Victoria," 58.
39 *Col.,* 27 Jan. 1872.
40 Walter B. Cheadle, *Cheadle's Journal of a Trip Across Canada, 1862–63* (Edmonton: Hurtig 1971), 268.
41 VG, 10 Aug. 1858.
42 *Col.,* 1 Mar. 1860.
43 Ibid., 8 Mar. 1860.
44 Ibid., 5 Apr. 1864.
45 James Morton, *In the Sea of Sterile Mountains: The Chinese in British Columbia* (Vancouver: J.J. Douglas 1974), 47.
46 Ibid., 48.
47 *Col.,* 4 Jan. 1872.
48 Ibid., 13 Sept. 1878.
49 Ibid., 18 Sept. 1878.
50 Ibid., 22 Sept. 1878.
51 Ibid., 18 Sept. 1878.
52 Ibid., 26 Oct. 1878.
53 Edgar Wickberg, ed., *From China to Canada* (Toronto: McClelland & Stewart 1982), 35.
54 *Col.,* 21 Jan. 1877, 5 Jan. 1879.
55 Chuen-yan David Lai, "Far from South China Shores," Victoria *Times*, 25 Jan. 1980, 2.
56 British Columbia, Land Registry Office, *Absolute Fees Book*, Vol. 5, Fol. 324, No. 182A. The property was purchased by Tsay Ching and Dong Sang.
57 *Col.,* 23 Jan. 1876.
58 S.S. Osterhout, *Orientals in Canada* (Toronto: United Church of Canada 1929), 73.
59 *Col.,* 13 Mar. 1874.
60 Ibid., 20 Mar. 1874.
61 Ibid., 17 Oct. 1874.
62 Ibid., 2 Nov. 1871.
63 Ibid., 27 Jan. 1872.
64 Ibid., 26 Feb., 9 May 1866.
65 Ibid., 15, 17 Nov. 1861.
66 Ibid., 11 Feb. 1871.
67 Ibid., 5, 6 May 1864.
68 Ibid., 1 May 1875.
69 *St.,* 15 May 1875.
70 Martin Segger, *Victoria: A Primer for Regional History in Architecture* (New York: American Life Foundation 1979), 59.
71 H.M. Lai and P.P. Choy, *Outlines of History [sic] of the Chinese in America* (San Francisco: Chinese-American Studies Planning Group 1973), 85.

72 *Col.*, 4 Oct. 1871.
73 Ibid., 14 Dec. 1871.
74 Chuen-yan David Lai, *Arches in British Columbia* (Victoria: Sono Nis 1982), 51.

NOTES TO CHAPTER NINE

1 *Col.*, 20 Jan. 1886.
2 *Royal Commission, 1885*, 366.
3 *Col.*, 6 Oct. 1883.
4 Ibid.
5 G.A. Sargison, "Victoria, B.C. Originals: Notebook by Sargison as Chief Census Officer, Victoria District, 1891,"; *Col.*, 9 Sept., 14 Oct. 1891.
6 *Royal Commission, 1902*, 13.
7 Ibid., and *Royal Commission, 1885*, 363.
8 *Col.*, 8 Oct. 1907.
9 Ibid., 10 Feb. 1909.
10 TV, 5 Oct. 1908.
11 *Col.*, 27 July 1919.
12 Ibid., 11 July 1880, 29 Apr. 1899.
13 *Royal Commission, 1885*, 41.
14 SBC, 1920, Chap. 27, 99, 101.
15 Chuen-yan David Lai, "Home County and Clan Origins of Overseas Chinese in Canada in the Early 1880's," *BC Studies*, 27 (1975), 3–29.
16 Chuen-yan David Lai, "An Analysis of Data on Home Journeys by Chinese Immigrants in Canada, 1892–1915, *Professional Geographer*, XXIX (1977), 359–65.
17 David T.H. Lee, *A History of Chinese in Canada* (Taipei: published by the author 1967), 204 (Chinese script).
18 *A Special Issue of the Golden Anniversary of Lim Sai Hor (Kow Mock) Benevolent Association* (Vancouver: Lim Sai Hor Benevolent Association 1980), Sect. 3, 2 (in Chinese script).
19 *Col.*, 26 Sept. 1886.
20 Ibid., 12 and 13 Nov. 1886.
21 Ibid., 2 and 4 Oct. 1898.
22 Ibid., 1 Dec. 1898.
23 British Columbia, Land Registry Office, The Indenture between Joshua Davies and Lee Yow Young, 3 July 1899, and The Indenture between Lee Yow Young and Chee Kong Tong Society, 5 Feb. 1910, DD 17399; and *Col.*, 25 Nov. 1900.
24 Jianwu Cao, "The Participation of the Hongmen in the Chinese Revolution" (n.d., unfolioed).
25 British Columbia, Land Registry Office, The Indenture between Chee Kong Tong and the British Columbia Land and Investment Agency, 27 Feb. 1911, File No. 17399.
26 Chuen-yan David Lai, "Contribution of the Zhigongtang in Canada to the Huanghuagang Uprising in Canton, 1911," Canadian Ethnic Studies, XIV, 3 (1982), 103.
27 Osterhout, *Orientals in Canada*, 74.
28 *Col.*, 8 Aug. 1888.
29 Ibid., 4, 5, 7 Sept., and 15, 17, 19 Dec. 1886.
30 *Col.*, 6 July 1888.
31 Osterhout, *Orientals in Canada*, 172.
32 M.M. Washington, "The Story of the Oriental Home, Victoria," *The Missionary Monthly* (July 1943), 294.
33 *Col.*, 13 July 1888.
34 Ibid., 10, 15 Feb., and 18 May 1898, 17 Mar. 1908.
35 Ibid., 27 Oct. 1891.
36 Ibid., 28 Aug. 1889.
37 Ibid., 9 Mar. 1897.
38 *Royal Commission, 1902*, 38.
39 TV, 20 Apr. 1908.

40 *Col.*, 3 Apr. 1908.
41 TV, 4 Nov., 31 Dec. 1908.
42 Karen Van Dieren, "The Response of the WMS to the Immigration of Asian Women, 1888–1942," in *Not Just Pin Money: Selected Essays on the History of Women's Work in British Columbia,* ed. Barbara K. Latham and Roberta J. Pazdro (Victoria: Camosun College 1984), 94.
43 *Col.,* 8 July 1882.
44 Ibid., 16 June 1885.
45 Ibid., 4 Feb. 1885.
46 British Columbia, Land Registry Office, *Absolute Fees Book*, Vol. 10, Fol. 504. The property at Lot 454 was bought for $3,500.
47 *Col.*, 15 Mar. 1891.
48 *The Chinese Presbyterian Church, Victoria, B.C., 1892–1983* (Victoria: Chinese Presbyterian Church 90th Anniversary Celebrations Committee 1983), 15.
49 Ibid., 16.
50 Ibid., 18.
51 M.E. Cameron, *The Reform Movement in China, 1898–1912* (New York: Octagon Books 1963), 21.
52 Ibid.
53 Lee, *History of Chinese in Canada*, 276.
54 *Col.*, 27, 28 Dec. 1899.
55 British Columbia, Land Registry Office, *The Indenture between Lee Folk Gay and Trustees of the Chinese Empire Reform Association, 2 February 1905*. Absolute Fees Book, Vol. 22, Fol. 229, No. 1084.
56 Lee, *History of Chinese in Canada*, 298–9.
57 Huang Huayi, Kuomintang member, interview, Jan. 1978.
58 *Col.*, 31 Oct., 13 Dec. 1916; TV, 21–4 Nov. 1916, 31 Jan. 1917.
59 *Col.,* 6 Sept. 1907.
60 Ibid., 31 Mar., 18 June 1908.
61 TV, 29 May 1909.
62 *Col.,* 30 Nov. 1911.
63 Ibid., 2, 4 Aug. 1911.
64 Chuen-yan David Lai, "The Chinese Consolidated Benevolent Association in Victoria: Its Origins and Functions, *BC Studies*, 15 (1972), 56–68.
65 *Col.,* 15 July 1885.
66 CCBA, "The Proposal for the Establishment of Taipingfang, 1884" (Chinese script).
67 British Columbia, Land Registry Office, Conveyance from W.H. Liver to Wong Soon Lim and Lee Mong Kow, 27 June 1899; and Conveyance from Wong Soon Lim and Lee Mong Kow to CCBA, 27 Sept. 1899, DD Roll 109 B 987.
68 CCBA, Annual Report, 1915, minutes of meeting, 23 Apr. 1915 (CCBA Archives).
69 *Col.,* 18 Jan. 1899.
70 CCBA, Annual Report, 1899, minutes of meeting, 8 Jan. 1899 (CCBA Archives).
71 *Col.,* 30 June 1899.
72 Ibid.
73 Ibid., 31 Aug. 1907.
74 Ibid., 27 Aug. 1908.
75 British Columbia, Land Registry Office. The CCBA bought Lot 606 from Lim Dat for $3,500 on 24 Aug. 1908, and the west half of Lot 607 from Lee Mong Kow for $2,000 on 8 Mar. 1909. DD 505, DD 10925.
76 TV, 6, 7 Aug. 1909; and *Col.,* 7 Aug. 1909.
77 After the downfall of the Manchu in 1911, the school was called Huaqiao Gongli Xuexiao (Chinese Public School).
78 City of Victoria, Victoria District Assessment Rooll for 1902, Vol. 2, 45–50.
79 Chuen-yan David Lai, "A Feng Shui Model as a Location Index," *Annals of the Association of American Geographers*, 64 (1974), 512.

80 Chuen-yan David Lai, "The Chinese Cemetery in Victoria," *BC Studies*, 75 (1987), 28.
81 British Columbia, Land Registry Office, Conveyance from CCBA to George Hick, 4 June 1902, DD Pocket, 25482; and CCBA, Minutes of Meeting, 30 May 1902.
82 Conveyance from Mary Williams to the CCBA, 3 Apr. British Columbia, Land Registry Office, 1903. DD Pocket 987.
83 SBC, 1899, Chap. 39.
84 SBC, 1903, Chap. 17.
85 *Col.,* 7 Feb. 1899.
86 Ibid., 28 Feb. 1904.
87 TV, 12 Dec. 1911.
88 *Col.,* 20 Sept. 1882.
89 Contract signed between the British Columbia Sugar Refining Company and the City of Vancouver, Aug. 1890. (PABC).
90 *Col.,* 22 Aug. 1909.
91 TV, 20 Oct. 1910.
92 Ibid., 19 Dec. 1911.
93 Ibid., 8 July 1912.
94 Ibid., 18 Aug. 1913.
95 Ibid., 11 Nov. 1913.
96 *Royal Commission, 1885*, 394.
97 *Col.,* 11 Nov. 1884.
98 British Columbia, Land Registry Office, *Absolute Fees Book*, Vol. 9, Fol. 845, No. 7492.
99 *Royal Commission*, 1902, 212.
100 *Col.,* 11 Aug. 1886.
101 R.D. Harvey, *A History of Saanich Peninsula Railways* (Victoria: Department of Commercial Transport, Railway Branch 1960), 6; and BCD, 1892, 1133.
102 George Hearn, and David Wilkie, *The Cordwood Limited: A History of the Victoria and Sidney Railway* (Victoria: British Columbia Railway Historical Association 1976), 48.
103 Ibid., 51.
104 Canada, House of Commons, Sessional Papers, 1907–8, Paper No. 36b, 7.
105 *Col.,* 5 Aug. 1887.
106 Ibid., 9 Sept. 1988.
107 Ibid., 8 May 1888.
108 Ibid.
109 David Edward Owen, *British Opium Policy in China and India* (Hamden, CN: Archon Books 1968), 334.
110 Frederick T. Merrill, *Japan and the Opium Menace* (New York: Institute of Pacific Relations and the Foreign Policy Association 1942), 112.
111 During his visit to Victoria in Aug. 1895, the Earl of Aberdeen, Governor-General of Canada, even toured the Tai Yune Opium Factory in Chinatown (*Col.,* 9 Aug. 1895).
112 Canada, House of Commons, Sessional Papers, 1907–8, Paper No. 10, 52–3.
113 *Col.,* 16 July 1964, 8 May 1888.
114 Canada, House of Commons, Sessional Papers, 1907–8, Paper No. 74f, 15.
115 Canada, Statutes of Canada, 7–8 Edward VII, Chap. 50, An Act to prohibit the importation, manufacture and sale of opium for other than medicinal purposes, 20 July 1908.
116 British Columbia, Land Registry Office, *Absolute Fees Book*, Vol. 8, Folio 895.
117 Data were compiled from *Victoria City Directory* and *Vancouver Island Gazetteer*, 1912.
118 G.A. Sargison, "Victoria, B.C. Originals: Notebook by Sargison as Chief Census Officer, Victoria District, 1891," *Col.,* 9 Sept., 14 Oct. 1891.
119 CCBA, Annual Report, 1893, minutes of meeting on 3 Nov. 1893 (CCBA Archives).
120 *Royal Commission, 1902*, 14.
121 *Col.,* 22 May 1884.
122 Ibid., 28 May 1885.
123 Ibid., 6 Jan. 1886.
124 Ibid., 2 Feb. 1893.

125 C.E. Glick, *Sojourners and Settlers: Chinese Migrants in Hawaii* (Honolulu: University Press of Hawaii 1980), 227.
126 *Col.*, 17 Jan. 1900.
127 TV, 29 Oct. 1907.
128 *Col.*, 8 Dec. 1908.
129 Lai, *Arches in British Columbia*, 63.

NOTES TO CHAPTER TEN

1 *Col.*, 29 Jan. 1944.
2 TV, 22 Mar. 1945.
3 Ibid., 28 Dec. 1934.
4 *Col.*, 17 July 1945.
5 British Columbia, Capital Region Planning Board, Urban Renewal Study for Victoria, 1961, 35; and TV, 18 Feb. 1964.
6 Chuen-yan David Lai, "The Demographic Structure of a Canadian Chinatown in the Mid-Twentieth Century," *Canadian Ethnic Studies*, XI, 2 (1979), 54.
7 Ibid.
8 British Columbia, Capital Region Planning Board, *Social Characteristics of Victoria Census Metropolitan Area* (1961), 37.
9 British Columbia, Ministry of Industry and Small Business Development, Central Statistics Bureau, Demographic Data of Electoral District 23 By Enumeration Areas, 1971 Census.
10 Chuen-yan David Lai, "Socio-Economic Structures and Viability of Chinatown," in *Residential and Neighbourhood Studies in Victoria*, ed. C.N. Forward, Western Geographical Series, (Victoria: University of Victoria, Department of Geography 1973), 120.
11 Chuen-yan David Lai, *The Future of Victoria's Chinatown*, Vol. 2 Tabluation of Data (Victoria: City of Victoria 1979), 21.
12 Ibid., 24.
13 *Danger: The Anti-Asiatic Weekly*, 1 (Oct. 1921), 18; 9 (Dec. 1921), 23.
14 Hilda Glynn-Ward, *The Writing on the Wall*, Introduction by Patricia E. Roy (reprint, Toronto: University of Toronto Press 1974).
15 *Col.*, 29 Aug. 1922.
16 Ibid., 5 Dec. 1939.
17 TV, 14 Nov. 1939.
18 *Pro.*, 14 Nov. 1939.
19 TV, 15 Dec. 1939.
20 Charles P. Sedgwick, "The Context of Economic Change and Continuity in an Urban Overseas Chinese Community" (MA thesis, University of Victoria 1970), 145.
21 For a detailed study of this issue, see David Chuen-yan Lai, "The Issue of Discrimination in Education in Victoria, 1902–1923," *Canadian Ethnic Studies*, XIX, 3 (1987), 47–67.
22 *Col.*, 13 Mar. 1902.
23 Ibid., 24 Aug. 1907.
24 Victoria, Victoria School Board, *Minutes, 1905–1910*, minutes of meeting on 29 August 1907, 77.
25 Ibid., "Minutes of Meeting on 8 Jan. 1908," 89.
26 Ibid., "Minutes of Meeting on 27 March 1908," 99.
27 TV, 2 Nov. 1908.
28 W. Peter Ward, *White Canada Forever* (Montreal: McGill-Queen's University Press 1978), 63.
29 TV, 28 Nov. 1921.
30 *Col.*, 29 Nov. 1921.
31 Victoria, Victoria School Board, *Minutes, Jan. 1920–Dec. 1922*, "Minutes of Meeting on 11 Jan. 1922."
32 Ibid.
33 CCBA, Annual Report, 1922, minutes of meeting on 4 Sept. 1922 (CCBA Archives).
34 *Col.*, 6 Sept. 1922.

35 CCBA, Annual Report, 1922–3, minutes of meeting on 1 Sept. 1923 (CCBA Archives).
36 Ibid., Annual Report, 1924, minutes of meeting on 12 Apr. 1923 (CCBA Archives).
37 Ibid., Annual Report, 1924, minutes of meeting on 4 May 1924 (CCBA Archives).
38 *Col.*, 9 Oct. 1927.
39 Ibid., 8 Jan. 1959.
40 A.H. Fenwick, "For East Is East," Maclean's Magazine, 15 Jan. 1928, 45.
41 TV, 12 Aug. 1939.
42 Carol F. Lee, "The Road to Enfranchisement: Chinese and Japanese in British Columbia," *BC Studies*, 30 (1976), 46.
43 Patricia E. Roy, "The Soldiers Canada Didn't Want: Her Chinese and Japanese Citizens," *Canadian Historical Review*, LIX (1978), 342–3.
44 TV, 4 Feb. 1941; *Pro.*, 14 Aug. 1944; TV, 24 Aug. 1944.
45 TV, 11 and 13 May, 26 Oct. 1943; COL., 7 Oct. 1945.
46 Carol F. Lee, "The Road to Enfranchisement," 44.
47 *Col.*, 18 June 1947.
48 Ibid., 30 June 1949.
49 Ibid., 2 June 1963.
50 Chuen-yan David Lai, Socio-Economic Structures and Viability," 112.
51 Chuen-yan David Lai, *The Future of Victoria's Chinatown, Vol. 2*, 12.
52 Ibid., 16.
53 Ibid., 17.
54 British Columbia, Land Registry Office, *Absolute Fees Book*, Vol. 22, Fol. 229, No. 38112-I, and No. 107793-I.
55 Ibid., Certificate No. 82521.
56 TV, 7 Feb. 1928.
57 CCBA, Annual Report, 1923, minutes of meeting on 26 Sept. 1931 (CCBA Archives).
58 *Col.*, 6 Mar. 1938.
59 The Rev. Chow Ye Ching, Chinese Presbyterian Church, private interview, Nov. 1976.
60 British Columbia, Hospital Act, Chap. 178, 1 Aug. 1975. Part II. Private Hospitals, 1806.
61 Peter Wong, Administrator, Chinese Hospital, private interview, Nov. 1976.
62 Dr. Andrew Rose, Hospital Inspector, private interview, Nov. 1970.
63 Letter from Dr. Andrew Rose, Hospital Inspector, to Mr. Peter Wong, Administrator of the Chinese Hospital, 2 Dec. 1976.
64 British Columbia, Community Care Facilities Licensing Act, Chap. 4, 1969, 6.
65 Miss A. Liu, social worker, Long Term Care Program, private interview, July 1977.
66 *Col.*, 20 June 1979.
67 Ibid., 21 May 1950.
68 Ibid., 4 Aug. 1961.
69 TV, 14 July 1966.
70 British Columbia, Ministry of Recreation and Conservation, Heritage Conservation Branch. *Chinatown Victoria* (August 1977), pages unfolioed.
71 Alastair W. Kerr, "The Architecture of Victoria's Chinatown," *Datum*, 4 (1979), 11.
72 Martin Segger, *Victoria: A Primer for Regional History in Architecture* (New York: American Life Foundation, 1979), 75.
73 British Columbia, Capital Region Planning Board, *Victoria: Urban Renewal Study for Victoria* (1961), 22–4.
74 Ibid., 35.
75 Ibid., 36.
76 Ibid., 35.
77 A.W. Toone, "Victoria Rediscovered," *Western Wonderland*, Dec. 1964, 6.
78 *Col.*, 2 Aug. 1962.
79 TV, 18 Feb. 1964.
80 TV, 18 Feb. 1964; *Col.*, 19 Feb. 1964; and *Col.*, 4 Nov. 1969.
81 Milton Tisdelle, developer, private interview, Sept. 1972.
82 TV, 18 May 1977.
83 Sien Lok Society, Calgary, National Conference on Urban Renewal as It Affects Chinatown, 6–9 Apr. 1969, 59.

84 CCBA, Annual Report, 1971, minutes of meeting on 6 Nov. 1971 (CCBA Archives).
85 Ibid., 21 Nov. 1971.
86 Ibid., 7 Dec. 1975, 11 Jan. 1976.
87 Augustine Low, Chairman, Cultural Centre and Community Hall Project Committee, private interview, Apr. 1976.
88 City of Victoria, Planning Department, *The Central Area of Victoria: Neighbourhood Improvement Program* (1977), 5.
89 TV, 19 Mar. 1970; Victoria *Express*, 26 Mar. 1974; TV, 30 Jan. 1976; and *Monday Magazine*, Victoria, 28 Mar.–3 Apr. 1977.

NOTES TO CHAPTER ELEVEN

1 Letter, Peter Lou-Poy, President of CCBA to Mayor Michael Young, 11 Mar. 1979.
2 Chuen-yan David Lai, *Victoria's Chinatown*, Vol. 2, 4–45.
3 Chuen-yan David Lai, *The Future of Victoria's Chinatown: A Survey of Views and Opinions*, Vol. I. Recommendations (Victoria: City of Victoria 1979), 7.
4 Ibid., 8.
5 Ibid., 22–4.
6 Lai, *Victoria's Chinatown*, Vol. 2, 42.
7 Ibid., 37.
8 Ibid., 46.
9 Ibid., 48.
10 Letter, from B.D. Strongitharm, Planning Officer, to C.J. Greenhalgh, Director of Department of Community Development, City of Victoria, 5 July 1979.
11 City of Victoria, Chinatown Ad Hoc Committee, *A Plan for the Rehabilitation of Chinatown* (Aug. 1979), 3–8.
12 Ibid., 10.
13 Letter, from Alderman R.H. Wright to Mayor Michael D.W. Young, 29 Aug. 1979.
14 City of Victoria, Group B Committee's Report to Mayor and Board of Aldermen in Council, 4 Sept. 1979.
15 TV, 8 Sept. 1979.
16 Chuen-yan David Lai, *Arches in British Columbia*, (Victoria: Sono Nis 1982), 22, 51, 63, 101.
17 Chuen-yan David Lai, *The Gate of Harmonious Interest: From Concept to Reality*, 2nd printing (Victoria: City of Victoria 1982), 7.
18 TV, 1 Apr. 1980.
19 *Gate of Harmonious Interest*, 14.
20 Ibid., 15.
21 Chuen-yan David Lai, *A Care Home in Victoria's Chinatown: A Survey of the Need* (Victoria: Capital Regional Hospital and Health Planning Commission 1979), 24.
22 Ibid.
23 British Columbia, Department of Health, *The British Columbia Classification of Types of Health Care* (Sept. 1973), 5: "The intermediate care is required by patients whose primary need is for room and board, daily professional nursing supervision, assistance with some of the activities of daily living, and a planned programme of social and recreational activities."
24 Letter from J.W. Pennington, Assistant Director of Hospital Planning, Capital Regional Hospital and Health Planning Commission, to Mrs. Isabel Kelly, Assistant Deputy Minister, Ministry of Health, 29 Nov. 1979.
25 TV, 30 Apr. 1980.
26 Letter, from Jeremy V. Tate, Facility Planning and Development Co-ordinator, Ministry of Health, to the CCBA, 21 May 1980.
27 *Times-Colonist,* 17 Feb. 1982.
29 *Chinatown News,* 3 Mar. 1980, 11.
29 Letter, Hon. Hugh Curtis, Minister of Provincial Secretary to Peter Lou-Poy, President, CCBA, 25 Oct. 1979.
30 Chuen-yan David Lai, "Final Report on the Survey of the Need for Low-Rental Housing in Chinatown and Elsewhere in Downtown Victoria," 29 July 1982 (submitted to Victoria's Housing Advisory Committee), 1.

31 Chuen-yan David Lai, ''Interim Report on Chinatown's Vacant Floors and the Designated Heritage Buildings'', 5 July 1982 (submitted to Victoria's Housing Advisory Committee), 1.
32 Letter, from Augustine Low, President of CCBA, to Canada Mortgage and Housing Corporation, 2 Dec. 1982.
33 *Chinese Canadian Bulletin*, Vol. 25, No. 153, Dec. 1985, 15.
34 Lai, *Final Report on the Survey of the Need for Low-Rental Housing*, 1.
35 City of Victoria, Heritage Advisory Committee, regular meeting minutes, 23 Jan. 1984, 3.
36 Letter, from Mark Johnston, Administrative Assistant, City of Victoria, to Peter Lou-Poy, Advisor of CCBA, 17 May 1984.
37 *Times-Colonist*, 24 July 1984.
38 Betty Campbell, ed., *Royal Visit to Victoria* (Victoria: Campbell's), pages unfolioed.
39 Horst Meissner, owner of Capital City Auto Ltd., Victoria, interview, May 1979.
40 I attended the extraordinary general meeting on 1 Apr. 1987.

NOTES TO CHAPTER TWELVE

1 O.D. Duncan, ''Human Ecology and Population Studies,'' in *The Study of Population*, ed. P.M. Hauser and O.D. Duncan (Chicago: University of Chicago Press 1959), 678–716.
2 See, for example, B. Berry and F.E. Horton, *Geographical Perspectives on Urban Systems* (New York: Prentice-Hall 1970); Larry S. Bourne, *Internal Structure of the City* (Toronto: Oxford University Press 1971); E.W. Burgess and D.J. Bogue, *Contributions to Urban Sociology* (Chicago: University of Chicago Press 1964); and G.D. Suttles, *The Social Construction of Communities* (Chicago: University of Chicago Press 1972).
3 See, for example, Leonard Dinnerstein, Roger L. Nichols, and David M. Reimers, *Natives and Strangers: Ethnic Groups and the Building of America* (New York: Oxford University Press 1979), and John Norris, ed., *Strangers Entertained: A History of the Ethnic Groups of British Columbia* (Vancouver: British Columbia Centennial '71 Committee 1971).
4 Brinley Thomas, ''International Migration,'' in *The Study of Population*, ed. P.M. Hauser and O.D. Duncan (Chicago: University of Chicago Press 1959), 519–20.
5 D.Y. Yuan, ''Voluntary Segregation: A Study of New York Chinatown,'' *Phylon: The Atlanta University Review of Race and Culture*, XXIV (1963), 259–60.
6 *Chinese Canadian Community News*, Ottawa (Oct. 1987), 27.
7 Rose Hum Lee, ''The Decline of Chinatowns in the United States,'' *American Journal of Sociology*, 54 (Mar. 1949), 425.
8 See, for example, N. Smith and P. Williams, ed., *Gentrification of the City* (Winchester, MA: Allen & Unwin 1986); David Ley, ''Alternative Explanations for Inner-City Gentrification: A Canadian Assessment,'' *Annals of the Association of American Geography*, 76 (1986), 521–35; Michael H. Lang, *Gentrification and Urban Decline: Strategies for America's Older Cities* (Cambridge, MA: Ballinger 1982); C. Hartman, D. Keating, and R. LeGates, ed., *Displacement: How to Fight It* (Berkeley, CA: National Housing Law Project 1982).
9 Mickey Lam, Senior Planner (Urban Design), City of Victoria, private interview, June 1987.
10 Calvin Lee, *Chinatown, U.S.A.* (New York: Doubleday 1965), 143.
11 *South China Morning Post*, Hong Kong, 3, 10 Aug. 1986.
12 Peter Leung, Honorary Citizen of Monterey Park, private interview, May 1987.
13 *Chinatown News*, 18 Mar. 1986, 19.
14 Ibid., 3 Apr. 1986, 6, 42, 43.
15 Nancy M. Siew and Michael J. Maclean, ''Perceptions of Elderly Hong Kong Immigrants: Implications for 1997,'' paper presented at the annual meeting of the Canadian Asian Studies Association, Hamilton, Ontario, 5 June 1987, Table 1, 2.
16 Ibid., 8.
17 Some rich Hong Kong people have bought fully-equipped launches to be used for fleeing from Hong Kong if there is a crisis after 1997.

Bibliography

GOVERNMENT DOCUMENTS

British Columbia. Barkerville Historic Park. *Chi Kung Tong Research Contract, Schedule A*, May 1987

British Columbia. Capital Region Planning Board. *Urban Renewal Study for Victoria* 1961

British Columbia. Capital Region Planning Board. *Social Characteristics of Victoria Census Metropolitan Area* 1961

British Columbia. Consolidated Statutes of the Province of British Columbia 1877

British Columbia. Department of Health. *The British Columbia Classification of Types of Health Care*, Sept. 1973

British Columbia. Department of Mines and Resources. Annual Report 1947

British Columbia. Gazette. Vol. CXI, No. 7, 18 Feb. 1971

British Columbia. Heritage Conservation Branch. Parks and Outdoor Recreation Division. *Barkerville Historic Park Concept Plan* 1981

British Columbia. Land Registry Office. *Absolute Fees Book*, Vol. 2, Fol. 619; Vol. 5, Fol. 324, and Fol. 776; Vol. 8, Fol. 895; Vol. 9, Fol. 845; Vol. 10, Fol. 504; and Vol. 22, Fol. 229; DD Pocket 987 (3 Apr. 1903), DD Pocket 17399 (3 July 1899, and 5 Feb. 1910); DD Pocket 505 (24 Aug. 1908); DD Pocket 10925 (8 Mar. 1909); Certificate No. 12864-N; Certificate No. 14877-15 and Certificate No. 14877-3; File No. 83407-1 (17 June 1903), and File No. 17399 (27 Feb. 1911); DD roll 109 B 987 (27 June 1899, 27 Sept. 1899)

British Columbia. Legislative Assembly. Sessional Papers, 1899

British Columbia. Legislative Assembly. Journals 1871, 1872, and 1879

British Columbia. Legislative Assembly. Report on Oriental Activities within the Province 1927

British Columbia. Ministry of Health. Community Care Facilities Licensing Act 1969

British Columbia. Ministry of Health. Hospital Act. Chap. 178, 1 August 1975. Part II. Private Hospitals.

British Columbia. Ministry of Industry and Small Business Development, Central Statistics Bureau: Demographic Data of Electoral District 23 By Enumeration Areas, 1971 Census

British Columbia. Ministry of Lands, Parks and Housing. *Barkerville Provincial Historic Park* 1986

British Columbia. Minister of Mines, Annual Report 1893

British Columbia. Ministry of Recreation and Conservation. Heritage Conservation Branch. *Chinatown Victoria*, August 1977

British Columbia. Report of Chinese in Barkerville, Lightning Creek and Quesnellmouth Polling Divisions, submitted by J. Lindsay to provincial secretary, 30 Nov. 1879

British Columbia. Statutes of the Province of British Columbia, 1872, 121; 1876, 3; 1878, 30.; 1899, Chap. 39; 1900, Chap. 11; 1903, Chap. 17, and 1920, Chap. 27, 99

British Columbia. Statutes of the Province of British Columbia, 2nd Session, 4th Parliament, 1884, Chap. 2, 3; Chap. 3, 5–6; and Chap. 4, 7–12

Canada. Censuses of Canada 1881–1981

Canada. Department of Employment and Immigration, *Annual Report to Parliament on Future Immigration Levels*, Ottawa 1987

Canada. Department of Citizenship and Immigration, *Immigration Statistics*, Ottawa 1957

Canada. Department of Employment and Immigration, *Kaleidoscope: The Quarterly Review of Immigration and Ethnic Affairs* 1982–3

Canada. Department of External Affairs, *Documents on Canadian External Affairs*, Vol. 1, 1908–18, Despatch 565

Canada. Department of Immigration and Colonization. Annual Reports, 1918–25, 1931–2, 1934–35

Canada. Department of the Interior, Superintendent of Immigration, Annual Reports 1904–17

Canada. Department of Manpower and Immigration. *Immigration Statistics*, Ottawa 1975, 1976

Canada. Department of Manpower and Immigration. *Highlights from the Green Paper on Immigration and Population*, Ottawa 1975

Canada. Department of Mines and Resources, Annual Report 1947

Canada. Dominion Bureau of Statistics. *Canada Year Book*, Ottawa 1948–9, 1968

Canada. Employment and Immigration Commission. *Immigration* Vol. 1, No. 1, 1986

Canada. Canada Gazette, 14 July 1919, 3824

Canada. House of Commons. Debates. 12 May 1882, 1476; 8 May 1922, 1509; May 1947, 2644–6

Canada. House of Commons. Sessional Papers, 1907–8, Paper No. 74f; Paper No. 10; Paper No. 36b

Canada. Ministry of Supply and Services. *Canada Year Book*, Ottawa 1985

Canada. Ministry of Industry, Trade and Commerce. *Canada Year Book*, Ottawa 1980–1

Canada. Order in Council PC 1202, 9 June 1919; PC 2115, 16 Sept. 1930; PC 1378, 17 June 1931; PC 1962–86, 18 Jan. 1962

Canada. Report of the Task Force on Housing and Urban Development. Ottawa: Queen's Printer 1969

Canada. Royal Commission on Coal Mining Disputes on Vancouver Island. Ottawa: King's Printer 1913

Canada. Royal Commission on Chinese Immigration: Report and Evidence. Ottawa: Printed by Order of the Commission 1885

Canada. Royal Commission on Chinese and Japanese Immigration Report, 1902, *Sessional Papers of the Dominion of Canada*, No. 54, Vol. 13, Ottawa 1902

Canada. *Report on Losses Sustained by the Chinese Population of Vancouver, B.C. on the Occasion of the Riots in that City in September, 1907* (Royal Commission: W.L. Mackenzie King) Sessional Papers, No. 74f, 1908

Canada. Statutes of Canada. An Act to Restrict and Regulate Chinese Immigration into Canada, 1885. Ottawa: 48–9 Victoria, Chap. 71, 207–12

Canada. Statutes of Canada. An Act Respecting and Restricting Chinese Immigration, 1900. Ottawa: 63–4 Victoria, Chap. 32, 215–21

Canada. Statutes of Canada. An Act Respecting and Restricting Chinese Immigration, 1903. Ottawa: 3 Edward VII, Chap. 8, 105–11

Canada. Statutes of Canada, 7–8 Edward VII, Chap. 50, An Act to Prohibit the Importation, Manufacture and Sale of Opium for Other than Medicinal Purposes, 20 July 1908.

Canada. Statutes of Canada, 10–11 George V, Chap. 46, An Act Respecting the Election of Members of the House of Commons and the Electoral Franchise, 1 July 1920

Canada. Statutes of Canada. An Act Respecting Chinese Immigration, 1923, Ottawa, 13–14 George V, Chap. 38, 301–15

Canada. Statutes of Canada. An Act to Amend the Immigration Act and to Repeal the Chinese Immigration Act, 1947. Ottawa: 11 George VI, Chap. 19, 107–9

Canada. Statutes of Canada. An Act Respecting Immigration to Canada, 25–26 Elizabeth II, Chap. 52, 1976–7

City of Calgary. Planning Department. *Chinatown-Calgary*, Nov. 1972

City of Calgary. Planning Department. *Calgary Chinatown Design Brief*, Sept. 1976

City of Calgary. Planning Department. *Chinatown Workshop Proposal Call*, 16 Nov. 1982
City of Calgary. Planning and Building Department. *Chinatown Area Redevelopment Plan* 1986
City of Calgary. Planning and Building Department. *Chinatown Handbook of Public Improvements* 1985
City of Edmonton. Planning Department. *Boyle Street/McCauley Area Redevelopment Plan Bylaw 1981*
City of Edmonton. Planning Department. *Downtown Plan Working Paper No 1: The Future of Chinatown* 1978
City of Edmonton. Planning Department. Revised Chinatown Plan 1980
City of Edmonton. Mayor's Task Force on the Heart of the City. *Final Report* 1984
City of Lethbridge. Planning and Development Department. *Downtown Phase II Area Redevelopment Plan* 1979
City of Montreal. By-law 6513: Zoning of the Downtown Area Located South of Sherbrooke Street between Bleury and Sanguinet streets, 22 Oct. 1984
City of Montreal. Urban Planning Department. W. "A Study of Various Chinese Associations"
City of Ottawa. Planning Branch. *Dalhousie Study: Existing Conditions Report*, Nov. 1975
City of Regina. Economic Development Department. W. Henry K.C. Ng, *Discussion Paper Regarding the Possibility of a Chinatown in Regina*, Oct. 1985
City of Ottawa. Planning Branch. *Dalhousie North Redevelopment Plan*, July 1980
City of Scarborough. Planning Department. The Glen Watford Commercial Area Study, 20 Feb. 1985
City of Toronto. Planning Board. *Official Plan for the City of Toronto Planning Area, Part I*, Oct. 1969
City of Toronto. Planning Board. *South-East Spadina Appraisal* 1971
City of Toronto. Planning Board. *Towards a Part II Plan for South-East Spadina* 1972
City of Toronto. Planning Board. *Planning in South-East Spadina Status Report*, June 1973
City of Toronto. Planning Board. *Towards a Neighbourhood Plan: South Riverdale* 1977
City of Toronto. Planning Board. *Official Plan Proposals: South-East Spadina*, Jan. 1978
City of Toronto. Planning Board. *Final Recommendations: South-East Spadina*, Mar. 1979
City of Vancouver. Planning Department for the Housing Research Com-

mittee. *Vancouver Redevelopment Study*, Dec. 1957
City of Vancouver. Planning Department. *Urban Renewal Proposed Study Under Part V of the National Housing Act*, Aug. 1966
City of Vancouver. Planning Department. *Urban Renewal in Vancouver, Progress Report, No.* 7, 1966
City of Vancouver. Planning Department. Chinatown Planning Newsletter, Nov. 1976
City of Vancouver. Planning Department. *Urban Renewal Program Scheme 3: Sub-area 1—Strathcona, Appendices.* "Brief to City Council from a Citizen's Group," n.d., 1
City of Vancouver. Planning Department. *Reports to Council: North Park Development Concept*, 12 Mar. 1986, 1–12
City of Vancouver. Strathcona Rehabilitation Committee. *Strathcona Rehabilitation Program: Recommendations of the Strathcona Rehabilitation Committee*, July 1971
City of Vancouver. Strathcona Rehabilitation Committee. *Strathcona Rehabilitation Project, Stage II Evaluation* 1977
City of Victoria. Chinatown Ad Hoc Committee. *A Plan for the Rehabilitation of Chinatown*, Aug. 1979
City of Victoria. Planning Department. *The Central Area Victoria—Neighbourhood Improvement Program* 1977
City of Victoria. Victoria District Assessment Roll for 1902
City of Victoria, Victoria School Board, *Minutes, 1905–1910*
City of Victoria, Victoria School Board, *Minutes, Jan. 1920–Dec. 1922*
City of Winnipeg. Department of Housing and Urban Renewal. *Final General Report of Urban Renewal Plan No.* 2, Jan. 1968
City of Winnipeg. Planning Department. *Winnipeg Core Area Initiative Status Report of Program Activities*, 30 Sept. 1985
Saskatchewan. Statutes of the Province of Saskatchewan, 4th Session of the Second Legislature, 1912: Chap. 17. An Act to Prevent the Employment of Female Labour in Certain Capacities (assent given 15 Mar. 1912)
United Kingdom. *Blue Books of Statistics, Colony of Vancouver Island*. London: Colonial Office 1865, 1867
United Kingdom. *Blue Books of Statistics, Colony of British Columbia*. London: Colonial Office 1861, 1862, 1865, 1867, 1871
Village of Cumberland. *Application for Expo '86 Legacy Funding to Build and Develop a Restoration of Cumberland's Historic Chinatown*, June 1986

CITY DIRECTORIES

British Columbia Directory. Victoria: Williams 1882–99
British Columbia Directory. Victoria: Wrigley's 1918–32

British Columbia Directory and Victoria Directory. Victoria: Howard and Barnett 1863

British Columbia Gazetteer and Directory. Victoria: Henderson 1889–90, 1899–1905

British Columbia and Yukon Directory. Vancouver: Sun Publishing 1935–47

Calgary City Directory. Calgary: Henderson 1910

Directory of the City and District of Calgary. Calgary: Calgary Herald Co. 1902

First Victoria Directory and British Columbia Guide. Victoria: E. Malladaine 1874

Guide to the Province of British Columbia. Victoria: T.N. Hibbon & Co. 1877

Lethbridge City Directory. Lethbridge: Henderson 1909

Montreal City Directory. Montreal: Lovell 1910–11, 1921–3

Ottawa City Directory. Ottawa: Might 1931, 1941, 1961

Regina City Directory. Regina: Henderson 1907

Toronto City Directory. Toronto: Might 1910

Victoria City Directory and Vancouver Island Gazetteer. Victoria: Henderson 1912–14.

NEWSPAPERS, NEWSLETTERS, MAGAZINES

Asian Leader, Montreal. Dec. 1985–Jan. 1986

British Colonist, Victoria, British Columbia (11 Dec. 1858–28 July 1860); *Daily British Colonist* (31 July 1860–23 June 1866); *Daily British Colonist* and *Victoria Daily Chronicle* (25 June 1866–6 Aug. 1872); *Daily British Colonist* (7 Aug. 1872–31 Dec. 1886), and *Daily Colonist* (1 Jan. 1887–31 Oct. 1980).

Calgary Herald, 1945–82

Cariboo Sentinel, BC, 6 June 1865–Oct. 1875

Centretown News, Ottawa, 11 Apr. 1981

Chinatown News, Vancouver, 3 Nov. 1956, 16; 3 Mar. 1980, 11; 3 Feb. 1982, 3, 33; 18 Aug. 1982, 14; 3 Mar. 1983, 31; 3 Mar. 1984, 35; 3 June 1984, 6, 28; 3 May 1985, 21; 18 May 1985, 8; 18 June 1985, 34; 3 Sept. 1985, 29; 3 Mar. 1986, 16; 3 Aug. 1986, 30

Chinese Canadian Bulletin, Vancouver, Jan. 1967, 5; Jan. 1981, 42; Dec. 1985, 15

Citizen, Ottawa, Apr.–Sept. 1986

Comox District Free Press, 18 May 1966

Daily Globe, San Francisco, 15 May 1858

Daily Times, Victoria (1884–6 Nov. 1971); *The Times,* Victoria (8 Nov. 1971–30 Oct. 1980)

Danger: The Anti-Asiatic Weekly, Vol. 1, No. 3, Oct. 1921, 18, and No. 9 Dec. 1921, 23.
Downtowner, Winnipeg, 22 Oct. 1986
Edmonton *Journal,* 1960–80
Financial Post, Toronto, 17–23 Nov. 1986
Globe and Mail, Toronto, 1965–85
Hamilton *Spectator,* 1985
Hope *Standard,* 29 Sept. 1982
Kamloops *Daily Sentinel,* 1927–65
Lethbridge *News,* 7 May 1890
Maclean's Magazine, 1949
Mirror, Scarborough, 1984
Monday Magazine, Victoria, 1977
Montreal *Star,* 1969–80
Nanaimo *Free Press,* 1883–1920
Quartier Chinois de Montréal, 2 Mar. 1982
Regina *Leader-Post,* 1986, 1987
Similkameen Star, 10 Sept. 1915
South China Morning Post, Hong Kong, 1977, 1986
Standard, Victoria, 1870–89, also known as *Daily Standard* and *Evening Standard*
Telegram, Toronto, 21 June 1971
Times-Colonist, Victoria, 4 Sept. 1980
Toronto *Star,* 1984–85
Toronto *Sun,* 1984
Vancouver *Daily World,* 29 Sept. 1889–14 July 1900
Vancouver *Daily News-Advertiser,* 1886–1917, also known as *Vancouver News* and *Daily News-Advertiser*
Vancouver *News-Herald,* 1933–54
Vancouver *Province,* 1894–1986
Vancouver *Sun,* 1912–86
Vancouver *World,* 1892–9
Victoria *Express,* 1974
Victoria *Gazette* (25 June–24 July 1858); *Daily Victoria Gazette* (28 July –26 Oct. 1858); and *Victoria Gazette* (28 Oct. 1858–26 Nov. 1859)
Winnipeg *Free Press,* 1877–1986
Winnipeg *Tribune,* 1942

UNPUBLISHED DOCUMENTS

Brief of the Chinatown Property Owners Association of Vancouver presented to the Royal Commission on Expropriation Laws and Procedures, 17 July 1961

Brief on the Chinese Community in Toronto, prepared by the Chinese Canadian Association and presented to the Special Committee on the Chinese Community, 18 June 1970

Constitution of the Workingmen's Protective Association, A (Provincial Archives of British Columbia)

Contract, signed between the British Columbia Sugar Refining Company and the City of Vancouver, Aug. 1890 (Provincial Archives of British Columbia)

Agreement signed between Hop Kee & Co. and Allan Lowe & Co., 24 June 1858 (Provincial Archives of British Columbia)

Baureiss, Gunter and Julia Kwong, "The History of the Chinese Community of Winnipeg," Chinese Community Committee 1979

Bell, Larry I. and Richard Moore, "The Strathcona Rehabilitation Project: Documentation and Analysis." Vancouver: Social Policy and Research Department of United Way of Greater Vancouver Dec. 1975

Canada. Historic Sites and Monuments Board of Canada. "Program of the Unveiling Ceremony of the Plaque at Yale, 25 Sept. 1982"

Chinatown Ratepayers Association of Calgary, *Chinatown Development Proposal*, Oct. 1982

"Chinese Labour" (vertical file, Provincial Archives of British Columbia)

City of Victoria. Group B Committee's Report to Mayor and Board of Aldermen in Council, 4 Sept. 1979

City of Victoria. Heritage Advisory Committee. "Minutes of Regular Monthly Meeting," 23 Jan. 1984

Cumberland and District Historical Society. "Schedule of Events on Workers' Memorial Day at Cumberland," 21 June 1986

Cumberland Heritage Conservation Society. Presents a Legacy of the Comox Valley 1987

Day, J.P. "Wong Sing Fuen and the Sing Lee Laundry," Historical and National Science Services, Edmonton Parks and Recreation (Nov. 1978)

Ellis, L.G. "Love Thy Neighbour: Social and Racial Attitudes in Lethbridge, 1885–1945"

Eversole, Linda J. "Chinese Community in Yale," Yale, Jan. 1984 (mimeographed)

Hong, Jane, Josephine Li, Jim Mah, and Mark Tang. "Chinese Community in Newfoundland." Memorial University of Newfoundland: Chinese Student Society 1976

Hunter, Margaret. "The Danger of Canada's 'Open Door' Immigration Policy"

Joyner, J.B. "Lethbridge Chinatowns: An Analysis of the Kwong On Lung Co. Building, the Bow On Tong Co. Building, and the Chinese Free Masons Building, 1985"

Lai, Chuen-yan David. "A Care Home in Victoria's Chinatown: A Survey of the Need." Victoria: Capital Regional Hospital and Health Planning Commission 1979

—. Interim Report on Chinatown's Vacant Floors and the Designated Heritage Buildings, 5 July 1982

—. Final Report on the Survey of the Need for Low-Rental Housing in Chinatown and Elsewhere in Downtown Victoria," 29 July 1982

Lau, C. "Proposal for a Chinese Cultural Center and Development of a 'Chinatown' Project in Essex County, 2 Feb. 1982"

Laurier papers Series. Correspondence, Vol. 638–41, 174002 (NAC)

Letter, A.F. Buckham of Cumberland to W.E. Ireland, Provincial Archivist, 6 Apr. 1960

Letter, Alderman R.H. Wright to Mayor Michael D.W. Young, 29 Aug. 1979

Letters, Armin Preiksaitis, Manager, Area Planning Branch, City of Edmonton, to David C.Y. Lai, 18 Mar. 7, Apr. 1986

Letter, Augustine Low, President of CCBA, to Canada Mortgage and Housing Corporation, Victoria, 2 Dec. 1982

Letter, B.D. Strongitharm, Planning Officer to C.J. Greenhalgh, Director of Department of Community Development, City of Victoria, 5 July 1979

Letter, Commissioner of Planning, City of Toronto to Committee on Buildings and Development, 6 Feb. 1979

Letter, Commissioner of Planning, City of Toronto to Land Use Committee, 23 Mar. 1981

Letter, David C.Y. Lai, to Rose Kung, Project Co-ordinator, Community Development Department, City of Ottawa, 20 Jan. 1988

Letter, Dr. Andrew Rose, Hospital Inspector to Peter Wong, Administrator of the Chinese Hospital, 2 Dec. 1976

Letter, Dr. Joseph N.H. Du, Chairman of Winnipeg Chinatown Development (1981) Corporation to David C.Y. Lai, 10 Sept. 1986

Letter, Dr. Michael Carley, Field Representative of Historical Resources, Historic Sites Service to Mr. Felix Michna, Director of Planning Department, City of Lethbridge, 1 Mar. 1983

Letter, Foon Sien, Chairman of the CBA (NH) to Dr. W.G. Black, Regional Liaison Officer, Department of Citizenship and Immigration, 3 Mar. 1951

Letter, J.W. Pennington, Assistant Director of Hospital Planning, Capital Regional Hospital and Health Planning Commission to Mrs. Isabel Kelly, Assistant Deputy Minister, Ministry of Health, 29 Nov. 1979

Letter, Jeremy Tate, Facility Planning and Development Co-ordinator, Ministry of Health to the CCBA, 21 May 1980

Letter, L.O. Spencer, A/Director, Community Planning Branch, City of Ottawa, to David C.Y. Lai, 8 Jan. 1988

Letter, Mark Johnston, Administrative Assistant, City of Victoria to Peter Lou-Poy, Advisor to CCBA, 17 May 1984

Letters, William McKim, City Manager, City of Regina to Mayor and Council, 22 Dec. 1966, 2 Feb. 1987

Letters, William McKim, City Manager, City of Regina to Mayor and Council, 22 Dec. 1966, 2 Feb. 1987

Letter, Peter Lou-Poy, President of CCBA to Mayor Michael Young, 11 Mar. 1979

Letter, Hon. Hugh Curtis, Provincial Secretary to Peter Lou-Poy, President of CCBA, 26 Oct. 1979

Letter, William Moncrief, jr., Mayor of Cumberland, to David C.Y. Lai, 12 Jan. 1987

Mah, Valerie A. "An Indepth Look at Toronto's Early Chinatown, 1913–1933." 1977

— . "The 'Bachelor' Society: A Look at Toronto's Early Chinese Community from 1878 to 1924." 1978

Meares, John. "Voyages Made in the Years 1788 and 1789, From China to the North West Coast of America," London 1790; and "Authentic Copy of the Memorial to the Right William Wyndham Grenville, 30 April 1790," London 1790 (presented to the House of Commons, 13 May 1790)

Montreal Chinese Professional and Businessmen's Association. Report of the Public Hearings on the Impacts of Municipal Zoning By-law No. 6513 upon Chinatown in Montreal. n.d., page unfolioed

Ng, Henry K.C. "Preliminary Study/Report on Redevelopment and Revitalization of Montreal Chinatown, 14 Jan. 1982

Ottawa. Chinatown Development. "Minutes of Second Meeting, 2 February 1986"

Scarborough. CBL TV. Transcript by MediaReach: No. 1504. "Metro Morning," 28 May 1984; No. 1509, "For Your Information," 29 May 1984

Sien Lok Society, Calgary. "National Conference on Urban Renewal as It Affects Chinatown," 6–9 Apr. 1969

Siew, Nancy M. and Michael J. Maclean, "Perceptions of Elderly Hong Kong Immigrants: Implications for 1997." Paper presented at annual meeting of Canadian Asian Studies Association, Hamilton, Ont. 5 June 1987

Spearing, David N., W.W. Wood, and W.H. Birmingham, "Draft of a Report on 'Rehabilitation through Cooperation' in Strathcona," December 1970

Submission of the Chinatown Property Owners Association of Vancouver to Mr. J.V. Clyne, Royal Commission on Property Expropriation, Regarding Vancouver Redevelopment in Area A. n.d.

United Community Services of the Greater Vancouver Area. "Draft Report

of Committee on Redevelopment and Relocation," n.d.
United Community Services of the Greater Vancouver Area. "Urban Renewal Scheme III, Strathcona." July 1966
Victoria School Board. Minutes of meeting, 31 July 1922
Winnipeg Chinatown Development (1981) Corporation. Chinatown Redevelopment Project 1984
Wong, May, Juliana Huang, and Sally Chung. "A Report on the Development of the Chinese Community in Hamilton." Chinese Cultural Association of Hamilton 1984
Woodgreen Community Centre of Riverdale. "Riverdale: The Changing Community," n.d.
Yee, Tim; Leesa Ritchie, Kee Yee, May Yee, Verlin Gwin, Gwen Choi, and Tim Tze. "An Ethnic Study of the Chinese Community of Moose Jaw." Moose Jaw: Opportunities for Youth Project, 14 May–25 Aug. 1973

THESES

Aiken, Rebecca B. "Montreal Chinese Property Ownership and Occupation Change: 1881–1981." PH.D. thesis, McGill University 1984
Chan, Ellen Y.L. "Planning for Change: The Winnipeg Chinese Community and Its Responsiveness to Government Services." Master in Community Planning thesis, University of Manitoba 1962
Nann, Richard. "Urban Renewal and Relocation of Chinese Community Families." PH.D. thesis, University of California, Berkeley 1970
Robert, Percy A. "Dufferin District: An Area in Transition." MA thesis, McGill University 1928
Sedgwick, Charles P. "The Context of Economic Change and Continuity in an Urban Overseas Chinese Comminuty." MA thesis, University of Victoria 1970
Wargo, A.J. "The Great Coal Strike: The Vancouver Island Coal Miners' Strike, 1912–1914." BA graduating essay, University of British Columbia 1962
Winnicki, W.R. "Chinatown in Transition: The Impact of the New City Hall and Court House on Toronto's Chinatown." BA thesis, University of Waterloo 1969
Wynne, R.E. "Reaction to the Chinese in the Pacific Northwest and British Columbia, 1850–1910." PH.D. thesis, University of Washington 1964
Yee, Paul Richard. "Chinese Business in Vancouver, 1886–1914." MA thesis, University of British Columbia 1983
Young, Raymond Edger. "Street of Tongs: Planning of Vancouver Chinatown." MA thesis, University of British Columbia 1975

BOOKS AND ARTICLES

Adachi, K. *The Enemy that Never Was*. Toronto: McClelland & Stewart 1976

Allport, G. *The Nature of Prejudice*. Cambridge, MA: Addison-Wesley 1954

Andracki, Stanislaw. *Immigration of Orientals into Canada with Special Reference to Chinese*. New York: Arno 1978

Artibise, Alan F.J., "The Urban West: The Evolution of Prairie Towns and Cities to 1930," *Prairie Forum*, 4 (1979), 237–62

Balf, Mary. *Kamloops: A History of the District up to 1914*. Kamloops: Kamloops Museum 1969

Balf, Ruth. *Kamloops: 1914–1945*. Kamloops: Kamloops Museum 1975

Balakrishnan, T.R. "Ethnic Residential Segregation in the Metropolitan Areas of Canada," *Canadian Journal of Sociology,* 1 (1976), 481–98

Baureiss, Gunter. "The Chinese Community of Calgary," *Canadian Ethnic Studies,* II (1971), 43–55

— . "The Chinese Community of Calgary," *Alberta Historical Review,* 22 (1974), 1–8

— . and L. Driedger. "Winnipeg's Chinatown: Demographic, Ecological and Organizational Change, 1900–1980," *Urban History Review,* X (1982), 11–24

Berry, B. and F.E. Horton. *Geographical Perspectives on Urban Systems*. New York: Prentice-Hall 1970

Berton, Pierre. *The Last Spike*, Toronto: McClelland & Stewart 1971

Bo, Lao. "Hostages in Canada: Toronto's Chinese (1880–1947)," *The Asianadian,* 1 (1978), 11–13

Bourne, Larry S. *Internal Structure of the City*. Toronto: Oxford University Press 1971

Bowen, Lynne. *Boss Whistle: The Coal Miners of Vancouver Island Remember*. Lantzville: Oolichan 1982

Bowring, Philip "Directing the Elections," *Far Eastern Economic Review,* 7 July 1987, 30–1.

Brown, Forbes. "Mon Sheong Home for Elderly Chinese," *Living Places,* 11 (1975), 2–9

Burgess, E.W. and D.J. Bogue. *Contributions to Urban Sociology*. Chicago: University of Chicago Press 1964.

Cameron, M.E. *The Reform Movement in China, 1898–1912*. New York: Octagon Books 1963

Century Regina, 1882–1982. Regina: Southland Mall 1982

Chan, Anthony B. *Gold Mountain*. Vancouver: New Star Books 1982

Chan, Kwok B. "Ethnic Urban Space, Urban Displacement and Forced Relocation: The Case of Chinatown in Montreal," *Canadian Ethnic Studies,* XVIII (1986), 65–78

Cheadle, Walter B. *Cheadle's Journal of a Trip Across Canada, 1862–63*. Edmonton: Hurtig 1971

Chen, T. *Chinese Migrations with Special Reference to Labour Conditions*. Taipei: Ch'eng Wen 1967

Chinese Presbyterian Church 90th Anniversary Celebrations Committee, Victoria 1983. *Chinese Presbyterian Church, Victoria, B.C., 1892–1983*

Chinese Students Athletic Club, Victoria. *Chinese Students Athletic Club 20th Anniversary Celebration and Reunion, 1931–1950*

Da Roza, Gustavo. *A Feasibility Study for the Development of Chinatown in Winnipeg*. Winnipeg: Winnipeg Chinese Development Corporation Ltd. 1974

— . *Winnipeg Chinatown: A Proposal*. Winnipeg: Winnipeg Chinese Development Corporation Ltd. 1971

Dawson, J. Brian. "The Chinese Experience in Frontier Calgary: 1885–1910," in *Frontier Calgary: Twon, City and Region*, edited A.W. Rasporich and H.C. Klassen. Calgary: McClelland & Stewart West 1975

Dinnerstein, Leonard, Roger L. Nichols, and David M. Reimers. *Natives and Strangers: Ethnic Groups and the Building of America*. New York: Oxford University Press 1979

Duncan, O.D. "Human ecology and population studies." *The Study of Population*. Ed. P.M. Hauser and O.D. Duncan. Chicago: University of Chicago Press 1959

Elliott, Gordon R. *Barkerville, Quesnel & the Cariboo Gold Rush*. Vancouver: Douglas & McIntyre 1978

Fenwick, A.H. "For East Is East," *Maclean's Magazine,* 15 Jan. 1928, 45–7

Ferguson, T. *A White Man's Country: An Exercise in Canadian Prejudice*. Toronto: Doubleday 1975

Forward, C.N. *Land Use of the Victoria Area, British Columbia*. Ottawa: Geographical Branch, Department of Energy, Mines and Resources 1969

Gao, Wenxiong. "Hamilton: The Chinatown that Died," *The Asianadian,* 1 (1978), 15–17

Gibbon, J.M. *The Romantic History of the Canadian Pacific*. New York: Tudor 1937

Gibbon, John. *Steel of Empire*. New York: Bobbs-Merrill, 1935

Glick, C.E. *Sojourners and Settlers: Chinese Migrants in Hawaii*. Honolulu: University Press of Hawaii 1980

Glynn-Ward, Hilda. *The Writing on the Wall*. Introduction by Patricia E. Roy. Toronto: University of Toronto Press, 1974, reprint

Goodfellow, J.C. *The Story of Similkameen*. n.p., n.d.

Gregson, Harry. *A History of Victoria, 1842–1970*. Victoria: Victoria Observer 1970

Hartman, C., D. Keating, and R. LeGates, ed. *Displacement: How to Fight It*. Berkeley, CA: National Housing Law Project 1982

Hartwell, George E. "Our Work among the Japanese and Chinese in British Columbia," *Missionary Bulletin*, 9 (1913), 513–29

Harvey, R.D. *A History of Saanich Peninsula Railways*. Victoria: Department of Commercial Transport, Railway Branch 1960

Hearn, George and David Wilkie. *The Cordwood Limited: A History of the Victoria and Sidney Railway*. Victoria: British Columbia Railway Historical Association 1976

Hoe, Ban Seng, *Structural Changes in Two Chinese Communities in Alberta*. Ottawa: Canadian Centre for Folk Culture Studies 1976

— . "Chinese Community and Cultural Traditions in Quebec City," *Chinese Consolidated Benevolent Association 1985 Tri-Celebration Special Issue*. Victoria: Chinese Consolidated Benevolent Association, 1986, 131–43.

Hong, W.M. *And So That's How It Happened: Recollections of Stanley Barkerville, 1900–1975*. Quesnel: by the author 1978

Howay, F.W. *British Columbia from the Earliest Times to the Present*. Vancouver: S.J. Clarke 1914. Vol. II

Huang, Ten-ming, *The Legal Status of the Chinese Abroad*. Taipei: China Cultural Service 1954

Isard, Walter, ed. *Methods of Regional Analysis: An Introduction to Regional Science*. Cambridge, MA: MIT Press 1960

Johnson, A. and Andy A. den Otter. *Lethbridge: A Centennial History*. Lethbridge: City of Lethbridge 1985

Johnstone, Bill. *Coal Dust in My Blood: Heritage Record No. 9*. Victoria: Provincial Museum 1980

Kerr, Alastair W. "The Architecture of Victoria's Chinatown," *Datum*, 4 (1979), 8–11

— . "Barkerville in the Eighties: The Future of Barkerville's Past," *Datum*, 6 (1981), 17–19.

Kidder, L. *The Psychology of Intergroup Relations*. New York: McGraw-Hill 1975

Kwong, Julia. "Transformation of an Ethnic Community: From a National to a Cultural Community." *Asian Canadians and Multiculturalism: Selections from the Proceedings of the Asian Canadian Symposium IV, May 1980*. Ed. Victor Ujimoto and Gordon Hirabayashi

La Forest, G.V. *Disallowance and Reservation of Provincial Legislation*. Ottawa. Queen's Printer 1955

Lai, Chuen-yan. "The Chinese Consolidated Benevolent Association in Victoria: Its Origins and Functions." *BC Studies*, 15 (1972), 53–67

— . "Chinese Attempts to Discourage Emigration to Canada: Some Find-

ings from the Chinese Archives in Victoria.'' *BC Studies,* 18 (1973), 33–49
Lai, Chuen-yan David. ''Socio-economic Structures and Viability of Chinatown.'' *Residential and Neighbourhood Studies in Victoria.* Ed. Charles N. Forward. Department of Geography, Western Geographical Series, Vol. 5. Victoria: University of Victoria 1973
— . ''A Feng Shui Model as a Location Index.'' *Annals of the Association of American Geographers,* 64 (1974), 506–13
— . ''Home County and Clan Origins of Overseas Chinese in Canada in the Early 1880's.'' *BC Studies,* 27 (1975), 3–29
— . ''An Analysis of Data on Home Journeys by Chinese Immigrants in Canada, 1892–1915. *Professional Geographer,* 29 (1977), 359–65
— . ''Chinese Imprints in British Columbia.'' *BC Studies,* 39 (1978), 20–9
— . ''Ethnic Groups.'' In *Vancouver Island: Land of Contrasts.* Ed. by Charles N. Forward. Department of Geography, Western Geographical Series, Vol. 17. Victoria: University of Victoria 1979
— . *The Future of Victoria's Chinatown: A Survey of Views and Opinions, Vol. 1. Recommendations.* Victoria: City of Victoria 1979
— . *The Future of Victoria's Chinatown: A Survey of Views and Opinions, Vol. 2. Tabulation of Data.* Victoria: City of Victoria 1979
— . ''Far from South China Shores.'' *Victoria Times,* 26 Jan. 1980.
— . ''The Demographic Structure of a Canadian Chinatown in the Mid-Twentieth Century.'' *Canadian Ethnic Studies,* XI 2 (1979), 49–62
— . *The Gate of Harmonious Interest: From Concept to Reality, 2nd printing.* Victoria: City of Victoria 1982
— . *Arches in British Columbia.* Victoria: Sono Nis 1982
— . ''Contribution of the Zhigongtang in Canada to the Huanghuagang Uprising in Canton, 1911.'' *Canadian Ethnic Studies,* XIV, 3 (1982), 95–104
Lai, David Chuenyan. ''The Chinese Cemetery in Victoria.'' *BC Studies,* 75 (1987), 24–42
— . ''The Issue of Discrimination in Education in Victoria, 1902–1923,'' *Canadian Ethnic Studies,* XIX, 3 (1987), 47–67
Lai, H.M. and P.P. Choy. *Outlines of History of the Chinese in America.* San Francisco: Chinese-American Studies Planning Group 1973
Lang, Michael H. *Gentrification and Urban Decline: Strategies for America's Older Cities.* Cambridge, MA: Ballinger 1982
Lau, Emily, ''On to Greener Pastures,'' *Far Eastern Economic Review,* 18 June 1987, 22–5
Lee, Calvin. *Chinatown, U.S.A.* New York: Doubleday 1965
Lee, Carol F. ''The Road to Enfranchisement: Chinese and Japanese in British Columbia,'' *BC Studies,* 30 (1976), 44–76

Lee, Rose Hum. "The Decline of Chinatowns in the United States." *American Journal of Sociology,* 54 (1949), 422–32

Ley, David. "Alternative Explanations for Inner-City Gentrification: A Canadian Assessment," *Annals of the Association of American Geography,* 76 (1986), 521–35

Li, Peter S. "Chinese Immigrants on the Canadian Prairie, 1910–1947," *Canadian Review of Sociology and Anthropology,* 19 (1982), 527–40.

Lyman, Stanford M., W.E. Willmott, and B. Ho. "Rules of a Chinese Secret Society in British Columbia," *Bulletin of the School of Oriental and African Studies,* 27 (1964), 530–9

Ma, Ching. *Chinese Pioneers.* Vancouver: Versatile 1979

Macdonald, J.S. and L.D. Macdonald. "Chain Migration, Ethnic Neighborhood Formation and Social Networks." *The Milbank Memorial Fund Quarterly,* 42 (1964), 82–91

Macfie, Matthew. *Vancouver Island and British Columbia.* London: Longman, Roberts and Green 1865

MacNab, Frances. *British Columbia for Settler: Its Mines, Trade, and Agriculture.* London: Chapman & Hall 1898

MacNair, H.F. "The Relation of China to her Nationals Abroad," *Chinese Social and Political Science Review,* VII (1923), 23–43

Marsh, Leonard. *Rebuilding a Neighbourhood: Report on a Demonstration Slum Clearance and Urban Rehabilitation Project in a Key Central Area in Vancouver. Research Publication No. 1.* Vancouver: University of British Columbia 1950

Mather, Ken. "Barkerville: Then and Now," *Datum,* 7 (1982), 7

Mayse, Susan. "Coal Town, Boomtown, Ghost Town?" *Canadian Heritage* (Oct.–Nov. 1985), 17–19

McClellan, Robert. *The Heathen Chinee: A Study of American Attitudes toward China, 1890–1905.* Columbus: Ohio State University Press 1971

Merrill, Frederick T. *Japan and the Opium Menace.* New York: Institute of Pacific Relations and the Foreign Policy Association 1942

Morley, Alan. *Vancouver: From Milltown to Metropolis.* Vancouver: Mitchell 1961

Morton, James. *In the Sea of Sterile Mountains: The Chinese in British Columbia.* Vancouver: J.J. Douglas 1974

Nicol, Eric. *Vancouver.* Toronto: Doubleday 1970

Nipp, Dora. "The Chinese in Toronto." *Gathering Place: Peoples and Neighbourhoods of Toronto, 1834–1945.* Ed. Robert F. Harney. Toronto: Multicultural History Society of Ontario 1985

Norris, John, ed. *Strangers Entertained: A History of the Ethnic Groups of British Columbia.* Vancouver: British Columbia Centennial 71 Committee 1971

Ormsby, M.A. *British Columbia: A History.* Toronto: Macmillan, 1971

Osterhout, S.S. *Orientals in Canada.* Toronto: United Church of Canada 1929

Owen, David Edward. *British Opium Policy in China and India.* Hamden, CN: Archon 1968

Paterson, T.W. *Ghost Town Trails of Vancouver Island.* Langley: Stage Coach 1975

Paupst, K. "A Note on Anti-Chinese Sentiment in Toronto Before the First World War." *Canadian Ethnic Studies,* IX (1977), 54–9

Pemberton, J.D. *Facts and Figures Relating to Vancouver Island and British Columbia.* London: Longman 1860

Phillips, P.A. *No Power Greater: A Century of Labour in British Columbia.* Vancouver: BC Federation of Labour 1967

Plasterer, Herbert P. *Fort Victoria: From Fur Trading Post to Capital City of British Columbia, Canada.* Victoria: Colonist Printer, n.d.

Ramsey, A.B. *Barkerville: A Guide in Word and Picture to the Fabulous Gold Camp of the Cariboo.* Vancouver: Mitchell 1961

Ramsey, Bruce. *Ghost Towns of British Columbia.* Vancouver: Mitchell 1970

Rickard, T.A. "A History of Coal Mining in British Columbia." *The Miner* (June 1942), 30–5

Roy, Patricia E. "A Choice between Evils." In *The CPR West*, Ed. Hugh Dempsey. Vancouver: Douglas & McIntyre 1984

— . "Educating the 'East': British Columbia and the Oriental Question in the Interwar Years." *BC Studies*, 18 (1973), 50–9

— . "The Soldiers Canada Didn't Want: Her Chinese and Japanese Citizens." *Canadian Historical Review,* LIX (1978), 341–58

— . "The Preservation of the Peace in Vancouver: The Aftermath of the Anti-Chinese Riot of 1887," *BC Studies,* 31 (1976), 44–59

Sage, Walter. "Federal Parties and Provincial Groups, 1871–1903," *British Columbia Historical Quarterly,* 12 (1948), 151–69

Segger, Martin. *Victoria: A Primer for Regional History in Architecture.* New York: American Life Foundation 1979

Smith, N. and P. Williams, ed. *Gentrification of the City* Winchester, MA: Allen & Unwin 1986

Staunton, G.T. and N.B. Edmondstone. "A Letter to the Governor at Prince of Wales Island." *Journal of the Indian Archipelago and Eastern Asia,* 6 (1852), 147–8

Sung, B.L. *The Story of the Chinese in America.* New York: Collier 1967

Suttles, G.D. *The Social Construction of Communities.* Chicago: University of Chicago Press 1972

Thomas, Brinley. "International Migration." In *The Study of Population.* Ed. P.M. Hauser and O.D. Duncan. Chicago: University of Chicago Press 1959

Toone, A.W. "Victoria Rediscovered." *Western Wonderland,* Dec. 1964, 6–9

Tuchman, B.W. *The Proud Tower*. New York: Bantam 1966

University of British Columbia, Alma Mater Society, *The Ubyssey,* 28 Sept. 1978

Van Dieren, Karen. "The Response of the WMS to the Immigration of Asian Women, 1888–1942." In *Not Just Pin Money: Selected Essays on the History of Women's Work in British Columbia*. Ed. Barbara K. Latham and Roberta J. Pazdro. Victoria: Camosun College 1984

Vancouver. Chinese Cultural Centre. *Chinese Cultural Centre* (leaflet) n.d.

Vancouver, Chinese Cultural Centre, *Chinese Cultural Centre Reports*, Vol. 1, Nov. 1976

Vancouver, Committee to Democratize the CBA, *CBA Issue* (Jan. 1978)

Voisey, P. "Chinatown on the Prairies: The Emergence of an Ethnic Community." *Selected Papers from the Society for the Study of Architecture in Canada: Annual Meetings in 1975 and 1976*. Ottawa: Society for the Study of Architecture in Canada 1981

Ward, W. Peter. *White Canada Forever*. Montreal: McGill-Queen's University Press 1978

Washington, M.M. "The Story of the Oriental Home, Victoria," *The Missionary Monthly* (July 1943), 293–4

Wickberg, Edgar, ed. *From China to Canada*. Toronto: McClelland & Stewart 1982

Willmott, W.E. "Some Aspects of Chinese Communities in British Columbia Towns." *BC Studies,* 1 (1968–9), 27–36

Woodland, A. *New Westminster: The Early Years, 1858–1898*. New Westminster: Nunaga 1973

Wynne, R.E. "American Labor Leaders and the Vancouver Anti-Oriental Riot." *Pacific Northwest Quarterly,* 57 (1966), 172–9

Yee, Paul. *A Walking Tour of Vancouver's Chinatown*. Vancouver: Weller Cartographic Services Ltd. 1983

Yuan, D.Y. "Voluntary Segregation: A Study of New York Chinatown," *Phylon: The Atlanta University Review of Race and Culture,* XXIV (1963), 255–65

PUBLISHED AND UNPUBLISHED MATERIALS IN CHINESE

Commemorative Issue of the Opening of the Yue Shan Society Building in Vancouver, Vancouver 1949

Special Issue of the Golden Anniversary of Lim Sai Hor (Kow Mock) Benevolent Association, Vancouver, Lim Sai Hor Benevolent Association 1980

Cao Jianwu, "The Participation of the Hongmen in the Chinese Revolution" (unpublished)

Chinese Canadian Community News (Jiahua Qiaobao), Ottawa, May 1982, 4, 9; Jan. 1986, 1; July 1986, 9, 18.

Canadian Chinese Times (Jiahuabao), Calgary, 8 Oct. 1982

Chinese Press (Huaqiao Shibao), Montreal, 10 May 1976

Chinese Times (Dahan Gongbao), Vancouver, 14 Feb. 1986

Edmonton Chinese News (Aihua Bao), 4 Apr. 1986

Lee, David T.H. *A History of Chinese in Canada*. Taipei: By the author 1967

Manitoba Chinese Post (Miansheng Huabao), 1 Jan., 1 Nov. 1986

Sing Tao Daily, Vancouver, 18 Oct. 1983, 28 Jan. 1984

Tan, Rongxu, "The Sod-Turning Ceremony of Montreal's Chinese United Building," *Quartier Chinois de Montréal,* Dec. 1983, 1.

Tan, Zhenqiao, "A Preliminary Suggestion of the Development of Montreal's Chinatown," *Quartier Chinois de Montréal,* 2 Mar. 1982, 2

Victoria. Chinese Consolidated Benevolent Association. *Annual Reports, Minutes of Meetings, 1912–1986*

Victoria. Chinese Consolidated Benevolent Association. *The Proposal for the Establishment of Taipingfang, 1884* (unpublished)

Index